High temperature superconductors (HTS) for energy applications

Related titles:

Electricity transmission, distribution and storage systems
(ISBN 978-1-84569-784-6)
Advanced electricity transmission and distribution network technologies are required to reduce losses, sustain higher network efficiency, and maintain power quality and security. Research in this field must also consider the increasing role and penetration of renewable and distributed generation, for which advanced electricity storage systems are required. This book presents a comprehensive review of the materials, architecture and performance of electricity transmission and distribution networks, and the application and integration of electricity storage systems.

High temperature superconductors
(ISBN 978-1-84569-578-1)
High temperature superconductors reviews growth techniques, properties and applications of key high temperature superconducting films. The first part of the book provides an overview of high temperature superconductor properties such as optical conductivity and transport properties. Part II reviews the growth and properties of particular types of superconducting film, while Part III describes how they can be applied in practice in various areas of electronics.

Stand-alone and hybrid wind energy systems
(ISBN 978-1-84569-527-9)
Wind power generation is fast becoming one of the leading renewable energy sources worldwide, with increasing penetration of stand-alone and hybrid wind energy systems, particularly in distributed, isolated, and community power networks. Advanced energy storage and grid integration systems are required to provide secure, reliable power supply to the end user. This book provides an extensive reference on the development of stand-alone and hybrid wind energy systems, as well as energy storage and building-/grid-integration systems.

Details of these and other Woodhead Publishing materials books can be obtained by:

- visiting our web site at www.woodheadpublishing.com
- contacting Customer Services (e-mail: sales@woodheadpublishing.com; fax: +44 (0) 1223 832819; tel.: +44 (0) 1223 499140 ext. 130; address: Woodhead Publishing Limited, 80 High Street, Sawston, Cambridge CB22 3HJ, UK)
- contacting our US office (e-mail: usmarketing@woodheadpublishing.com; tel.: (215) 928 9112; address: Woodhead Publishing, 1518 Walnut Street, Suite 1100, Philadelphia, PA 19102–3406, USA)

If you would like to receive information on forthcoming titles, please send your address details to: Francis Dodds (address, tel. and fax as above; e-mail: francis.dodds@woodheadpublishing.com). Please confirm which subject areas you are interested in.

Woodhead Publishing Series in Energy: Number 27

High temperature superconductors (HTS) for energy applications

Edited by
Ziad Melhem

WOODHEAD PUBLISHING

Oxford Cambridge Philadelphia New Delhi

Published by Woodhead Publishing Limited,
80 High Street, Sawston, Cambridge CB22 3HJ, UK
www.woodheadpublishing.com

Woodhead Publishing, 1518 Walnut Street, Suite 1100, Philadelphia, PA 19102-3406, USA

Woodhead Publishing India Private Limited, G-2, Vardaan House, 7/28 Ansari Road, Daryaganj, New Delhi – 110002, India
www.woodheadpublishingindia.com

First published 2012, Woodhead Publishing Limited

British Library Cataloguing in Publication Data
A catalogue record for this book is available from the British Library.

Library of Congress Control Number: 2011919188

ISBN 978-0-85709-012-6 (print)
ISBN 978-0-85709-529-9 (online)
ISSN 2044-9364 Woodhead Publishing Series in Energy (print)
ISSN 2044-9372 Woodhead Publishing Series in Energy (online)

Typeset by RefineCatch Limited, Bungay, Suffolk

Contents

Contributor contact details

(* = main contact)

Editor

Dr Ziad Melhem
Oxford Instruments NanoScience
Tubney Woods
Abingdon
Oxon
OX13 5QX
UK
E-mail: ziad.melhem@oxinst.com

Chapter 1

Dr Ken Marken
Superconductivity Technology Center
Los Alamos National Laboratory
Mail Stop K763
PO Box 1663
Los Alamos
New Mexico 87545
USA
E-mail: kmarken@lanl.gov

Chapter 2

Professor Venkat Selvamanickam
Department of Mechanical Engineering and Texas Centre for Superconductivity
University of Houston
4800 Calhoun
Houston
TX 77204-4006
USA
E-mail: selva@uh.edu

Chapter 3

Dr Sonja I. Schlachter*
Karlsruhe Institute of Technology (KIT)
Institute for Technical Physics
Hermann-von-Helmholtz-Pl. 1
76344 Eggenstein-Leopoldshafen
Germany
Email: sonja.schlachter@kit.edu

Dr Wilfried Goldacker
Karlsruhe Institute of Technology (KIT)
Institute for Technical Physics
Hermann-von-Helmholtz-Pl. 1
76344 Eggenstein-Leopoldshafen
Germany
E-mail: wilfried.goldacker@kit.edu

Chapter 4

Dr Tim Coombs
University of Cambridge
Department of Engineering
Electrical Engineering Division
9 JJ Thomson Ave
Cambridge
CB3 0FA
UK
E-mail: tac1000@cam.ac.uk

Chapter 5

Professor Harry Jones
Head of Magnet Group
Clarendon Laboratory
Department of Physics
University of Oxford
Parks Road
Oxford
OX1 3PU
UK
E-mail: h.jones1@physics.ox.ac.uk

Chapter 6

Dr Probir K. Ghoshal
Principal Engineer – Magnets Design
Consultancy Group, Oxford Instruments NanoScience
Tubney Woods, Abingdon
Oxfordshire
OX13 5QX
UK
E-mail: pk_ghoshal@yahoo.com; Probir.Ghoshal@oxinst.com

Chapter 7

Dr Fedor Gömöry
Institute of Electrical Engineering
Slovak Academy of Sciences
Dúbravská cesta 9
841 01 Bratislava
Slovak Republic
E-mail: fedor.gomory@savba.sk

Chapter 8

Dr William Hassenzahl
Advanced Energy Analysis
USA
E-mail: advenergy1@aol.com

Chapter 9

Professor Pascal Tixador
Grenoble INP/Institut Néel – G2Elab
CNRS Batiment E.B.P. 166
38 042 Grenoble Cedex 09
France
E-mail: pascal.tixador@grenoble.cnrs.fr

Chapter 10

Dr Ruben Fair
Superconducting Projects Manager
Advanced Technology Group
Converteam UK Ltd
Leicester Road
Rugby
Warwickshire
CV21 1BD
UK
E-mail: ruben.fair@talktalk.net

Chapter 11

Dr Jean-Luc Duchateau
CEA
IRFM
F-13108 St-Paul-les-Durance Cedex
France
E-mail: jean-luc.duchateau@cea.fr

Chapter 12

Dr Nikolaj Zangenberg
Centre for Materials Testing
Danish Technological Institute
Kongsvang Allé 29
DK-8000 Aarhus C
Denmark
E-mail: nikolaj.zangenberg@teknologisk.dk; nzg@teknologisk.dk; nzg@dti.dk

Woodhead Publishing Series in Energy

1 **Generating power at high efficiency: Combined cycle technology for sustainable energy production**
Eric Jeffs
2 **Advanced separation techniques for nuclear fuel reprocessing and radioactive waste treatment**
Edited by Kenneth L. Nash and Gregg J. Lumetta
3 **Bioalcohol production: Biochemical conversion of lignocellulosic biomass**
Edited by K. W. Waldron
4 **Understanding and mitigating ageing in nuclear power plants: Materials and operational aspects of plant life management (PLiM)**
Edited by Philip G. Tipping
5 **Advanced power plant materials, design and technology**
Edited by Dermot Roddy
6 **Stand-alone and hybrid wind energy systems: Technology, energy storage and applications**
Edited by J. K. Kaldellis
7 **Biodiesel science and technology: From soil to oil**
Jan C. J. Bart, Natale Palmeri and Stefano Cavallaro
8 **Developments and innovation in carbon dioxide (CO_2) capture and storage technology Volume 1: Carbon dioxide (CO_2) capture, transport and industrial applications**
Edited by M. Mercedes Maroto-Valer
9 **Geological repository systems for safe disposal of spent nuclear fuels and radioactive waste**
Edited by Joonhong Ahn and Michael J. Apted
10 **Wind energy systems: Optimising design and construction for safe and reliable operation**
Edited by John D. Sørensen and Jens N. Sørensen
11 **Solid oxide fuel cell technology: Principles, performance and operations**
Kevin Huang and John Bannister Goodenough
12 **Handbook of advanced radioactive waste conditioning technologies**
Edited by Michael I. Ojovan
13 **Nuclear safety systems**
Edited by Dan Gabriel Cacuci

14 **Materials for energy efficiency and thermal comfort in buildings**
Edited by Matthew R. Hall
15 **Handbook of biofuels production: Processes and technologies**
Edited by Rafael Luque, Juan Campelo and James Clark
16 **Developments and innovation in carbon dioxide (CO_2) capture and storage technology Volume 2: Carbon dioxide (CO_2) storage and utilisation**
Edited by M. Mercedes Maroto-Valer
17 **Oxy-fuel combustion for power generation and carbon dioxide (CO_2) capture**
Edited by Ligang Zheng
18 **Small and micro combined heat and power (CHP) systems: Advanced design, performance, materials and applications**
Edited by Robert Beith
19 **Advances in clean hydrocarbon fuel processing: Science and technology**
Edited by M. Rashid Khan
20 **Modern gas turbine systems: High efficiency, low emission, fuel flexible power generation**
Edited by Peter Jansohn
21 **Concentrating solar power (CSP) technology: Developments and applications**
Edited by Keith Lovegrove and Wes Stein
22 **Nuclear corrosion science and engineering**
Edited by Damien Féron
23 **Power plant life management and performance improvement**
Edited by John E. Oakey
24 **Direct-drive renewable energy systems**
Edited by Markus Mueller and Henk Polinder
25 **Advanced membrane science and technology for sustainable energy and environmental applications**
Edited by Angelo Basile and Suzana Pereira Nunes
26 **Irradiation embrittlement of reactor pressure vessels (RPVs) in nuclear power plants**
Edited by Naoki Soneda
27 **High temperature superconductors (HTS) for energy applications**
Edited by Ziad Melhem
28 **Infrastructure and methodologies for the justification of nuclear power programmes**
Edited by Agustín Alonso
29 **Waste to energy (WtE) conversion technology**
Edited by Marco Castaldi
30 **Polymer electrolyte membrane and direct methanol fuel cell technology Volume 1: Fundamentals and performance of low temperature fuel cells**
Edited by Christoph Hartnig and Christina Roth
31 **Polymer electrolyte membrane and direct methanol fuel cell technology Volume 2: *In situ* characterisation techniques for low temperature fuel cells**
Edited by Christoph Hartnig and Christina Roth
32 **Combined cycle systems for near-zero emission power generation**
Edited by Ashok Rao
33 **Modern earth buildings: Materials, engineering, construction and applications**
Edited by Matthew R. Hall, Rick Lindsay and Meror Krayenhoff

34 **Handbook of metropolitan sustainability: Understanding and improving the urban environment**
Edited by Frank Zeman
35 **Functional materials for energy applications**
Edited by John Kilner, Stephen Skinner, Stuart Irvine and Peter Edwards
36 **Nuclear decommissioning: Planning, execution and international experience**
Edited by Michele Laraia
37 **Nuclear fuel cycle science and engineering**
Edited by Ian Crossland
38 **Electricity transmission, distribution and storage systems**
Edited by Ziad Melhem
39 **Advances in biodiesel production: Processes and technologies**
Edited by Rafael Luque and Juan Antonio Melero
40 **Biomass combustion science, technology and engineering**
Edited by Lasse Rosendahl
41 **Ultra-supercritical coal power plant: Materials, technologies and optimisation**
Edited by Dongke Zhang
42 **Radionuclide behaviour in the natural environment: Science, impacts and lessons for the nuclear industry**
Edited by Horst Geckeis and Christophe Poinssot
43 **Calcium and chemical looping technology for power generation and carbon dioxide (CO_2) capture: Solid oxygen- and CO_2-carriers**
P. Fennell and E. J. Anthony
44 **Materials ageing and degradation in light water reactors: Mechanisms, modelling and mitigation**
Edited by K. L. Murty
45 **Structural alloys for power plants: Operational challenges and high-temperature materials**
Edited by Amir Shirzadi, Rob Wallach and Susan Jackson
46 **Biolubricants: Science and technology**
Jan C. J. Bart, Emanuele Gucciardi and Stefano Cavallaro
47 **Wind turbine blade design and materials: Improving reliability, cost and performance**
Edited by Povl Brøndsted and Rogier Nijssen
48 **Radioactive waste management and contaminated site clean-up: Processes, technologies and international experience**
Edited by William E. Lee, Michael I. Ojovan, Carol M. Jantzen
49 **Probabilistic safety assessment for optimum nuclear power plant life management (PLiM)**
Gennadij V. Arkadov, Alexander F. Getman and Andrei N. Rodionov
50 **Coal utilization in industry**
Edited by D. G. Osborne
51 **Coal power plant materials and life assessment: Developments and applications**
Edited by Ahmed Shibli
52 **The biogas handbook: Science, production and applications**
Edited by Arthur Wellinger and David Baxter
53 **Advances in biorefineries: Biomass and waste supply chain exploitation**
Edited by K. W. Waldron

54 **Geoscience of carbon dioxide (CO_2) storage**
Edited by Jon Gluyas and Simon Mathias
55 **Handbook of membrane reactors Volume I: Fundamental materials science, design and optimisation**
Edited by Angelo Basile
56 **Handbook of membrane reactors Volume II: Industrial applications and economics**
Edited by Angelo Basile
57 **Alternative fuels and advanced vehicle technologies: Towards zero carbon transportation**
Edited by Richard Folkson
58 **Handbook of microalgal bioprocess engineering**
Christopher Lan and Bei Wang
59 **Fluidized-bed technologies for near-zero emission combustion and gasification**
Edited by Fabrizio Scala
60 **Managing nuclear projects: A comprehensive management resource**
Edited by Jas Devgun
61 **Handbook of process integration: Energy, water, waste and emissions management in processing and power industries**
Edited by Jiří Klemeš
62 **Membranes for clean and renewable power applications**
Edited by Annarosa Gugliuzza and Angelo Basile

This book is dedicated to my wife Salam and my boys Nabil and Tariq

Preface

Superconductivity was discovered in 1911 by H. Kamerlingh Onnes, and the scientific community celebrated the 100th year of the discovery in 2011. However, it is only in the last 50 years that superconducting materials have been used in industrial applications, and these applications have been dominated by superconducting magnets operating at low temperatures (i.e. usually below 20 K or −253°C). Since the discovery of high temperature superconductivity by J. Georg Bednorz and K. Alex Müller (1987), the last 25 years have seen the development of novel materials with much stronger market potential.

Almost all superconducting products today use low temperature superconductor (LTS) materials in their magnet systems, from compact magnets used in research laboratories, to large-scale systems in particle physics and fusion experiments, and also including superconducting magnets for scientific, pharmaceutical and materials research applications. Magnetic resonance imaging (MRI) for medical diagnosis dominates the sector. The majority of these LTS applications operate at liquid helium temperatures. High temperature superconductors (HTS) have the potential to facilitate a more fundamental change to many of these industrial applications and to electric power technologies.

In many ways, the transmission and distribution of electricity is poised for advancement via HTS materials in the way that the Internet was poised for its takeoff in the 1990s. Just as fibre optics enabled the 'information superhighway', by supplanting lower-capacity copper, superconductivity is enabling an 'energy superhighway' by supplanting copper electrical conductors with a ceramic superconducting alternative that has higher capacity while eliminating resistive losses.

There are four challenges that HTS materials have to overcome to be widely used: cost, refrigeration, reliability and acceptance. Modest growth rates are expected for the well-established businesses of superconductivity, which may reach about €5 billion in 2014. The contribution from HTS applications is anticipated at first to be in niche areas. This will continue until HTS see improved performance at about 65–75 K operating temperature, at which point the use of the liquid nitrogen cryogen will neutralise the impact of the cooling cost barrier.

First-generation (1G) HTS wires are now being produced in industrial lengths, and can potentially offer the basis for the development of ultra-high-field magnets of >25 T and reduce the need for high-field magnets using resistive coils. Production of second-generation (2G) materials such as YBCO tape and MgB2 wires has also increased during the last few years. A significant improvement in long-length performance and price is still required to meet application and commercial requirements.

I undertook this book project with the aim of providing a reference source for HTS materials and applications. The book provides a structured account of the fundamental principles of HTS materials and the range of applications that will benefit directly in the electric power sector. The book presumes a familiarity with the fundamentals of design and analysis of conventional superconducting applications such as conductor architecture, electromagnets, rotating machinery, power cables and superconducting magnetic energy storage. The intended audience for the book is scientists, engineers and managers in superconductor companies, the energy industry, and research laboratories, as well as academics and students from graduate level up, and anyone else interested in research and development and the industrial developmental challenges in this fascinating field.

The first part of the book provides detailed coverage of the science and engineering of HTS materials and components. Chapter 1 covers fundamental issues in HTS materials science and engineering, in particular the critical current density of superconductors in enabling applications and the impact of price and performance on adoption of HTS materials. Chapter 2 provides an overview on HTS wire and tape development, scale up and manufacturing, with a close look at progress in increasing the critical current densities and the increase in lengths of wires and tapes with high-performance current densities. Chapter 3 gives an overview of cabling techniques for HTS wires and tapes. HTS cables with high operating currents, in excess of 5000 A, are needed for applications such as rotating machinery, transformers and magnets with low inductance and AC losses. Chapter 4 introduces HTS bulk materials, another form of superconductor with the promise of providing high magnetic fields in a very compact package. Chapter 5 is concerned with electromagnets made wholly or partly with HTS in wire form. These magnets are primarily used for research, and measurements at specified magnetic fields. Chapter 6 reviews cryogenics for HTS, wet and dry system cooling methods, issues, challenges and developments. Chapter 7 covers electromagnetic modelling of HTS materials and applications. Such modelling is essential in the design of electric power applications.

The second part of the book reviews the application of HTS materials and components in energy systems. Chapter 8 deals with superconducting power cables and fault current limiters. These are two examples critical for the electric power industry, in particular in power transmission and distribution. Chapter 9 covers another power application: the emerging technology of superconducting magnetic energy storage (SMES). SMES is a particularly promising technology for high power and quick response applications. Chapter 10 introduces the application of HTS to rotating machinery for the electrical sector. An overview is provided together with a comparison between machine types, issues and challenges, in particular relating to the design, build, test and commissioning of such machines using HTS materials. Chapter 11 deals with HTS for fusion application. Currently, HTS use in fusion reactors is confined to current leads, reducing the cryogenic cost. Future reactors will benefit even further from using HTS for coils, which will simplify the cryogenic system requirements and reduce significantly the overall cost of future fusion machines. Finally, Chapter 12 presents HTS for accelerators and the challenges concerning the use of HTS for various accelerator components.

I hope that the contents presented in this book will help the reader to appreciate the benefits that HTS technology can provide to various industrial, scientific, energy and medical applications. The adoption and acceptance of new technologies such as HTS is usually slow, and can be frustrating due to user concerns over technological risks; and hence the need for fully proven characterisation and reliability. It is important that scientists and engineers continue to develop HTS technology, to improve materials performance and to demonstrate cost-effectiveness, as HTS has the potential to enable a step change in many applications that will affect our quality of life.

This idea behind this book originated following a workshop on HTS materials and applications held in January 2010 at the Institute of Physics (IOP), London, which was sponsored by Oxford Instruments, SuperPower Inc. and the Applied Superconductivity group of the IOP. Many of the chapter contributors gave invited talks at that workshop, and I would like to thank all of my colleagues who made that workshop a successful event. I would also like to thank all of the chapter authors, who have provided excellent contributions in their respective areas of expertise. Finally, though most important, thanks to Ian Borthwick, who invited me to take on board this project and shared many of his thoughts and advice, Sarah Lynch for her patience, to Cathryn Freear and to the staff at Woodhead Publishing for their support and guidance in bringing this important book to market.

Ziad Melhem

Part I

Fundamental science and engineering of high temperature superconductor (HTS) materials and components

1

Fundamental issues in high temperature superconductor (HTS) materials science and engineering

K. MARKEN, Los Alamos National Laboratory, USA

Abstract: A key factor for enabling high temperature superconductor (HTS) applications is the critical current density (J_c). The basic physics determining J_c is presented, along with basic concepts of the materials science of engineered flux-pinning defects, which provide tools to increase J_c in a superconductor. A history of the development of HTS materials concludes with comparison of processing and performance in three practical materials. Other key properties that enable applications are discussed, including economic considerations. Present status and price/performance trends indicate a wane in importance of BSCCO materials while YBCO-coated conductors seem to be growing in importance. The significant challenges that remain for the technical and commercial success of YBCO are outlined.

Key words: critical current density, flux pinning, BiSrCaCuO, YBaCuO, dollars per kiloamp-meter.

1.1 Introduction

The field of superconductivity is rich and deep both theoretically and experimentally, and there is vast detail beyond the scope of a brief introduction. This chapter focuses on an understanding of those characteristics and challenges in these materials that are important for enabling practical applications. Foremost of these is the critical current density (J_c), which is a key factor for almost all energy applications. Section 1.2 describes the basic properties of superconductors that influence and determine J_c. Section 1.3 provides a brief history of the various chemical families of HTS and moves quickly to a discussion of J_c in practical materials, including the importance of this key performance parameter for economic viability of HTS. A short list of other properties important for applications is discussed to provide context for further explanation in the applications chapters. Section 1.4 outlines the predominant manufacturing methods for the HTS materials that have become important for the range of applications explored in this book.

1.2 Fundamental superconductor physics

The twin super properties of superconductors – perfect conductivity and perfect diamagnetism – have entertained and employed physicists, chemists and materials scientists for nearly a century. Perfect conductivity, discovered in mercury by

Kammerlingh Onnes at Leiden University in 1911 (Onnes, 1911), prompted the name for this class of materials. In 1933 at the Physikalisch-Technische Bundesanstalt in Berlin, Walther Meissner and Robert Ochsenfeld observed the complete expulsion of magnetic flux from superconducting mercury and lead samples in an external field (Meissner and Oschenfeld, 1933), a phenomenon referred to as the Meissner Effect. While both properties have been exploited for practical uses, it must be said that perfect conductivity has provided the basis for the commercial success of superconductors. There are numerous superconductivity textbooks that cover these fundamental physical properties in detail and several are listed in Section 1.7. In this section we take a practical approach and describe characteristics of superconductors with the aim of understanding their application.

1.2.1 Condensed phase state and critical surface

Figure 1.1 illustrates a superconducting critical surface in temperature–magnetic field–current density phase space. When (T, H, J) are less than (T_c, H_c, J_c) the material is in the superconducting state, but if any of these parameters exceeds a critical value (denoted by subscript c), the material enters the normal state. The transition from normal to a superconducting state is a thermodynamic one, with a difference in energy equal to $\mu_0 H_c^2/2$, which is the condensation energy. H_c is the thermodynamic critical field. Strictly speaking the phase transition along the current density axis is not a thermodynamic one, but we will treat that case later

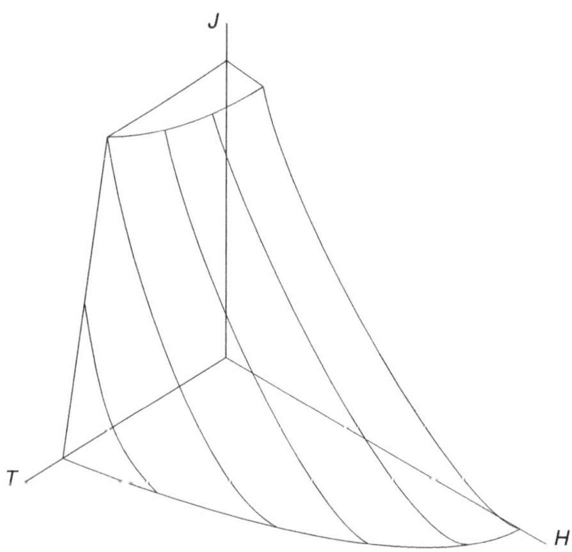

1.1 Critical surface of the superconducting phase state in temperature (*T*), current density (*J*) and applied field (*H*); the superconducting state exists below the surface and the normal state outside.

in Section 1.2.4. In practice the critical current density is a key property for many applications.

Superconductivity is a quantum mechanical phenomenon on a macroscopic scale. The BCS theory, which earned the 1972 Nobel Prize in Physics for Bardeen, Cooper and Schrieffer (BCS), describes a condensed state of electrons with long range order (Bardeen *et al.*, 1957). This correlation was described in terms of Cooper pairs, two electrons coupled through their mutual interaction with phonons (atomic lattice vibrations). BCS theory is accepted as an accurate description of superconductivity in metals, which are also called low-temperature superconductors (LTS). While the electron-phonon mechanism does not explain high temperature superconductivity, and there is not yet agreement on the mechanism, high temperature superconductivity certainly arises from a strongly correlated electronic state and has much in common with LTS including a critical surface in temperature, field and current density.

1.2.2 Perfect conductivity

Figure 1.2 compares resistivity as a function of temperature in a normal metal and a superconductor. Electron scattering from phonons in a normal metal decreases with temperature, since thermal lattice vibration decreases. Normal state resistivity saturates at a non-zero value determined by scattering from impurities in the

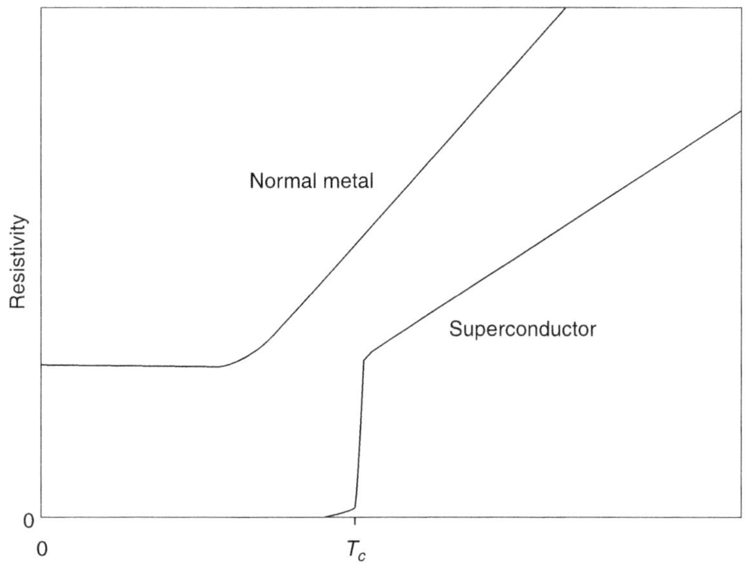

1.2 Comparison of the temperature dependence of resistivity in normal and superconducting metals.

metal. Similar behavior occurs in a superconductor in the normal state, that is for $T > T_c$. However at T_c the material enters the condensed electron state (superconducting phase state) which enables truly zero resistance for direct current. We should note that this is not strictly true for alternating currents. The reasons for dissipation from AC currents in superconductors will be explained in Section 1.2.4.

1.2.3 Perfect diamagnetism

The magnetization (M) of a material is the induced magnetic moment per unit volume in response to an applied magnetic field (H). Figure 1.3 illustrates the magnetic response of a Type I superconductor: the magnetization is equal and opposite to the applied field. As a consequence the magnetic flux density B inside the superconductor, $B = \mu_0(H + M)$, is truly zero. This expulsion of all magnetic field, called the Meissner Effect, is perfect diamagnetism. For many years after the discovery of the Meissner Effect this was considered a fundamental property of all superconductors.

Shubnikov *et al.* (1937) succeeded in making superconducting alloys that did not respond as shown in Fig. 1.3, but it was not until Abrikosov (1957) theorized there could exist a different magnetic response in a superconductor that the Shubnikov observations were explained. Abrikosov coined the terms Type I and Type II to distinguish two different behaviors. Figure 1.4 illustrates Type II behavior. A Type II superconductor remains a perfect diamagnet up to its lower critical field (H_{c1}). It remains superconducting up to a much higher field called the

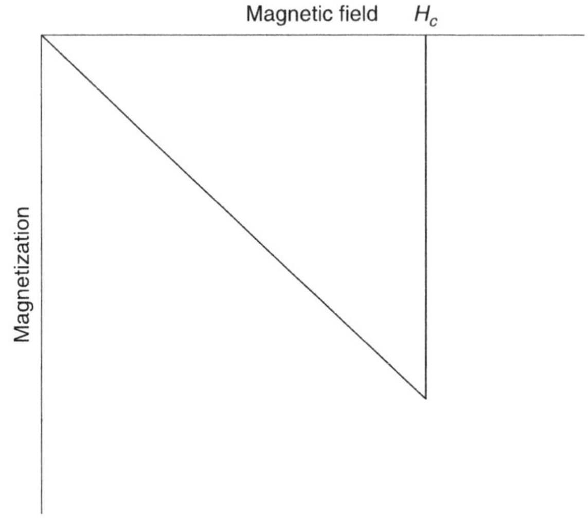

1.3 Magnetization as a function of applied magnetic field in a Type I superconductor.

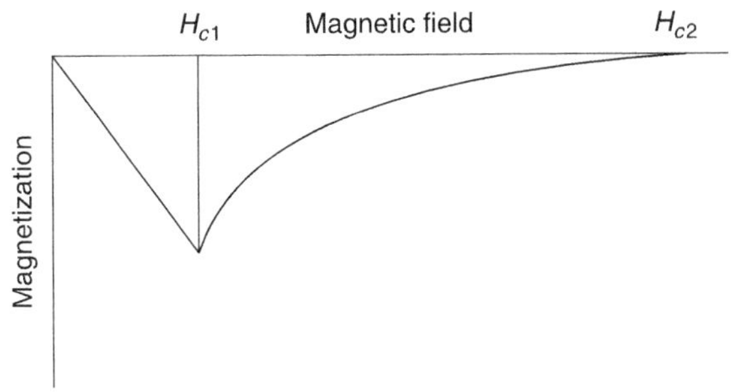

1.4 Magnetization as a function of applied magnetic field in a Type II superconductor.

upper critical field (H_{c2}). Between H_{c1} and H_{c2} it is still a diamagnet, but not a perfect one. Some magnetic flux is shielded, and some penetrates. As the applied field value increases, the density of fluxons increases until at H_{c2} there is full field penetration and the material transitions to the normal state.

Only pure elements have been identified as Type I superconductors. Approximately 30 elements display this behavior, all with T_c values below 7.2 K (the T_c for pure Pb), and most below 3 K. Type I superconductors have little practical value because their thermodynamic critical fields are quite small, typically less than 0.1 T. Type II superconductors can have quite high values for H_{c2}, making them much more practical materials. All commercially useful superconductors are Type II. An exhaustive table of properties for several thousand superconducting alloys and compounds discovered prior to 1976, including T_c, H_c, H_{c1} and H_{c2} where known, was compiled by Roberts (1976). An extensive compilation of T_c and H_{c2} values for HTS materials can be found in Poole (2000).

1.2.4 Mixed state, vortices and critical current

Originally called the Shubnikov phase, the regime between H_{c1} and H_{c2} where there is partial flux penetration is now conventionally referred to as the mixed state. Figure 1.5 illustrates this mixed state in a slab of Type II superconductor. Abrikosov's theory showed that flux penetrates in discrete units called fluxons, which are threaded through single cores of normal material. These cores are called vortices due to associated circulating currents. As predicted, a regularly spaced vortex lattice forms, which has since been observed experimentally by a variety of techniques. One example is shown in Fig. 1.6, which clearly demonstrates a lattice spacing inversely proportional to H; the vortex lattice spacing goes to zero at H_{c2}, which coincides with full flux penetration and the transition to the normal state.

When a transport current is applied to a Type II superconductor in the mixed state, there is a resultant Lorentz force $F_L = J \times B$, where J is the current density and

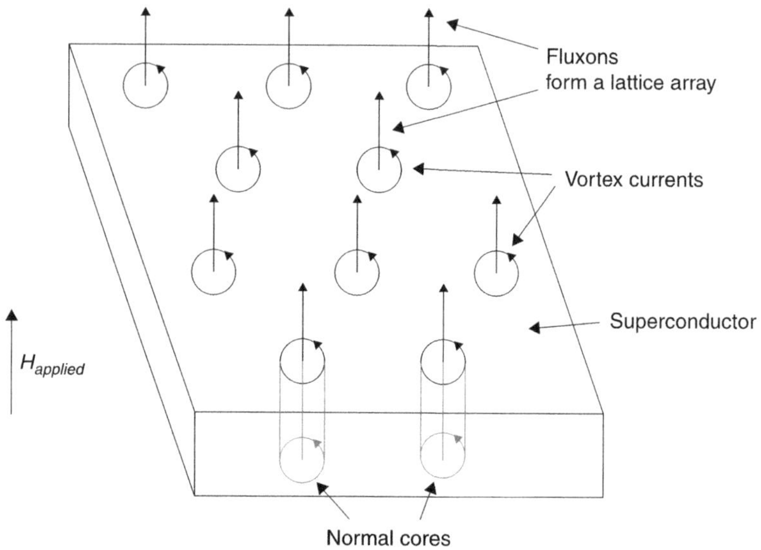

1.5 The mixed state in a Type II superconductor showing normal cores surrounded by vortex currents and threaded by fluxons.

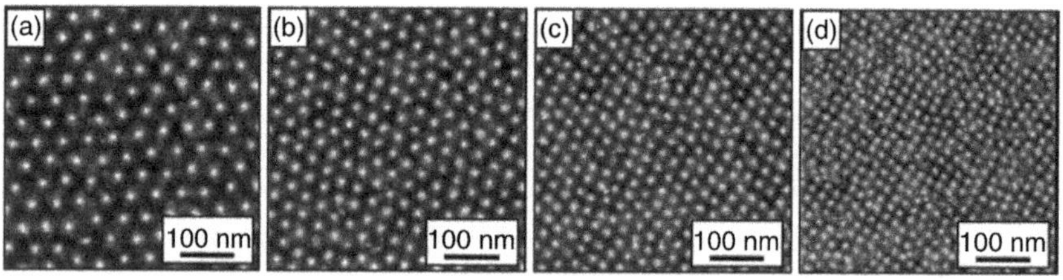

1.6 Images of the vortex lattice in V_3Si at 2.3 K taken at applied field values of (a) 1 T, (b) 2 T, (c) 3 T, (d) 5 T. (Source: Sosolik *et al.* 2003. Copyright (2003) American Physical Society.)

B is the magnetic flux density. This vector force is perpendicular to both the current and the vortices, acting transversely on the vortices. If there is no restraining force on the vortices they will move through the material, a phenomenon called flux flow, which causes dissipation. It is the onset of flux flow that determines J_c, the critical current density. Figure 1.7 shows a typical voltage-current characteristic with a sudden onset of voltage due to flux flow. The critical current I_c is typically determined using an electric field criterion, for example 1 μV/cm across the current carrying sample. J_c is then simply I_c/A where A is the cross-sectional area of the superconductor.

During the synthesis of practical materials there are defects deliberately added for the purpose of restraining, or pinning, the vortex lattice. The stronger the interaction between the defects and the vortices, and the closer the match between

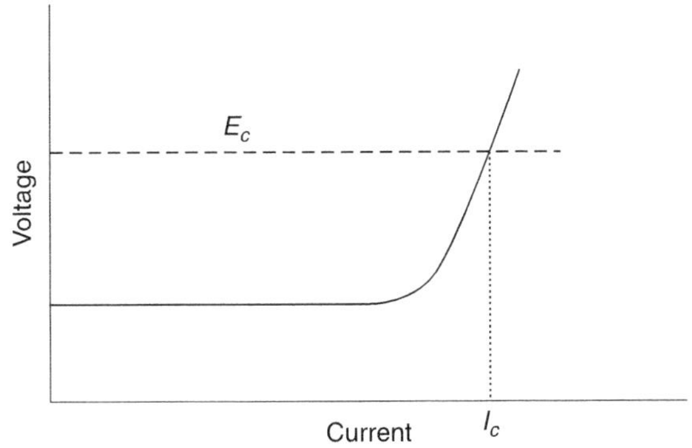

1.7 Voltage–current characteristic in a Type II material showing an electric field criterion (E_c) that defines I_c.

the defect spacing and the vortex spacing, the higher is the resultant pinning force (F_p). J_c then is determined by F_p, for flux flow begins when F_L exceeds F_p. Flux pinning is the key to engineering practical superconductors with high J_c, and enormous effort continues to be devoted to the science and engineering of understanding and controlling flux pinning in Type II superconductors.

Figure 1.8 shows a full magnetic hysteresis loop for a Type II superconductor. As H is increased or decreased, vortices move in or out of the material. The

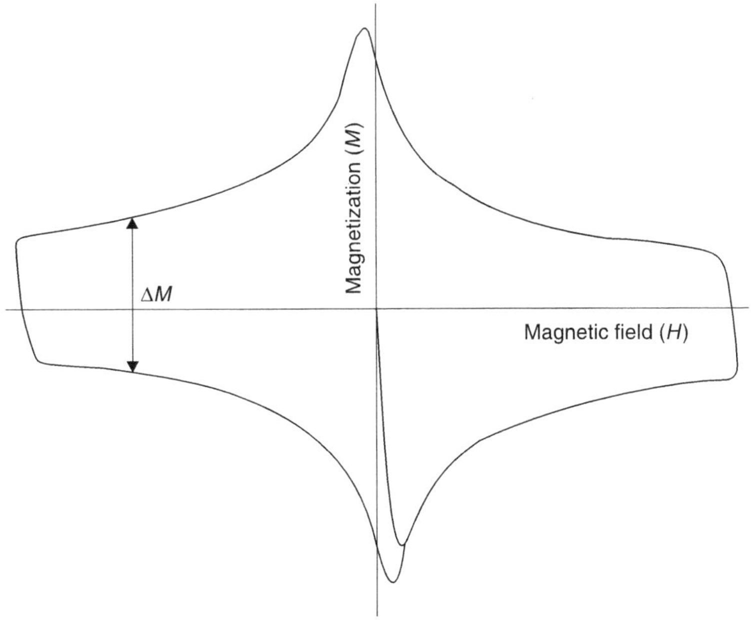

1.8 Full *M*(*H*) hysteresis loop in a Type II superconductor with substantial flux-pinning defects; ΔM is proportional to J_c.

higher the pinning force, the greater the energy required to move them. In fact the difference in M between the field increasing and field decreasing arms of the loop is proportional to J_c, and measuring $M(H)$ in a superconducting sample is sometimes used to determine J_c. This figure is shown here however to illustrate the total energy loss, represented by the area enclosed by the M(H) loop, in a time-varying field. This dissipation is the origin of the AC loss referred to in Section 1.2.2.

1.2.5 HTS vortex phase diagram

Although called 'high temperature superconductors' because of the enormously higher values of T_c that have been obtained in these materials, the high values of H_{c2} that many of these new materials reach is equally as exciting and perhaps just as important for applications of these materials. However there are key differences between LTS and HTS materials in behavior of vortices, particularly at high temperatures and in fields. Figure 1.9 compares the vortex phase diagrams for LTS and HTS materials. LTS materials remain in the mixed state with J_c determined by flux pinning at fields all the way to H_{c2}. In contrast, HTS materials have a vortex liquid phase at high H and T, partly as a result of the much greater thermal energy at these very high T_c values. The boundary between the vortex solid and vortex liquid states is called the irreversibility field (H_{irr}) (or sometimes the melting field H_m), and rather than H_{c2}, it is H_{irr} that determines the upper limit for J_c in HTS materials. It is interesting that the type of defect engineering that has gone on to increase F_p and J_c has been shown to also affect H_{irr} (Miura *et al.*, 2010). This is an active field of study that holds promise for future improvements in HTS materials.

1.3 Types, properties and performance of HTS materials

1.3.1 A brief history of high temperature superconductivity

In the middle 1980s applied superconductivity research and development was a healthy field focused on optimizing performance in the low temperature superconductors NbTi and Nb_3Sn. The main applications pull was from the planned Superconducting Supercollider's need for higher current density in NbTi to achieve 6.6 T dipole magnets, and the fusion energy science demand for higher current density in Nb_3Sn to enable fields above 12 T for magnetic confinement reactors. The proceedings of the 1986 Applied Superconductivity Conference (IEEE, 1987) runs to 1420 pages and bear witness to the vigor of LTS research and development at the time. They include the record of a session commemorating the 75th anniversary of superconductivity, but make no mention

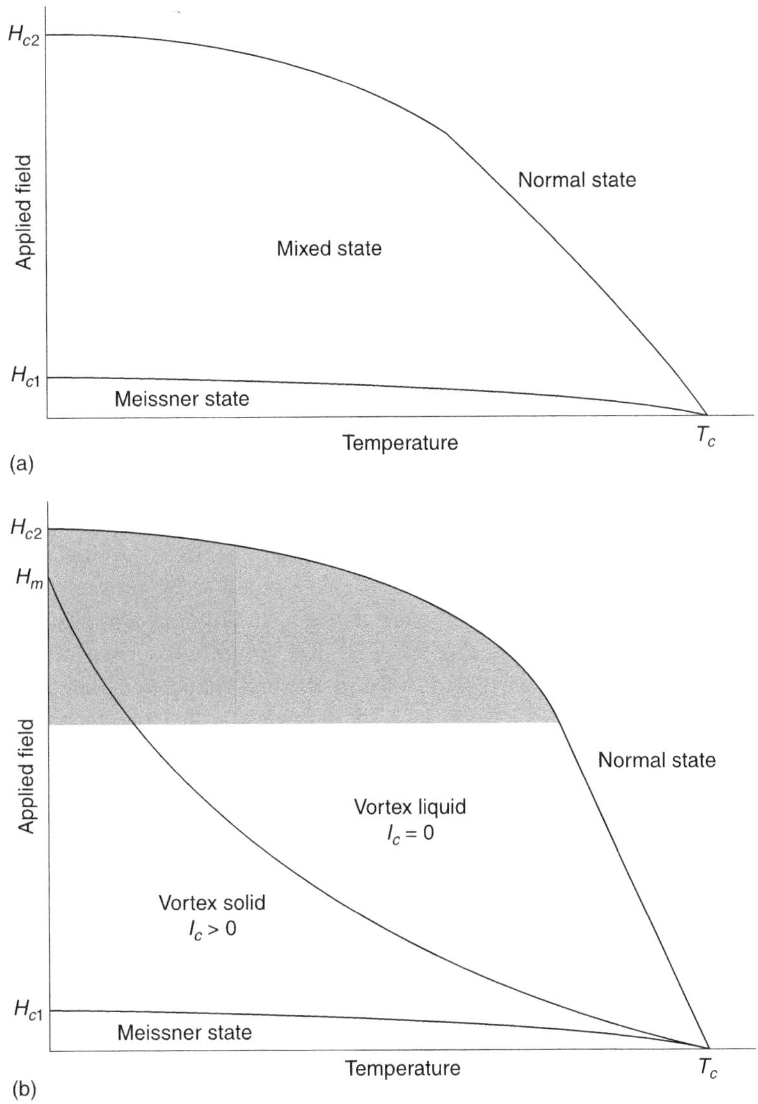

1.9 Comparison of field–temperature vortex phase diagrams in (a) LTS and (b) HTS materials. The shaded region at high field for high temperature superconductivity is speculative since this field regime cannot be experimentally reached.

of an obscure June 1986 publication with the intriguing title 'Possible high T_c superconductivity. . .' (Bednorz and Muller, 1986).

The start of the HTS era is dated to that publication, which announced a newly discovered material LaBaCuO with a T_c near 30 K, and which garnered the 1987 Nobel Prize in Physics for IBM scientists Georg Bednorz and Alex Muller. After a lag of a few months this report commanded the world's attention and there was a literal explosion in the level of effort and in published work exploring the new field of oxide superconductors with high values of T_c. At IBM alone there were 38 papers submitted within a year of the June 1986 announcement (Malozemoff and Grant, 1987). The March 1987 meeting of the American Physical Society in New York City added a special late-breaking evening session devoted to the new materials. An estimated 2000 scientists (including the author) tried to squeeze into a single ballroom at the New York Hilton; many ended up watching on closed circuit television in nearby rooms, while many were unable to view the talks because of the unanticipated numbers of attendees. The session, later dubbed the 'Woodstock of Physics', ended up lasting most of the night; the palpable buzz of excitement and anticipation on that night was indeed akin to a rock concert. The proceedings of the 1988 Applied Superconductivity Conference contain more than 150 papers devoted to high temperature superconductivity, and the attendance at that August 1988 meeting in San Francisco was 50% higher than the 1986 meeting. Clearly a revolution was underway.

Paul Chu of the University of Houston is widely credited with discovery of the first material with a T_c above the 77 K boiling point of liquid nitrogen. Collaborative work between Chu's group and M. K. Wu's group at the University of Alabama was published in a March 1987 paper that described the onset of superconductivity in YBaCuO above 90 K (Wu *et al.*, 1987). There are several accounts of the secrecy and controversy regarding the review and publication of this paper, due to the wildly competitive work on HTS discovery at the time (see, for example, Blundell, 2009). A related fierce patent battle ensued between the University of Houston, IBM and Bell Labs regarding applications all filed very early in 1987. The US patent for composition and structure of YBaCuO was not granted until October 2003 (Battlog *et al.*, 2003) to Lucent Technologies, the successor to AT&T Bell Labs, although IBM did obtain as early as 1995 a US patent regarding processing of YBCO and other HTS materials (Chaudhari *et al.*, 1995). It is interesting to note that IBM did obtain a European patent regarding bulk YBCO composition and processing in 1994 (Beyers *et al.*, 1994).

Figure 1.10 illustrates both the revolutionary leap in T_c that occurred around 1987 and the explosion in number of new materials over a very short period of time. In fact most of the materials on the chart are each but one of a family of related compounds. For example the so-called '123 compound' shown as YBaCuO forms not only with Y but with many other rare earth elements including Nd, Sm, Eu, Gd, Dy, Er, Tm and Yb. All of the HTS materials shown are called cuprate materials and share a common feature of one or more CuO_2 layers in the atomic cell. There are many excellent treatments of the structures of HTS cuprate materials, including Chapter 8 in Poole *et al.* (2007). While non-cuprate families of HTS materials such as the pnictides have since been discovered, the highest T_c values and the only practical HTS materials to date remain among the cuprates. Much of the excitement and motivation toward applications of high temperature superconductivity was generated by the large number of materials with T_c

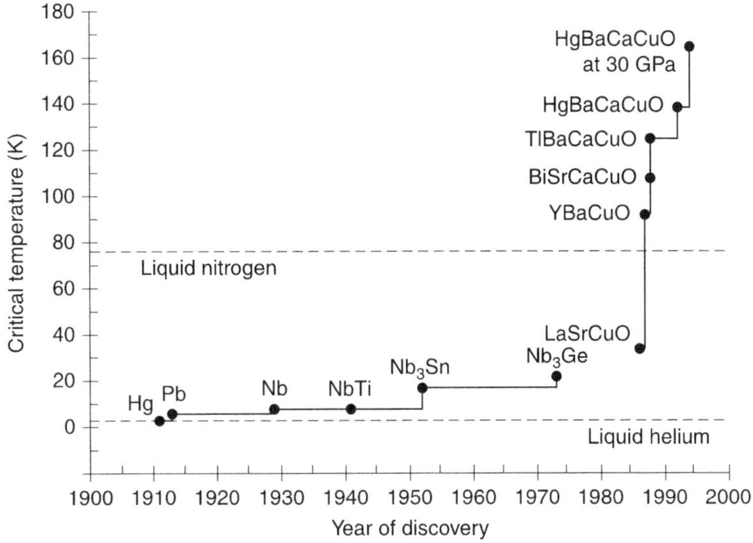

1.10 Critical temperature in selected LTS and HTS materials plotted against the year of discovery.

exceeding the 77 K boiling point of liquid nitrogen, as shown in Fig. 1.10. Prior to discovery of high temperature superconductivity, nearly all applications of superconductivity involved cooling with liquid helium. In 2011 the price of liquid helium was 25–100 times higher than liquid nitrogen (depending on quantities and delivery arrangements). Helium is limited in supply, with the commercial supply extracted from natural gas wells, and for the past 10 years the price has been rising. On the other hand 78% of Earth's atmosphere is nitrogen, which can be economically liquefied from air; it is reasonable to expect liquid nitrogen to be available at low cost indefinitely. Realizing the advantages of superconducting devices at liquid nitrogen temperatures is indeed a revolutionary opportunity.

About one year after the discovery of the LaBa- and YBa-cuprates, Hiroshi Maeda and co-workers at the National Research Institute for Metals in Japan reported discovery of multiple high temperature superconductivity phases in BiSrCa-cuprates (Maeda *et al.* 1988, Matsui *et al.* 1988). Both $Bi_2Sr_2Ca_2Cu_3O_x$ (BSCCO-2223) with $T_c = 105$ K and $Bi_2Sr_2CaCu_2O_x$ (BSCCO-2212) with $T_c = 80$ K quickly became the most practical conductors for applications development, a point discussed further in Section 1.3.3. At about the same time Alan Hermann at the University of Arkansas reported superconductivity at 120 K in TlBaCa-cuprates (Sheng and Hermann, 1988). Later it was shown that under pressure the Tl-cuprates could reach $T_c = 132$ K (Berkley *et al.*, 1993). In 1993 HgBa-cuprates were first shown to have transition temperatures at 94 K (Putilin *et al.*, 1993), followed by a rapid proliferation of different HgBaCa-cuprate phases with higher temperatures. The record critical temperature for any superconductor is in

$HgBa_2Ca_2Cu_3O_x$ (also called Hg-1223) under 30 GPa of hydrostatic pressure, with reported $T_c = 164$ K (Gao *et al.*, 1994).

As a practical note about names, the IUPAC (International Union of Pure and Applied Chemistry) conventions often result in quite cumbersome sequences of characters for these complex multi-element superconducting cuprates. Common usage in the community is a minimalist shorthand to denote elements and sometimes chemical stoichiometry. Common designations for $YBa_2Cu_3O_x$ are 'YBCO' and 'Y-123'; $Bi_2Sr_2CaCu_2O_x$ and $Bi_2Sr_2Ca_2Cu_3O_x$ are frequently BSCCO-2212 and BSCCO-2223, or even simply Bi-2212 and Bi-2223. RE denotes a rare earth element and Ln a lanthanide element.

1.3.2 Anisotropy and grain connectivity

All of the cuprate superconductors exhibit a high degree of anisotropy in crystal structure (Poole *et al.*, 2007) and in fundamental physical properties including superconducting critical current and critical field (Saxena, 2010). For example, Poole (2007) reported that YBCO crystal lattice spacing in the a-b plane is 3.85 and along the c-axis is 11.68, an aspect ratio greater than 3. (These values are for stoichiometric oxygen and vary somewhat with oxygen content.) The structural anisotropy is even greater in BSCCO, with a lattice spacing aspect ratio of 8 in Bi-2212 and 9.7 in Bi-2223. Figure 1.11 shows the large difference in YBCO J_c

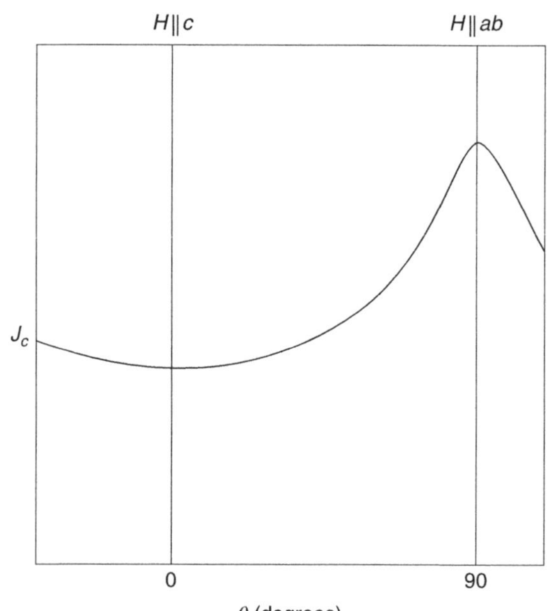

1.11 Typical J_c variation with the angle θ between applied field and sample orientation in YBCO; $H \parallel ab$ denotes field parallel to the plane of the film.

between external field orientations parallel to the *a-b* crystal plane and perpendicular to the plane. This anisotropy in current has enormous implications for applications of these materials, which in many cases involve large external fields.

A serious consequence of the anisotropy in J_c is that, in a polycrystalline superconductor, there can be significant limitations on current flow if all of the grains are not aligned with each other, and if they are not well aligned with the current flow direction. Figure 1.12 from Feldmann *et al.* (2008) is a clear example of this problem in YBCO. There is a precipitous drop in J_c for mis-orientation angles greater than 2°. In practical terms this means that to take advantage of the intrinsically high J_c in YBCO, any manufacturing technique must produce long lengths of conductor with grains aligned to within a few degrees. In YBCO bi-axial texture is required, i.e. not only must the crystallographic c-axes be aligned, but also the axes within the *a-b* plane. Chapter 2 will address this challenge in some detail.

It is interesting to note that in BSCCO materials the texture required for significant transport current is uniaxial (Scanlan *et al.*, 2004), a fact that enabled success in making conductors fairly quickly using powder-in-tube methods.

1.3.3 Practical conductor materials

Very soon after the discovery of BSCCO HTS materials, scientists at Vacuumschmelze GmbH in Germany made a Bi-2212 monofilament wire by

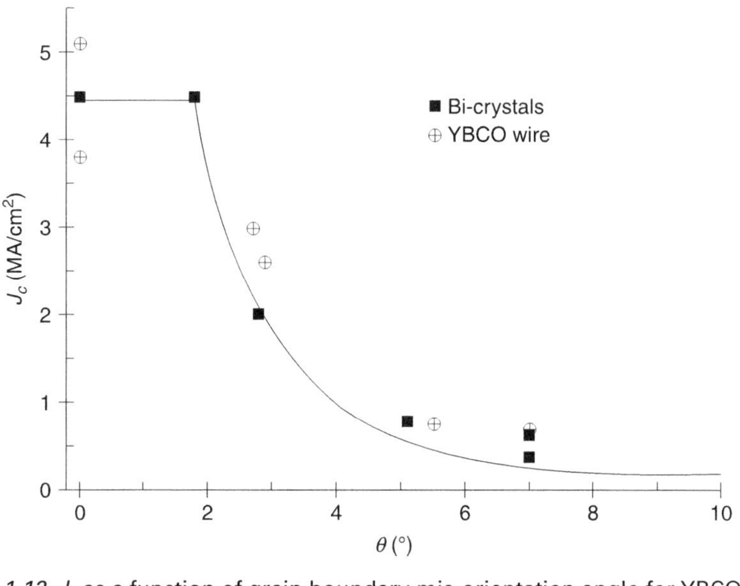

1.12 J_c as a function of grain boundary mis-orientation angle for YBCO deposited on bi-crystal substrates and for single grain boundaries in YBCO wire. The data are at 77 K self field (Feldmann *et al.*, 2008).

filling a silver tube with a powder of the appropriate stoichiometry followed by conventional wire drawing processes and a final heat treatment in air (Heine *et al.*, 1989). This first experimental wire resulted in the world's highest critical current density conductor in applied magnetic fields above 18 T, a quite extraordinary result. Figure 1.13 shows the original data from Heine *et al.* (1989), comparing the new material performance at 4.2 K with the best high-field LTS materials of the time. Not only did this report greatly accelerate research and development in BSCCO conductors, but it demonstrated the advantages of HTS materials for very-high-field applications.

Although this seminal demonstration was with Bi-2212, a significant worldwide effort to develop BSCCO conductors focused on Bi-2223 from 1989 until around 2005, primarily because of the 108 K T_c in the 2223 phase. (The 2212 phase T_c varies with oxygen content but is typically in the 80 K range, not sufficiently above the 77 K nitrogen boiling point to be an effective conductor with nitrogen cooling.) In the mid 1990s there were more than ten companies worldwide developing and producing Bi-2223 tape, all using powder-in-silver tube and mechanical deformation. The mechanical deformation process included substantial tape rolling, followed by carefully controlled heat treatment, which enabled the uniaxial texture needed for good superconducting performance. It is important to note that this process works only in a silver matrix, for two reasons: the oxide superconductor releases and takes-up oxygen during heat treatment, requiring a

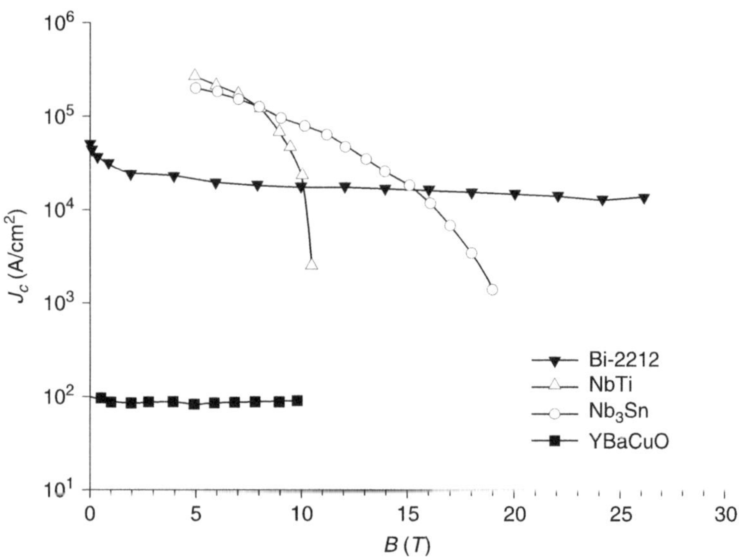

1.13 Early high-field J_c results in B-2212 wire at 4.2 K compared with the best LTS materials and YBCO wire of the time (source: Heine *et al.*, 1989, American Institute of Physics).

matrix with high oxygen diffusion rate; second, the BSCCO material is reactive with most metals, and silver is one of the few metals that do not degrade the BSCCO. Kellers and Masur (2002) provide a useful summary of commercial considerations of this process. A significant quantity of 2223 tape has gone into applications and demonstration projects: more than a dozen fault current limiter projects, more than twenty cable projects, and numerous magnets have been built or are in construction using this material. In 2011, however, only two companies remain that manufacture B-2223, and one of those has announced that it will exit that business when existing purchase agreements have been completed (Dietrich, 2011).

Two main reasons exist for the demise of the Bi-2223 business: the price of silver and the relative performance of Bi-2223 and YBCO. BSCCO composite wires contain 60–75% silver by volume, thus the cost to manufacture depends heavily on that of silver. Figure 1.14 presents the historical price of silver from 1992 through May 2011, showing an order of magnitude increase between 2004 and mid-2011. Even at the relatively stable price of $4/oz throughout the 1990s, there were many who projected the price of BSCCO wire and tape to be too high to compete with copper (a topic for further discussion in Section 1.3.5); at the time of writing the price is expected to continue to rise for the foreseeable future and continue to be high for many years. For many users of HTS, this motivates the interest in YBCO. The performance comparison will be considered in the next section.

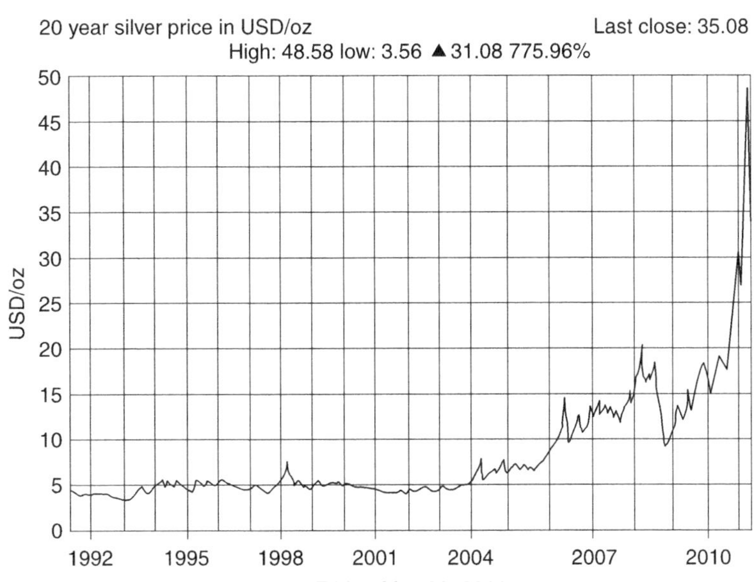

1.14 Historical price of silver (silverprice.org, 2011).

The earliest efforts to make YBCO wires using a simple powder-in-tube method (as used for BSCCO) failed miserably. The reasons are now well understood – the anisotropy and the need for biaxial texture discussed above. Mechanical deformation followed by heat treatment are not sufficient to enable the biaxial texture required for good performance in YBCO conductors. From very early after the discovery of YBCO, high J_c samples were made for electronics applications by thin-film epitaxial deposition on single crystal substrates. Building on this success the first two reports of epitaxial YBCO on polycrystalline metal substrates appeared in 1992 (Iijima *et al.*, 1992; Reade *et al.*, 1992); both of these groups used an ion beam assisted deposition (IBAD) method to impart good grain alignment to a buffer layer between the metal substrate and the YBCO. Within a few years continuous processing of epitaxial YBCO films was accomplished using a variety of methods, which will be described in Chapter 2. In 2011 more than a dozen companies worldwide are in various stages of development and commercialization of so called 'coated conductors', and the competition to achieve longer lengths and higher performance remains strong.

Although the interest in Bi-2223 for higher temperature applications has been largely supplanted by the development of coated conductors, there remains a focused effort to develop Bi-2212 for low temperature, very-high-field applications. High-energy physics accelerator magnets typically are low inductance and operated at very high currents, which requires a highly compacted cable conductor. Magnet designers greatly prefer a round wire for this purpose; neither Bi-2223 nor YBCO-coated conductors can be made round. Since Bi-2212 can be manufactured as round wire and has been successfully cabled, it is high on the list of potential conductors for accelerator magnets that generate fields higher than that possible with LTS (Scanlan *et al.*, 2004).

So despite the large number of HTS materials discovered, and despite considerably higher values of T_c in some of them, only three practical HTS have been developed: Bi-2223 tapes, YBCO-coated conductor tapes, and Bi-2212 round wires. The reasons come down to performance and economics. By far the best critical current in temperatures and fields of interest has been obtained in these three materials. In addition the thallium- and mercury-based cuprates are more hazardous, more difficult to work with safely, and thus more costly. Scanlan *et al.* (2004) present a good review of these three practical conductors and compare them with LTS conductors in the context of applications.

1.3.4 Critical current density

For many applications, including magnets and rotating machinery that rely on field generation, one of the key performance specifications is critical current density. For other applications, including cables, a more important parameter is critical current, which of course depends upon the current density. Bulk HTS materials are primarily used for applications requiring large values of trapped magnetic flux, effectively making them strong permanent magnets. The field trapped in these materials is directly proportional to critical current. Thus for nearly all applications the J_c in the material is paramount.

As pointed out in Section 1.2.4, adding flux-pinning defects is the key to engineering practical superconductors with high J_c, and a significant part of HTS development today is devoted to understanding and controlling pinning in these materials. J_c as determined by the superconducting materials' microstructure and chemistry, including these engineered defects, is often referred to as the intrinsic J_c. The intrinsic J_c is frequently suppressed by extrinsic limitations such as voids, cracks, delamination or unwanted 2nd phases that dilute the superconducting phase or the array of pinning defects. Minimizing such extrinsic limitations through careful process control is another important development activity.

Figure 1.15 provides a comparison of engineering current density (J_e) in the three practical HTS types at 4.2 K. J_e is the measured critical current of a conductor normalized to the entire cross-section of the wire or tape. The figure is adapted from Lee (2010), a highly cited and frequently updated comparison of state-of-the-art performance in commercial conductors, and the reference supplies details of the wires compared. The volume fraction of superconductor in a conductor varies widely, from about 1% in coated conductors to about 50% in some NbTi/Cu composites; for applications such as magnets, J_e is as important as J_c. The figure includes the two important commercial LTS types of wire (NbTi

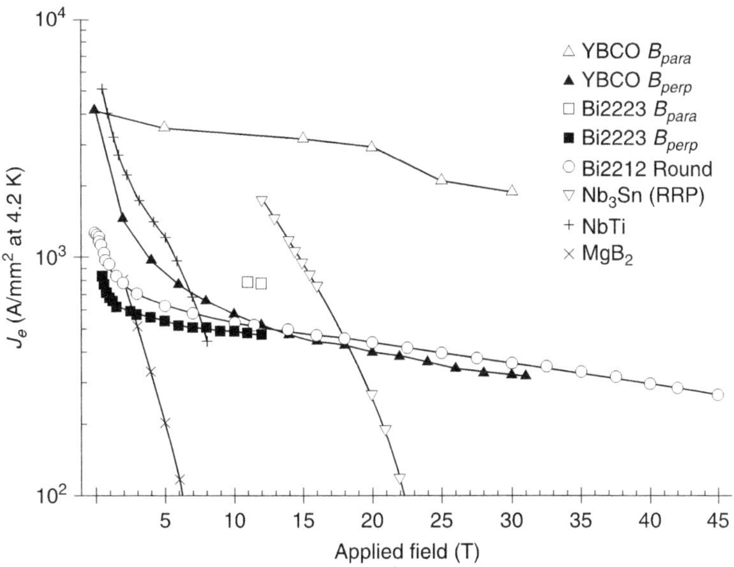

1.15 Comparison of engineering current density J_e as a function of magnetic field at 4.2 K in commercial superconductors (adapted from Lee, 2010).

and Nb_3Sn) for comparison; also included is MgB_2 wire, which some people categorize as an HTS. The comparison at 4.2 K is useful because it allows a direct comparison between HTS and LTS conductors. One immediate conclusion from this comparison is the vast superiority of the in-field current density in YBCO when the applied field is parallel to the tape; equally apparent is the very large anisotropy in YBCO, with J_e for B perpendicular to the tape (B_{perp}) only 15% of the B parallel value (B_{para}) at 20 T. It is also interesting to note that the B_{perp} results for YBCO are nearly as good as the isotropic J_e values reported in round-wire Bi-2212. The average YBCO critical current performance in fields above 10 T is clearly superior to every other practical conductor available; however there remain considerations of cost, quantity available, and the feasibility of making cables with high current for magnet applications that counter the high J_e advantage. Also apparent from the Fig. 1.15 comparison is the vastly superior field dependence of the HTS at low temperature, demonstrating the great potential that high temperature superconductivity holds for very high-field magnet applications. It is interesting to note that in LTS wires the fill factor (volume fraction of superconductor in the wire) is relatively high; in NbTi it is in the range 35–50%, and in Nb_3Sn 15–25%. In YBCO the fill factor is typically 1–2%, partly because of the fact that in most YBCO films there is a strong decrease in J_c with film thickness. Foltyn *et al.* (2007) indicate the likelihood that this decrease can be mitigated, raising the potential for improving the low fill factors in YBCO tapes.

As will be discussed in greater detail in Chapter 2, YBCO has proven to be quite amenable to improving J_e(B) performance through engineered flux-pinning microstructures; a good general review of this fruitful field of study has been provided by Foltyn *et al.* (2007). The prospects are very good for improving J_c, I_c and thus J_e, as well as for reducing the anisotropy in current density with applied field orientation. In contrast, the LTS conductors are quite technologically mature, and large increases in J_c and J_e are much less likely.

1.3.5 Price/performance ratio

Although not a physical property, the selling price of HTS wire is a key factor for commercial success of applications that use these materials. A commonly accepted economic metric is the price/performance ratio in units of dollars per kiloamp-meter ($/kA-m). Although sometimes used in relation to cost, the more relevant metric is the selling price normalized to the critical current performance of the wire. To calculate and compare one needs to know a price in dollars per meter. This is sometimes a great challenge since selling prices for high-value-added goods are typically commercially sensitive, and indeed they vary with a wide range of factors including quantity ordered, delivery time required, market conditions for raw materials, and even the credit rating of the purchaser. Selling price obviously includes a range of overhead costs which vary widely by geographic location and prevailing economic conditions. Although fraught with peril, the comparison of different types of wire, and especially an understanding of the trends with time, can provide valuable input to a discussion of applications of these materials.

Cooley *et al.* (2005) provided a systematic method to compare prices of superconducting wire specifically for high-field magnet applications. This treatment took the approach of independently calculating raw material costs and comparing them with historical selling prices of wires; this determined a ratio P between the price actually paid and the raw materials cost. This production scaling factor (P) includes processing costs, overhead, product yield, profit margin, and all factors other than materials that influence price. If known accurately this factor allows one to convert raw materials cost to purchase price. In the case of NbTi and N_3Sn wires for accelerator magnets there exists a fairly substantial history of government orders, which enabled P values to be determined with some confidence. Cooley *et al.* (2005) also made some reasonable assumptions and estimated P for Bi-2212 and MgB_2 for comparison.

Using this method we here update and revise material costs and P factors published by those authors. Between 2004 and 2011 metal prices have increased significantly. For example, the copper price took big jumps in 2005 and 2006 and at the time of writing averages about 2.6 times the 2004 price (Papp, 2011a); niobium, a chief component in LTS wires, had price increases in 2007–8 and has remained relatively stable since, at 1.9 times the 2004 price. Silver, a large component in BSCCO conductors, was stable in price for most of the years that these conductors were under development. As shown in Fig. 1.14, since 2004 the price has steadily increased and today is about 5.5 times the value used by Cooley (2005). In the case of a raw material cost at this high level, especially a component that comprises 60–75% of the product, it does not make sense to apply the P factor. Rather, the data from Cooley (2005) was reanalyzed in terms of the difference in per meter materials costs and per meter selling price. The analysis shows that for high-volume and high-maturity conductors (NbTi and bronze-process Nb_3Sn) the difference is between \$0.5 and \$2; for lower volume and lower maturity conductors (RRP Nb_3Sn and Bi-2212) the difference is between \$4 and \$8. The BSCCO materials at this stage are judged to be at the higher end of this range.

Table 1.1 summarizes these updated comparisons. There is naturally uncertainty in the factors listed, and the table is ordered with the most certain value at the top with increasing uncertainty toward the bottom. The NbTi values are based on large quantities and the P factor is well established. RRP-type Nb_3Sn production processing has matured significantly since Cooley *et al.* (2005), and worldwide competition for Nb_3Sn products has increased, both of which tend to decrease P, so it is here adjusted down from 10.1 to 8. The YBCO value of current price is taken from a recent public presentation from SuperPower, Inc. (Selvamanickam, 2010); recent purchases of development-scale quantities of coated conductor at Los Alamos National Laboratory have found similar pricing from other

Table 1.1 Factors used for price/performance comparison

Wire type	Diameter (mm)	Raw materials cost ($/kg)		Materials ($/m)	P factor	Price ($/m)	Comments
NbTi LHC-type	0.8	NbTi alloy	150	0.36	3.3	1.19	Based on Cooley (2005)
		Cu	13				Nb cost increased 1.9x (Papp, 2011b)
		Barrier	420				Cu cost increased 2.6x (Papp, 2011a)
							P factor well established
Nb_3Sn RRP-type	0.8	NbTa alloy	550	0.82	8	6.50	Based on Cooley (2005)
		Cu	13				Nb cost increased 1.9x (Papp, 2011b)
		Barrier	420				Cu cost increased 2.6x (Papp, 2011a)
							P factor reduced for increased product maturity
YBCO						40	Selvamanickam (2010)
Bi-2223 40% superconductor	4.2 × 0.2	Ag alloy	965	7.50		15.5	Not in Cooley (2005)
		BSCCO powder	700				Materials cost calculated from wire type details
							P factor discussed in text, uses +$8/m
Bi-2212	0.8	Ag alloy	965	4.40		12.4	Based on Cooley (2005)
		BSCCO powder	1000				Ag cost increased 5.5x (silverprice.org)
							Powder price unchanged
							P factor discussed in text, uses +$8/m
MgB_2	0.8	Precursor powder	285	0.30	10	3	Based on Cooley (2005)
							Powder cost discussed in text
		Cu	13				Cu cost increased 2.6× (Papp, 2011a)
		Barrier	20				Barrier cost increased 1.9× (Papp, 2011b)

manufacturers. No effort was made to calculate materials per meter costs in this case, since the various manufacturing processes differ widely. There is a published cost model for coated conductor which examines one process method in detail (reactive co-evaporation) and predicts total production costs (materials, labor, capital and indirect) can be less than $10/meter (Matias *et al.*, 2007), a cost that is consistent with the $40/m selling price. The BSCCO-2223 case was not included in Cooley (2005); the raw materials cost calculations are based on the size of tape shown in the table and 40% fully dense superconductor filaments in the cross section. The present silver price is well known, the powder price less so but is based on the author's experience purchasing BSCCO powder from various suppliers for the last twenty years. Note that the Bi-2223 precursor powder supply is well established and stable, which is not the case for Bi-2212 powder. Partly due to much lower demand it is often difficult to obtain, and sometimes much more expensive than shown. At the time of writing it is still usually available at the $1000/kg price shown. Rather than use P factors, the new analysis of Cooley's data point toward adding $8 per meter to the raw materials cost. As with the J_e comparison shown above, MgB_2 is included for general interest, although this estimate is a bit more challenging. This estimate deviates from Cooley (2005) significantly, again partly due to generally higher metal prices. Another big difference is the powder cost shown here. Work over the last several years on MgB_2 conductors has shown that performance depends strongly on precursor powder purity, and high-purity boron powder is over $500/kg. The quoted $285/kg may still be low but is believed to be a better estimate of the cost required to produce wire with the performance shown in Fig. 1.15. The higher P factor reflects the fact that these multi-component composites contain materials with widely varying mechanical deformation behavior (Cu, Nb, monel, low density powders), which requires more care than the simpler two-component BSCCO composites.

Figure 1.16 uses the price estimates in Table 1.1 together with the J_e performance (converted to I_c) shown in Fig. 1.15 to compare price/performance in this range of conductors at 4.2 K. It is clear from the figure that a single value of $/kA-m for a particular conductor can be misleading, due to the field dependence of the current density. For example the ratio in YBCO with B_{perp} varies from 40 to nearly 600 $/kA-m between 0–30 T. There is similar strong temperature dependence of the price/performance ratio, so caution is needed to obtain the value for a particular temperature and field of intended application. At 77 K self field, which is a typical point of comparison for HTS, the values corresponding to the examples in the table and figure are $400/kA-m for YBCO ($40/m and 100 A) and $88/kA-m for Bi-2223 ($15.5/m and 176 A). None of the other materials in Table 1.1 carry significant current at 77 K.

It is useful to consider the trends with time for the price/performance ratio. As already discussed, the LTS materials are mature in both processing and in performance. It is unlikely that J_c values or P factors will change much; fluctuating raw material prices will have an effect, but these affect all conductors to varying

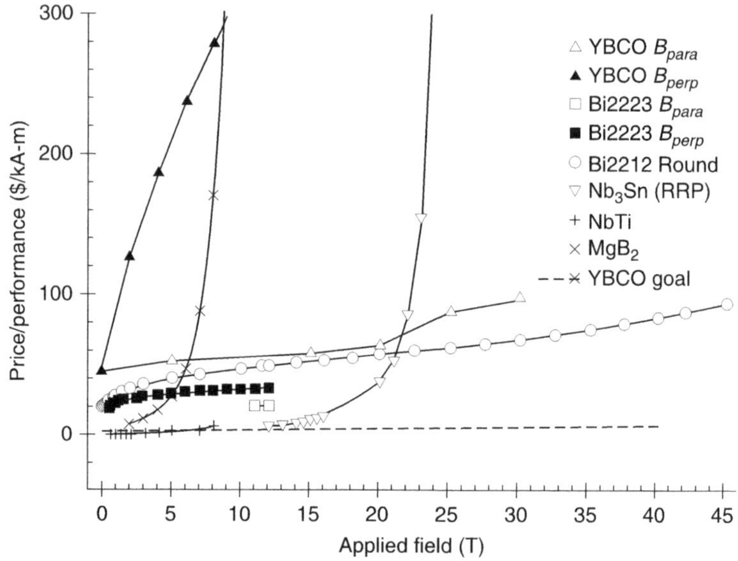

1.16 Comparison of price/performance ratio at 4.2 K in the same commercial superconductors compared in Fig. 1.15; the dashed line shows targeted values for YBCO.

degrees. During the period 1990–2005 there was a steady decrease in price/performance in Bi-2223, while the cost of raw materials was constant and the performance kept improving. That changed in 2004 when the silver price began to climb; the raw material costs for 2223 have increased from \$2.65/m to \$7.50/m in that period, with no sign of a reversal in trend. Both BSCCO conductor types show increases in price/performance due to the price of silver. YBCO is trending strongly down, due to continued strong improvements in both performance and manufacturing processing, as well as increasing production volumes. Selvamanickam (2010) plots this trend, which shows two orders of magnitude reduction in \$/kA-m at 77 K, self field between 2004–10. This strong decline is likely to continue for some time, and commercial manufacturers are targeting \$25/kA-m (77 K, self field), which is 6% of the current \$400/kA-m. At 4.2 K this goal would translate into \$3–6/kA-m, as indicated in Fig. 1.16, and make YBCO competitive with all existing conductors except NbTi.

1.3.6 Other key properties

The importance of other characteristics of HTS vary somewhat with specific applications. In most cases the strength of the material is an issue; the related electromechanical performance, in particular the dependence of J_c on strain, can limit the operating conditions for a magnet, cable or other device. The AC loss characteristics of a particular conductor strongly influence the amount of refrigeration required for a device and thus the overall system cost as well as the operating cost. For fault current limiters the homogeneity of the wire performance is an issue, since a region that has I_c well below the average could prematurely limit current before the desired threshold is reached. The normal state resistance of a conductor is also an issue for some applications, in particular fault current limiters. Also, some applications naturally demand long piece lengths of superconductor to minimize splices and joints. For reference, LTS conductors can be manufactured in lengths exceeding 10 km, and are frequently cut to lengths needed for particular applications. While BSCCO conductors are fairly easily manufactured in kilometer lengths, there has been a long struggle to manufacture coated conductor longer than a few hundred meters. In general there is a need for further development of HTS materials in this regard. These and other key parameters for HTS will be explored in subsequent sections of this book.

1.4 Manufacturing methods for HTS materials

1.4.1 Bulk materials

Most HTS materials were first synthesized in bulk form. Blocks or pucks of BSCCO and YBCO were widely made by grinding mixed oxide precursors followed by powder pressing and sintering. However, because of the fundamental anisotropy in these materials (discussed in Section 1.3.2), polycrystalline material with randomly oriented grains has very low J_c. Melt texturing can be used to achieve grain alignment. Aligned polycrystalline, BSCCO-2212 in particular, was made early on, and continues to be made, by melt casting precursor powder mixtures into near-final shape bars, rods or tubes (Bock *et al.*, 1995). Applications of such objects include current leads and fault current limiters (Elschner *et al.*, 2002).

Melt textured growth of aligned polycrystalline YBCO bulk samples was demonstrated in 1988 (Jin *et al.*, 1988), followed by several variants over the years. Murakami *et al.* (1989) were known for significant development of a quench melt growth method which achieved high J_c and trapped fields. However, the best bulk YBCO has been obtained by adding a single crystal seed to the surface of a pressed pellet followed by a melt process, which nucleates and grows YBCO with the orientation of the seed. This method and variants on it first appeared in the early 1990s in several groups, and an excellent review of the thermodynamics, processing details and properties can be found in Krabbes *et al.* (2006). Referred to generically as large-grain (RE)BCO superconducting materials, the large values of trapped field obtained in these materials have enabled interesting applications including magnetic bearings, levitated flywheels and trains, and trapped flux magnets that are 'permanent' as long as they are kept cold. Chapter 4 addresses these in more detail.

1.4.2 Thermomechanical processing

LTS wires have long been made by conventional mechanical deformation processes (extrusion, wire drawing, rolling, etc.) coupled with appropriate heat treatments for annealing and controlling microstructure (Marancik and Marken, 1995). Such manufacturing methods are referred to generically as 'thermomechanical processing'. These methods were quickly adapted to HTS processing of wires and tapes. Rather than solid metal precursor components, the HTS wire manufacturing started with mixed oxide powders and are referred to as powder-in-tube (PIT) or oxide powder-in-tube (OPIT).

Heine *et al.* (1989) achieved the first high-current OPIT wire using Bi-2212 monocores. The key to their success is now understood to be the partial melt heat treatment they applied to the final wire; when the 2212 phase solidifies and grows on a silver interface, the crystallographic c-axes preferentially aligned perpendicular to the interface (Hasegawa *et al.*, 1994). This achieves sufficient uniaxial texture along the conductor length to promote significant values of J_c. Multifilament wire was first reported by Motowidlo *et al.* (1994), and Rutherford cables made from round wire by Scanlan *et al.* (1999). Although Bi-2212 never received the level of development effort devoted to Bi-2223 or YBCO, there has been steady progress in 2212 OPIT wires for more than 20 years, as demonstrated in Fig. 1.15 by the competitive levels of J_e attained in round multifilament wire. In 2011 there remains a small but active program in the United States for development of Bi-2212 round wire, cables and coils. A large motivation for research and development in this material is that Bi-2212 in round-wire form is the only HTS material available that can be cabled by conventional means.

For more than 15 years an enormous worldwide effort was devoted to development of Bi-2223 tapes manufactured by OPIT plus thermomechanical processing. Although the basic processing was similar to that for 2212, the specific parameters required were much more demanding. The difference is that the 2223 phase of BSCCO is much less thermodynamically stable than the 2212 phase; in fact early efforts to melt-process 2223 invariably resulted in crystallization of the 2212 phase plus non-superconducting second phases. The tricks learned over the years for achieving a substantial volume fraction of grain-oriented 2223 phase in a conductor include higher density of the filaments, additions of controlled amounts of Pb to help stabilize the phase, and a narrow band of heat treatment temperatures that resulted in a process better described as liquid assisted sintering (rather that true melt processing). The filament density required for oriented 2223 filaments is higher than can be attained using wire drawing alone. For this reason Bi-2223 processing included substantial tape rolling, a process that achieves substantially greater filament density as well as contributing to grain alignment, and which explains why round-wire 2223 conductors are not feasible. The review by Vase *et al.* (2000) provides a good overview of Bi-2223 tape processing and issues. Continued recent advances in performance of B-2223 have been achieved by controlled overpressure heat treatments that substantially increase the filament density over what is attainable by rolling and conventional heat treatment, coupled with improved process control (Osabe *et al.*, 2010). These improvements have pushed 77 K, self field I_c performance to over 400 A/cm-width in a 2 km length of tape, and record J_c performance of 663 A/mm^2 (Osabe *et al.*, 2010).

1.4.3 Thick-film processing

Although considered by some a technological dead end, and although there is not an active effort today, there was sufficient development work and success with thick-film ceramic processing methods for HTS tapes that a brief description is warranted. In fact the highest performance Bi-2212 conductors have been made using these techniques, with dense, highly aligned 2212 layers obtained by a so-called pre-annealing and intermediate rolling (PAIR) process achieving J_c = 5000 A/mm^2 at 4.2 K, 10 T (Miao *et al.* 1998). For comparison, the best round-wire B-2212 has achieved about 2500 A/mm^2 at this temperature and field (corresponding to ~600 A/mm^2 J_e as shown on Fig. 1.15).

A good review of the different methods studied and the relative advantages of each was provided by Alford *et al.* (1997), who also point out that such manufacturing methods are extensively used in the electronics industry. Perhaps the most commonly used method for HTS was dip-coating from a ceramic slurry, but a wide range of other techniques used include electro-deposition, sol-gel coating, spray coating and the doctor-blade method. The most effort and the most success was achieved with Bi-2212, but results have also been reported for Bi-2223, YBCO and TBCCO-2212 and -2223 (TlBaCaCuO).

The TBCCO system was initially explored as a conductor because the T_c values are significantly greater than either YBCO or BSCCO, and the flux melting line $H_m(T)$ is significantly higher than that for BSCCO. One immediate challenge with this system is the toxicity of thallium; the general approach to synthesis was to fabricate conductor from precursors that omitted the thallium, and to heat treat the final conductor in a closed system that supplied thallium vapor. Obviously such a method requires good vapor exchange between the Tl source and the precursor, which favors an open approach such as film processing. Some initial modest levels of current density were obtained, but these efforts were abandoned because of the much better J_c performance in BSCCO and the rapid advances in YBCO-coated conductor.

A fair number of applications demonstrations were made with Bi-2212 thick-film conductors, including high-field inserts and MRI magnet coils (for example Dai *et al.*, 1995), an HTS transformer (Schwenterly *et al.*, 1999) and current leads (Ting *et al.*, 1997). The chief obstacle to further development proved to be the metal substrate for the coating. A primary requirement is chemical compatibility with the BSCCO melt, a very stringent criterion only met by silver and gold. The substrate must also be sufficiently strong to enable required applications, a requirement that is incompatible with the need to use silver for chemical compatibility. Although some progress was made in composite substrates (for example silver-plated nickel alloy), the progress demonstrated in multi-filamentary round 2212 wires combined with a general lack of research and development resources cut short the development of thick-film Bi-2212.

1.4.4 Thin-film processing

Within a year of the publication of the discovery of YBCO there were reports of high J_c values greater than 1 MA/cm^2 at 77 K in thin films grown on single crystal substrates (see for example Roas *et al.*, 1988). Already in mid-1988 high-quality thin films had been grown by a wide range of deposition methods including pulsed laser deposition, electron beam co-evaporation, molecular beam epitaxy and sputtering. This range of processing methods was soon extended to include chemical vapor deposition and solution deposition methods. As discussed earlier in this chapter, a key requirement for high J_c is achieving a high degree of biaxial grain alignment, and thin films on single crystals grew epitaxially to meet this requirement. A major step toward making lengths of conductor came when both Iijima *et al.*, (1992) and Reade *et al.*, (1992) reported high J_c on metal substrates by using IBAD to achieve an aligned buffer layer and then depositing epitaxial YBCO. Since that time a variety of other methods for achieving good grain alignment have been devised, most notably the rolling assisted biaxially textured substrate (RABiTS) method pioneered by Goyal *et al.* (1996). All manufacturing methods in use today employ either IBAD or RABiTS as the means of achieving biaxial texture.

A wider range of film deposition methods for buffer layers and superconductor are, however, in use. As the technology has evolved since the mid-1990s, the successful methods for growing high J_c films have narrowed to fairly traditional thin-film methods. Parans and Izumi (2004) as guest editors for the August 2004 MRS bulletin put together a very useful review of various methods of coated conductor manufacturing. This issue included three more detailed treatments of IBAD and RABiTS methods for texturing, plus detailed discussion of different YBCO deposition technologies: metalorganic solution deposition, metalorganic chemical vapor deposition, and high-rate pulsed laser deposition (PLD). One additional important technique for high-rate and potentially low-cost YBCO film deposition is reactive co-evaporation (Matias *et al.*, 2010). Chapter 2 provides a more detailed treatment of coated conductor processing.

1.5 Challenges and future trends

There remain active projects and installations using Bi-2223 conductor, and these will likely continue for the foreseeable future; one clear trend however, driven by the steep increases in the price of silver, is the diminishing role for such silver-based conductors. The predominance of 2223 for high temperature superconductivity applications over the last 20 years is in rapid decline, a trend that is unlikely to reverse unless the price of silver declines, which at the time of writing seems unlikely. There may be a continued use for B-2212 round wire, based mainly on the ability to make dense, high-current cables; for this to happen, however, will require substantial increases in the J_e performance of the wire. There is serious interest in round 2212 for high-field magnets based on the preference of many magnet designers for wire over tape; however the increasing cost of this predominantly silver wire will drive serious consideration of YBCO-coated conductors for high-field applications.

The trend apparent today is for a rapid increase in the use of YBCO for a range of applications. Significant challenges remain for this trend to continue and grow. The price of YBCO remains high, and significant effort will address manufacturing issues leading to simpler and more efficient processes, increased rates and throughput and higher yields. Recent years have brought enormous improvements in J_c performance in coated conductors through engineered flux-pinning defects, and it seems likely this trend will continue. Related to improvements in J_c at all temperatures and fields is the need to reduce the anisotropy in J_c(B), and the expectation is this will happen with better understanding and control of the flux-pinning landscape in these materials. Another key challenge is to increase the lengths of YBCO-coated conductor. The state-of-the-art today is 500–1000 m lengths, which will certainly need to be increased. Fledgling efforts to find ways to make high-current cables from highly aspected coated conductor are likely to grow and expand.

On a longer-term horizon there remains the distinct possibility for a new family of HTS materials to become available for applications. There are various efforts around the world to discover higher T_c materials, and there are significant efforts to develop processing methods to achieve high J_c in known materials such as the pnictides. Though difficult to predict when this might happen, the number and type of useful HTS materials are likely to grow.

1.6 Conclusions

A key factor for enabling applications of high temperature superconductivity is the critical current density. Because of the unique magnetic properties in Type II superconductors, discrete flux lines penetrate and form a vortex lattice; when a transport current is applied, a Lorentz force $J \times B$ acts on the vortex lattice. J_c is determined by the onset of flux flow under this Lorentz force, and can be increased substantially by engineering flux-pinning defects of the appropriate size and density in the superconductor. Although a large number of HTS materials have been discovered, only three practical materials have been developed for applications. The basic manufacturing methods of these three were described. The J_c and J_e performance in these materials was compared, and put in the context of

LTS materials. The economic considerations were represented by a price/performance figure of merit in dollars per kA m. The present status and the trends with time of the price/performance measure in these materials was discussed, with present conditions indicating a wane in importance of BSCCO materials while YBCO-coated conductors seem to be growing in importance. Significant challenges remain for the technical and commercial success of YBCO, but these are likely to be met.

1.7 Sources of further information and advice

Bennemann K and Ketterson J editors, (2008), *Superconductivity Conventional and Unconventional Superconductors*, Berlin, Springer Verlag.

Buckel W and Kleiner R, (2004), *Superconductivity Fundamentals and Applications*, 2nd ed., Weinheim, Wiley VCH.

De Gennes P, (1989), *Superconductivity of Metals and Alloys*, Reading, Addison-Wesley.

Fossheim K and Sudbo A, (2004), *Superconductivity Physics and Applications*, Chichester, John Wiley & Sons.

Orlando T and Delin K, (1991), *Foundations of Applied Superconductivity*, Reading, Addison-Wesley.

Rose-Innes A and Rhoderick E, (1978), *Introduction to Superconductivity*, Oxford, Pergamon.

Tinkham M, (1996), *Introduction to Superconductivity*, 2nd ed., New York, Dover.

1.8 References

Abrikosov A, (1957) 'On the magnetic properties of superconductors of the second groups', *Soviet physics, JETP*, **5**, 1174–1183.

Alford N, Penn S and Button T, (1997), 'High-temperature superconducting thick films', *Supercond Sci Techn*, **10**, 169–185.

Bardeen J, Cooper L and Schrieffer J, (1957), 'Theory of superconductivity', *Phys Rev*, **108**, 1175–1204.

Battlog B, Cava R and Van Dover R, (2003), US Patent No. 6,630,425 B1.

Bednorz J and Muller K, (1986), 'Possible high T_c superconductivity in the Ba-La-Cu-O system', *Z Phys B*,**64**, 189–193.

Berkley D, Skelton E, Moulton N, Osofsky M, Lechter W *et al.*, (1993), 'Pressure dependence of the superconducting transition temperature in single crystals of TlBaCaCuO', *Phys. Rev. B* **47**, 5524–5527.

Beyers R, Engler E, Grant P, Lim G and Parkin S, (1994), European Patent Publication No. 281753B1.

Blundell S, (2009), *Superconductivity, a Very Short Introduction*, Oxford, Oxford University Press.

Bock J, Elschner S and Herrmann P, (1995), 'Melt cast processed (MCP)-BSCCO 2212 tubes for power applications up to 10 kA', *IEEE Trans Appl Supercon*, **5**, 1409–1412.

Chaudhari P, Gambino R, Koch R, Lacey J, Laibowitz R *et al.*, (1995), US Patent No. 5,447,906.

Cooley L, Ghosh A and Scanlan R, (2005), 'Costs of high-field superconducting strands for particle accelerator magnets', *Supercond Sci Technol*, **18**, R51–65.

Dai W, Marken K, Hong S, Cowey L, Timms K *et al.*, (1995), 'Fabrication of high T_c coils from BSCCO 2212 powder-in-tube and dip coated tape', *IEEE Trans Appl Supercon*, **5**, 516–519.

Dietrich R (2011), 'Bruker HTS', presentation at International Energy Agency-HTS committee meeting, Santa Fe, NM, May 10–12.

Elschner S, Breuer F, Noe M, Wolf A and Bock J, (2002), 'Qualification of MCP BSCCO 2212 bulk material for use in resistive current limiters', *Physica C*, **372–376**, 1668–1672.

Feldmann D, Holesinger T, Feenstra R and Larbalestier D, (2008), 'A review of the influence of grain boundary geometry on the electromagnetic properties of polycrystalline YBaCuO films', *J Am Ceram Soc*, **91**, 1869–1882.

Foltyn S, Civale L, Macmanus-Driscoll J, Jia Q, Maiorov B *et al.*, (2007) 'Materials science challenges for high-temperature superconducting wire', *Nat Mat*, **6**, 631–642.

Gao L, Xue Y, Chen F, Xiong Q, Meng R *et al.*, (1994), 'Superconductivity up to 164 K in HgBaCaCuO under quasihydrostatic pressures', *Phys Rev B*, **50**, 4260–4263.

Goyal A, Norton D, Budai J, Paranthaman M, Specht E *et al.*, (1996) 'High critical current density superconducting tapes by epitaxial deposition of $YBa_2Cu_3O_{7-x}$ thick films on biaxially textured metals', *Appl Phys Lett*, **69** 1795–1797.

Hasegawa T, Kobayashi T, Kumakura K and Togano K, (1994), 'Formation and growth mechanisms of Bi-2212 on an Ag substrate in the partial melting method', *Supercond Sci Technol*, **7**, 579–582.

Heine K, Tenbrink J and Thoner M, (1989), 'High-field critical current densities in $Bi_2Sr_2CaCu_2O_{8+x}$/Ag wires', *Appl Phys Lett*, **55**, 2441–2443.

IEEE, (1987) *IEEE Transactions on Magnetics*, MAG-**23**, 343–1763.

Iijima Y, Tanabe N, Kohno O and Ikeno Y, (1992), 'In-plane aligned YBCO thin films deposited on polycrystalline metallic substrates', *Appl Phys Lett*, **60**, 769–771.

Jin S, Tiefel T, Sherwood R, Davis M, Van Dover R *et al.*, (1988), 'High critical currents in YBaCuO superconductors', *Appl Phys Lett*, **52**, 2074–2076.

Kellers, J and Masur, L (2002), 'Reliable commercial HTS wire for power applications', *Physica C*, **372–376**, 1040–1045.

Krabbes G, Fuchs G, Canders W, May H and Palka R, (2006), *High Temperature Superconductor Bulk Materials*, Weinheim, Wiley VCH.

Lee P, (2010), 'Engineering Critical Current Density vs. Applied Field', available from http://magnet.fsu.edu/~lee/plot/plot.htm, [Accessed 28 May 2011].

Maeda H, Tanaka Y, Fukutomi M and Asano T, (1988), 'A new high-T_c oxide superconductor without a rare earth element', *Jap J Appl Phys* **27**, L209–210.

Malozemoff A and Grant P, (1987), *Z Phys B*, **67**, 275–283.

Marancik G and Marken K, (1995) 'Manufacture of low T_c superconductor wire', in *Nonferrous Wire Handbook Volume 3*, Guilford CT, Wire Association International.

Matias V, Hänisch J, Sheehan C, Ugurlu O and Storer J, (2007), 'Reactive co-evaporation of YBCO for coated conductors', *Korea Inst Appl Supercon Cryo*, **9**, 1–6.

Matias V, Rowley J, Coulter Y, Maiorov B, Holesinger T *et al.*, (2010), 'YBCO films grown by reactive co-evaporation on simplified IBAD-MgO coated conductor templates', *Supercond Sci Technol*, **23**, 1–5.

Matsui Y, Maeda H, Tanaka Y and Horiuchi S, (1988), 'High-resolution electron microscopy of modulated structure in the new high-T_c superconductors of the Bi-Sr-Ca-Cu-O system', *Jap J Appl Phys* **27**, L361–364.

Meissner W and Ochsenfeld R (1933), 'Ein neuer Effekt bei Eintritt der Supraleitfähigkeit', *Naturwissenschaften*, **21**, 787–788.

Miao H, Kitaguchi H, Kumakura H, Togano K, Hasegawa T *et al.*, (1998), '$Bi_2Sr_2CaCu_2O_x$/Ag multilayer tapes with J_c > 500 000 A/cm^2 at 4.2 K and 10 T by using pre-annealing and intermediate rolling process', *Physica C*, **303**, 81–90.

Miura M, Baily S, Maiorov B, Civale L, Willis J *et al.*, (2010), 'Vortex liquid-glass transition up to 60 T in nanoengineered coated conductors grown by metal organic deposition', *Appl Phys Lett*, **96**, 072506-1 to -3.

Motowidlo L, Galinski G, Ozeryansky G, Zhang W and Hellstrom E, (1994), 'Dependence of critical current density on filament diameter in round multifilament Ag-sheathed Bi-2212 wires processed in O_2', *Appl Phys Lett*, **65**, 2731–2733.

Murakami M, Morita M, Doi K and Miyamoto K, (1989), 'A new process with the promise of high J_c in oxide superconductors', *Jpn J Appl Phys*, **28**, 1189–1194.

Onnes K, (1911), 'Further experiments with liquid helium. C. On the change of electric resistance of pure metals at very low temperatures, etc. IV. The resistance of pure mercury at helium temperatures', *Comm Phys Lab Univ Leiden*, No. 120b.

Osabe G, Ayai N, Kikuchi M, Tatamidani K, Nakashima T *et al.*, (2010), 'Recent progress of high performance Ag-sheathed Bi2223 wire', *Physica C*, **470**, 1365–1368.

Papp J, (2011a), *US Geological Survey Mineral Commodity Summaries 2011*, 110–111.

Papp J, (2011b) *US Geological Survey 2009 Minerals Yearbook: Niobium (Columbium) and Tantalum.*

Parans M and Izumi T, (2004), 'High performance YBCO-coated superconductor wires', *MRS Bulletin*, **29**, 533–541.

Poole C, editor, (2000), *Handbook of Superconductivity*, San Diego, Academic press.

Poole C, Farach H, Creswick R and Prozorov R, (2007), *Superconductivity*, 2nd ed., London, Elsevier.

Putilin S, Anipov, O, Chmaissem O and Marezio M, (1993), 'Superconductivity at 94 K in HgBaCuO', *Nature*, **362**, 226–228.

Reade R, Berdahl P, Russo R and Garrison S, (1992), 'Laser deposition of biaxially textured yttria-stabilized zirconia buffer layers on polycrystalline metallic alloys for high critical current YBaCuO thin films', *Appl Phys Lett*, **61**, 2231–2233.

Roas B, Schultz L and Endres G, (1988), 'Epitaxial growth of $YBa_2Cu_3O_{7-x}$ thin films by a laser evaporation process', *Appl Phys Lett*, **53**, 1557–1559.

Roberts B, (1976), 'Survey of superconductive materials and critical evaluation of selected properties', *J Phys Chem Ref Data*, **5**, 581–821.

Saxena, A, (2010), *High Temperature Superconductors*, Berlin, Springer-Verlag.

Scanlan R, Dietderich D, Higley H, Marken K, Motowidlo L *et al.*, (1999), 'Fabrication and test results for Rutherford-type cables made from BSCCO strands', *IEEE Trans Appl Supercond*, **9**, 130–133.

Scanlan R, Malozemoff A and Larbalestier D, (2004), 'Superconducting materials for large scale applications', *Proc IEEE*, **92**, 1639–1654.

Schwenterly S, McConnell B, Demko J, Fadnek A, Hsu J *et al.*, (1999), 'Performance of a 1-MVA HTS demonstration transformer', *IEEE Trans Appl Supercon*, **9**, 680–684.

Selvamanickam V, (2010) 'Progress in SuperPower's 2G HTS Wire Development and Manufacturing', US DOE 2010 HTS Peer Review, available from http://www.htspeerreview.com/agenda.html [accessed 28 May 2011].

Sheng Z and Hermann A, (1988), 'Bulk superconductivity at 120 K in the Tl-Ca/Ba-Cu-O system', *Nature*, **332**, 138–139.

Shubnikov L, Khotkevich V, Shepelev Y and Ryabinin Y, (1937), 'Magnetic Properties Of Superconducting Metals And Alloys', *Ukrainian J Physics, Shubnikov*, Yr: 2008 Pg: 42–52 (Translated And Reprinted From *Zh Eksper Teor Fiz* **7**, 221–237, 1937).

Sosolik C, Stroscio J, Stiles M, Hudson E, Blankenship S *et al.*, (2003), 'Real-space imaging of structural transitions in the vortex lattice of V3Si', *Phys Rev B*, **68**, 140503-1–140503-4.

Ting S, Marken K, Cowey L, Dai W, Nelson S *et al.*, (1997), 'Development of current leads using dip coated BSCCO-2212 tape', *IEEE Trans App Supercon*, **7**, 700–703.

Vase P, Flükiger R, Leghissa M and Glowacki B, (2000), 'Current status of high-T_c wire', *Supercond Sci Technol*, **13**, R71–R84.

Wu M, Ashburn J, Torng C, Hor P, Meng R *et al.*, (1987), 'Superconductivity at 93 K in a new mixed-phase Y-Ba-Cu-O compound system at ambient pressure', *Phys Rev Lett,* **58**, 908–910.

2
High temperature superconductor (HTS) wires and tapes

V. SELVAMANICKAM, University of Houston, USA

Abstract: High temperature superconductor (HTS) wires and tapes have been an area of intense research, development and scaling up for manufacturing. Remarkable success has been achieved in overcoming problems with grain boundaries, anisotropy, and poor mechanical properties of ceramic HTS materials to fabricate wires and tapes. First-generation HTS tapes based on thermomechanical processing have been the workhorse of the superconductor industry, and several prototypes have been made for over a decade. Second-generation (2G) HTS tapes based on epitaxial thin-film deposition on highly-textured templates on flexible alloy substrates have been developed with numerous innovations in materials science, deposition processes and manufacturing equipment.

Key words: wires, tapes, critical current, pinning, thin film, epitaxy, deposition.

2.1 Introduction

Ever since the discovery of superconductivity above 77 K in high temperature superconductors (HTS), several techniques were examined for fabrication of superconducting wires and tapes. In the first few years, the immense challenge with critical current degradation across grain boundaries, strong sensitivity of critical current to magnetic fields, and the weak mechanical properties of the superconductor were identified. It became apparent that composite techniques had to be developed for superconductor wire/tape fabrication. Thermomechanical processing techniques successfully used for low-temperature superconductor such as Nb-Ti were adopted for HTS with good success with the first generation (1G) Bi-based superconductors. In order to achieve higher critical currents, improved performance in magnetic fields, stronger wires, as well as potential for low-cost manufacturing, a second-generation (2G) tape fabrication approach based on thin-film deposition was developed. Numerous challenges were overcome in biaxial texture development, material compatibility, epitaxial growth, high throughput processing, process stability over long durations, engineering the wire to practical applications and scaling the technology to pilot-scale manufacturing. This chapter will elaborate on the pathway in the development of HTS wire/tape technology over the last 20 years.

2.2 Challenges with high temperature superconductor (HTS) wires and tapes

Several of the initial tape/wire processing approaches employed conventional ceramic processing techniques such as tape casting and wire extrusion, where the superconductors were fabricated using a slurry consisting of superconducting powder, binder, dispersants and plasticizers (Jin and Graebner 1991). After ceramic processing, the tapes/wires were heat-treated to remove organic constituents and sintered at a high temperature (~900 °C). $YBa_2Cu_3O_{7-x}$ (YBCO) was the commonly used superconductor in these wires and tapes and, in spite of improved processing to assure densification, reduced impurities and fewer secondary phases, the critical current density (J_c) was low, typically less than 1000 A/cm^2. Additionally, the current density of sintered YBCO was found to drop rapidly even in small magnetic fields, typically by two orders of magnitude at 100 G (Ekin 1987). A comparison of the magnetic field dependence of J_c of sintered YBCO and single crystal YBCO is shown in Fig. 2.1. It can be observed from the figure that the decay of J_c is much more rapid in low magnetic fields in sintered YBCO. The magnetic field dependence of J_c in sintered materials has been described by a model based on an array of strongly-superconducting grains connected by Josephson junctions (Peterson and Ekin 1989). It was postulated that the superconductor is transformed into a dissipative regime when the magnetic field exceeds a certain decoupling magnetic field, which was calculated to be as low as 5 G (D'Ovidio *et al.* 1991). The weak link behaviour of grain boundaries in sintered materials has also been evidenced in hysteresis effects in the magnetic field dependence of J_c (McHenry *et al.* 1989, Evetts and Glowacky 1988), multiple superconducting transitions (Goldschmidt 1989), sample-size dependence of J_c (Bransky *et al.* 1989, Shimizu and Ito 1989, Kumakura *et al.* 1987), and the inter-grain peak in AC susceptibility measurements (Kupfer 1988).

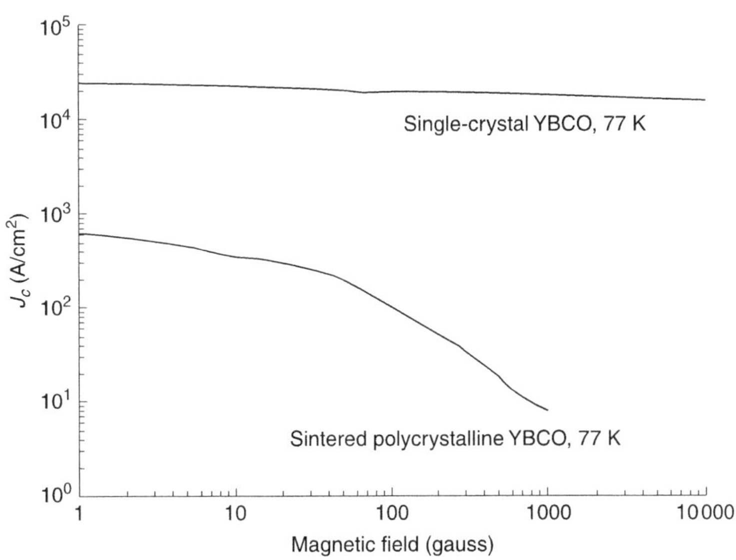

2.1 Magnetic field dependence of critical current density of sintered YBCO superconductor compared with that of single crystal YBCO.

The problem with low current density was traced to misorientation of grains across the multiple high-angle grain boundaries present in sintered HTS. A better understanding of the influence of grain-to-grain misorientation on J_c was gained from measurements across single grain boundaries of bicrystals of thin-film YBCO (Dimos *et al.* 1990, Heinig 1996, Verebelyi 2000). Figure 2.2 shows the grain boundary J_c value at 5 K normalized to the J_c value of the grain boundaries of laser-ablated films deposited on $SrTiO_3$ bicrystals with a tilt angle between (100) planes. It can be observed from the figure that the normalized J_c across 001 tilt grain boundaries decreases rapidly with increasing misorientation angle, particularly at values higher than 5°. It is seen that the critical current density decreases by a factor of 200 when the misorientation angle is increased to about 35°. The misorientation between the two crystals was found to be accommodated by dislocations, whose spacing was seen to vary inversely with the misorientation angle. The low J_c value, even at small misorientation angles, has been attributed to the structural order at the dislocation cores which suppresses the superconducting order parameter at the grain boundaries.

It was clear from the early results that the grains of HTS materials have to be aligned along the strongly conducting basal plane direction (Gyrogy *et al.* 1989) and several techniques were evaluated to achieve such an alignment such as magnetic (Chen *et al.* 1989, Ekin *et al.* 1990, Kawahara *et al.* 1990), mechanical (Poeppel 1987, Johnson *et al.* 1987, Chen *et al.* 1987, Lo and Glowacki 1991) and

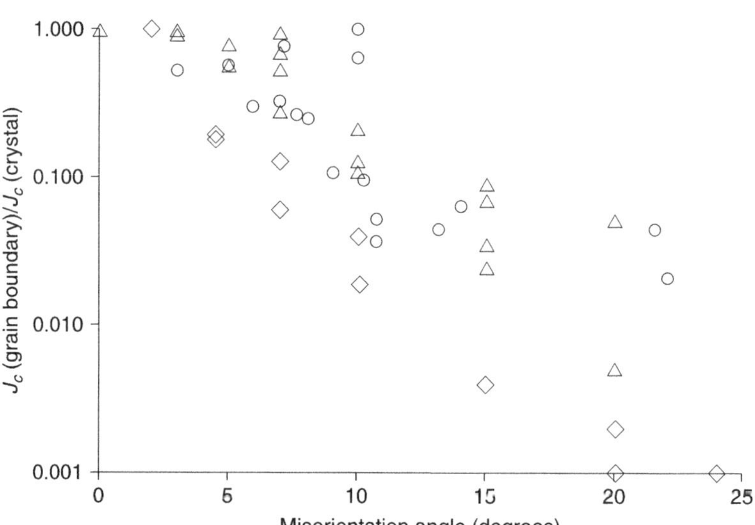

2.2 Critical current density across boundaries of superconductor bicrystal films at different misorientation angles normalized to the critical current density of individual films. (Data compiled by S. Foltyn from Dimos *et al.* 1990, Heinig *et al.* 1996, Verebelyi *et al.* 2000.)

melt-texturing (Jin *et al.* 1988, Salama *et al.* 1989, Murakami *et al.* 1989). The magnetic alignment method is based on the paramagnetic anisotropy exhibited by HTS at room temperature. When HTS grains are subjected to a strong magnetic field at room temperature, the c-axis of the grains tends to align along the magnetic field direction because of the susceptibility being maximum along this crystallographic direction. Some improvement in critical current density was achieved with magnetic alignment (Chen *et al.* 1989, Kawahara *et al.* 1990). However, magnetic alignment tends to align only the c-axis of grains, leaving a- and b-axes randomly aligned. Also, deficiencies of the magnetic alignment technique require the necessity of having to embed the superconductor grains in epoxy or suspend them in a solution during the alignment process, which makes it difficult to achieve high densities.

Melt-texturing was a popular technique in the early 1990s to achieve highly-aligned YBCO superconductors. The pioneering work on melt-texturing of HTS was done by Jin *et al.*, where the YBCO melt was solidified under a temperature gradient to yield a highly aligned microstructure that resulted in a critical current density above 10 000 A/cm^2 (Jin *et al.* 1988). Soon after, an excellent grain alignment was achieved in YBCO fabricated using a slow cooling process without using a temperature gradient, which led to critical current density close to 100 000 A/cm^2 (Salama *et al.* 1989). Further work on melt-textured YBCO was focused on improving pinning by the introduction of fine-scale precipitates (Murakami *et al.* 1989). While melt-texturing is still used by several groups today to fabricated highly-aligned (RE)BCO (*RE* = rare earth) with excellent J_c, the product is typically in the form of a disk rather than wires or tapes. The need to achieve an excellent grain orientation placed an additional burden on an already challenging problem of fabricating brittle ceramic superconductors as flexible wires or tapes in the long lengths required for commercial applications.

2.2.1 1G HTS tape/wire types, fabrication, properties

The brittle nature of HTS necessitated the use of composite structures to fabricate wires and tapes. The powder-in-tube (PIT) technique was first developed with YBCO where the superconducting powder was packed in a silver billet followed by drawing, rolling and a heat treatment to form a tape geometry (Jin *et al.* 1987). Due to the reactivity of most metals with superconductors, silver has been the only choice for encapsulation of HTS materials fabricated by PIT techniques. In spite of the ability to fabricate tapes, the PIT technique did not work out for (RE) BCO because of the grain boundary weak link problem. No significant alignment of (RE)BCO grains was achieved using this technique. Conversely, this technique has been successfully employed with $Bi_2Sr_2Ca_2Cu_3O_{10-x}$ (BSCCO). The reason for the success has been the formation of micaceous grains of BSCCO with a high aspect ratio and the ability to align these grains along the direction of tape rolling. Critical current density over 50 000 A/cm^2 have been achieved in short samples of

BSCCO fabricated by the PIT technique (Ueyama *et al.* 1991). It is remarkable that weak-link behaviour is not prominent in BSCCO grain boundaries, even though only c-axis alignment and no significant alignment of a- and b-axes is achieved in tapes fabricated by the PIT technique. A brick-wall model such as that shown in Fig. 2.3 has been used to explain the ability of transport current to travel through blocking boundaries by current flow along the c-axis of adjacent grains (Mannhart and Tsuei 1989). Because of the high aspect ratio of BSCCO grains, the grain boundary area will be much larger in the direction perpendicular to, rather than parallel to, the current flow direction. If the grains are elongated by a ratio r along the current direction and overlap on the side by half their length, the current can flow sideways so that the effective area of the current path is increased by a factor r. In BSCCO, the aspect ratio r of the grains is about 100 due to their mica-like morphology, whereas in YBCO, it is about five (Imayev 2000). Due to the large aspect ratio, the area available for current transfer perpendicular to the length direction is much higher in BSCCO.

The silver sheath in PIT-BSCCO wires also provides the benefit of highly aligned grains at the interface, presumably due to liquid-phase enhanced growth. Interfacial regions of BSCCO-Ag have been found to sustain high current densities (Holesinger *et al.* 2005). Silver-sheathed BSCCO wires exhibit poor mechanical properties because of the low yield strength of the metal. Therefore, silver alloy matrices or laminated wires are used to meet application requirements of mechanical properties.

BSCCO wires have been scaled up successfully to lengths of about two kilometres with critical current values of nearly 200 A over the entire length (Osabe *et al.* 2010). It has been the workhorse superconducting wire during 1995–2007, used to fabricate a number of superconducting devices, especially power transmission cables. However, due to the high content of high-purity silver, limited mechanical properties, as well as weak retention of critical current in high magnetic fields, alternative techniques to PIT BSCCO became a subject of intense investigation by several research groups.

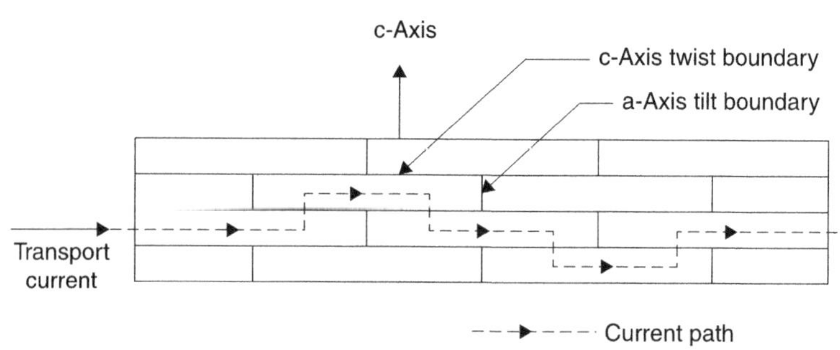

2.3 Brick wall model used to explain current transport in 1G HTS tapes.

2.2.2 2G HTS tape/wire types, fabrication and properties

Even four years after the discovery of superconductivity in YBCO, no technique that had promise to fabricate this material in the form of tapes or wires with a high current density was developed. At the same time, high critical current densities of 5 MA/cm^2 were routinely achievable in high-quality epitaxial thin films on single crystal substrates (Roas *et al.* 1990). Use of thin-film YBCO to fabricate tapes was initially explored using nickel alloy substrate and pulsed laser deposition (PLD) (Narumi *et al.* 1991). Employment of expensive silver was nearly eliminated and diffusion of cations from the metal substrate was avoided by use of buffer layers such as yttria-stabilized zirconia (YSZ) between the substrate and the superconductor film. Tapes made by the thin-film deposition technique exhibited good out-of-plane texture, i.e. c-axis orientation normal to the film plane because of the anisotropic growth characteristics of (RE)BCO. However, within the plane of the film, no orientation was present and, as a result, the critical current density was only 10^5 A/cm^2, which was more than a factor of 10 less than that achieved in epitaxial films on single crystal substrates. Nevertheless, this demonstration paved the path toward applying thin-film deposition to coat (RE)BCO films on nickel alloy substrates to fabricate superconducting tapes.

2.2.3 Textured template fabrication

Ion beam assisted deposition

The importance of in-plane texture to achieve high current densities in (RE)BCO tapes on nickel alloy substrates was first demonstrated by Iijima *et al.* (1991) when they showed that in-plane texture can be achieved in YSZ deposited by ion beam assisted deposition (IBAD). In the 1980s, simultaneous bombardment of growing films by low-energy ion beams was found to result in in-plane texture, where there is a preferential orientation in the plane of the film in addition to out-of-plane orientation (Yu *et al.* 1985). Such a texture is termed biaxial texture, since grains are aligned in the plane along two crystallographic axes (in fact, this phenomenon results in alignment of grains about all three axes including the out-of-plane axis). A specific form of biaxial texture, namely cube texture with (001) planes in the plane of the film and a texture of <100> and <010> axes within the film plane has been successfully demonstrated by IBAD. Fig. 2.4 displays the IBAD process using low-energy ion bombardment during electron-beam evaporation.

Earlier work using fluorites such as YSZ (Iijima *et al.* 1991) or pyrochlores such as $Gd_2Zr_2O_7$ (GZO) yielded good biaxial texture but required an extremely slow process (Iijima *et al.* 2005, 2004, 1991). The reason for the prolonged process was the evolutionary nature of texture development in these materials (Iijima *et al.* 1998). Biaxial texture develops through dominance of grains preferentially oriented with respect to the ion beam incidence direction. This evolutionary

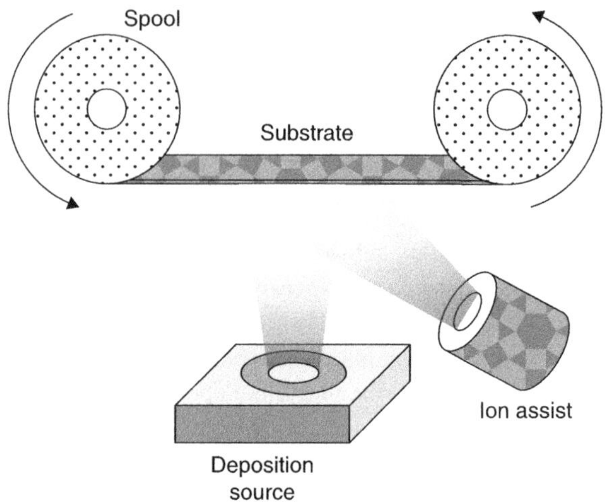

2.4 Schematic of the IBAD process.

growth typically requires a thickness of approximately 1 μm to reach a biaxial texture of about 10°, i.e. grain-to-grain misorientation within the plane of the film of about 10°. This requirement of a thick film coupled with a slow rate of deposition due to the necessity of an appropriate ion-to-atom ratio render this evolutionary type of texture development process unsuitable for practical use. A reason for the slow development of texture in materials like fluorites, pyrochlores and bixbyites during IBAD has been that the close-packed plane in these materials is (111). Therefore, in absence of an ion beam assist, they prefer to orient with a fibre texture of (111), i.e. with an out-of-plane texture of (111) and a random in-plane texture. In order to achieve a cube texture, the out-of-plane texture has to be reoriented towards (001) and, *additionally*, an in-plane texture of (100) needs to be achieved. Therefore, ion beam assist is encumbered with achieving the appropriate out-of-plane and in-plane textures resulting in a slow development of texture.

Since the YSZ or GZO layers are about 1 μm in thickness, these layers themselves act as diffusion barriers to cations in the substrate. Since the texture achieved in the IBAD template is independent of the substrate, and since the IBAD process is conducted near room temperature, there is wide choice for the substrate material. Typically, nickel alloys such as Hastelloy C-276 (Arendt and Foltyn 2004) as well as stainless steel (Usoskin *et al.* 2004) have been used as substrates. These substrates have high temperature resistant capabilities and can withstand the high deposition temperatures (750–850 °C) used in (RE)BCO film processing. Additionally, these substrates have excellent mechanical strength (yield stress > 700 MPa), are non-magnetic and highly resistive. Non-magnetic substrates are preferred from the point of view that no ferromagnetic losses can occur when a superconductor is used in an AC environment. Also, highly resistive substrates such as Hastelloy (room temperature resistivity of 125 μΩ cm) results in negligible eddy current losses. IBAD-based (RE)BCO thin films benefit from yet another advantage: small grain size. Typically, the grain size of the IBAD film is sub-micrometre in size. Filamentization of the superconductor film is the most commonly used approach to reduce hysteretic AC losses, with the filament size in the range of 50–200 μm. If the grain size of the superconductor film is of the order of 1 μm, numerous grains would still be present in a single filament and so, current flow through a filament will not be affected even if a few block current flow. On the other hand, if the grain size of the superconductor film itself is about 50 μm, then only one or two grains would be present in a filament, and current flow along a filament could be completely blocked even with the presence of one problematic grain. Hence, the small grain size afforded by IBAD film is very beneficial for the fabrication of multifilamentary tapes with low AC losses.

The superconductor film can be grown epitaxially on top of IBAD YSZ or GZO. The lattice parameter of YSZ is 5.14 Å and that of GZO is 10.52 Å. Even if the superconductor grows epitaxially with a 45° rotation directly on IBAD YSZ or IBAD GZO (i.e.) the [100] of the (RE)BCO aligned along [110] of YSZ or GZO, a 6% or 3.5% lattice mismatch would exist respectively. This can result in misoriented grains of the superconductor, which can be deleterious to the flow of current. Hence, a lattice-matching layer such as CeO_2 is typically used between the IBAD film and the superconductor film. The lattice parameter of CeO_2 is 5.41 Å, which matches the lattice parameter of the basal plane of YBCO within less than 1% after a 45° rotation. Since the IBAD YSZ or GZO process is very slow, some institutions had adopted the approach of using a thinner IBAD film with a broader texture and depositing CeO_2 by pulsed laser deposition at a high deposition rate (Yamada *et al.* 2005a). Still, the IBAD process was a bottleneck in the fabrication of (RE)BCO tapes using this technique.

A breakthrough in IBAD technology for (RE)BCO tape fabrication occurred with the demonstration of a strong biaxial texture using rock-salt-structured materials such as MgO (Wang *et al.* 1997) compared to the already proven fluorites, bixbyites and pyrochlores. In the case of rock-salt materials like MgO, the close-packed planes are along (001), and hence they tend to naturally orient with an out-of-plane texture of (001) even without ion assist. Hence, ion assist is needed only to orient the grains along the in-plane to achieve a cube texture of (001) (100). Consequently, a good degree of cube texture is developed within the first 10 nm of film during IBAD in MgO. *In such a case, the development of cube texture is determined during the nucleation process itself, unlike the evolutionary texture development in IBAD YSZ and GZO*. Hence, biaxial texture with IBAD MgO can be achieved about 100 times faster than previously accomplished and the bottleneck was eliminated.

Since the MgO film was only 10 nm thick, however, it was not thick enough to provide a diffusion barrier like that provided by 1 μm-thick YSZ or GZO. Therefore, an additional barrier layer is required with the IBAD MgO architecture. It was found that an amorphous alumina layer deposited on the metal substrate provided excellent diffusion barrier characteristics (Usov *et al.* 2004). Being amorphous, no grain boundary paths are available for easy diffusion. Additionally, the alumina layer prevented diffusion of oxygen in the processing atmosphere to interact with the substrate. Previously, it was found that C contamination on the substrate could lead to CO evolution and subsequent delamination of thin-film layers if oxygen were to interact with the carbon (Xiong 2007). Hence, just 80 nm of alumina provided an excellent diffusion barrier to both cations as well as oxygen.

Since the biaxial texture in the MgO layer develops in the nucleation stage itself, the surface on which it nucleates can detrimentally affect the nucleation process. The original work with IBAD MgO was done on an amorphous surface of silicon nitride. It was found that the [001] axis of MgO tended to deviate from perpendicular to the film surface with increasing film thickness (Groves *et al.* 2003). A layer of amorphous or nanocrystalline yttria was found to be a superior nucleation surface for IBAD MgO resulting in no tilt of [001] axis.

Since the texture development in IBAD MgO occurs fast and within a limited range of conditions, it is important to monitor development of texture *in situ* during IBAD processing. Reflection high energy electron diffraction (RHEED) is used for monitoring texture development and an example of a RHEED pattern obtained from a growing IBAD MgO film is shown in Fig. 2.5. Texture in the cubic film is clearly seen. RHEED is used during manufacture of kilometre-long IBAD MgO tapes as a quality assurance tool, wherein the processing conditions are adjusted throughout the process to maintain a sharp RHEED pattern.

The alumina, yttria and IBAD MgO layers are all deposited at a room temperature that is beneficial for manufacturing. The deposited IBAD MgO film is highly stressed due to the ion bombardment, and the texture of the IBAD layer itself is not good enough to achieve high critical current densities. Therefore, a thin (~30 nm) layer of MgO is homoepitaxially grown on the IBAD MgO layer to sharpen the texture. Also, since the homoepitaxial MgO deposition is conducted at a temperature of about 700°C, it serves to provide stress relief to the architecture.

The final cap layer in the IBAD-based template architecture is a lattice-matching layer. The lattice constant of MgO is 4.22 Å and that of the basal plane of YBCO is 3.85 Å. This nearly 9% lattice mismatch between the two can result in the formation of misoriented YBCO grains, such as those that are rotated 45° in the plane of the film. In order to transition the lattice parameters, a cap layer, typically a perovskite, which has a lattice parameter close to YBCO, is epitaxially grown on the MgO layer. $LaMnO_3$ (Paranthaman *et al.* 2003), $SrTiO_3$ (Wang *et al.* 2007), and $SrRuO_3$

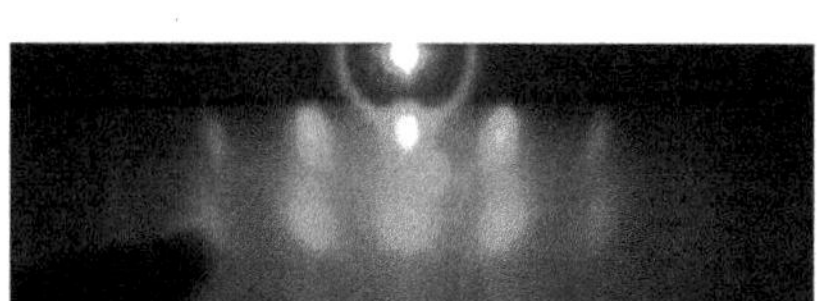

2.5 RHEED pattern obtained *in situ* from growing MgO film during IBAD.

(Jia *et al.* 2002) are some of the perovskites that have been used as a cap layer in the IBAD MgO template-based architectures. $SrTiO_3$ has a lattice parameter of 3.91 Å that matches well with that of YBCO and is a commonly used substrate in a single crystal form for high-quality epitaxial YBCO films (Roas 1990). $LaMnO_3$ has a lattice parameter of a = 5.53 Å, and b = 5.71 Å, but YBCO can grown epitaxially on it with a 45° rotation, i.e. the [100] of YBCO aligned with the [110] of $LaMnO_3$. $LaMnO_3$ is now commonly used in long 2G HTS tape manufacturing since it provides certain practical advantages that will be discussed in the next section.

A schematic of the architecture of 2G HTS tapes enabled by IBAD MgO templates is shown in Fig. 2.6. In Fig. 2.7, the cross-sectional microstructure of

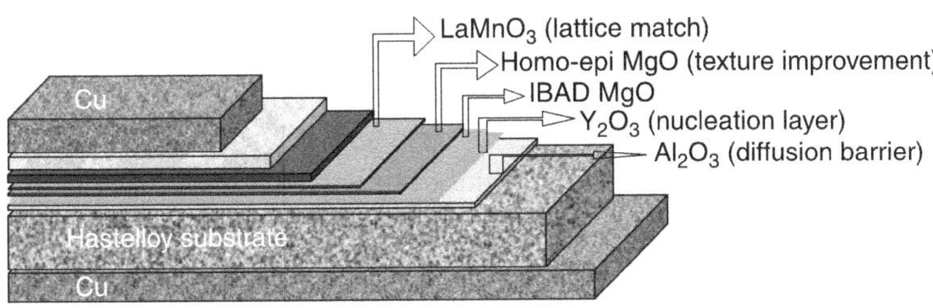

2.6 Architecture of 2G HTS tapes using textured templates.

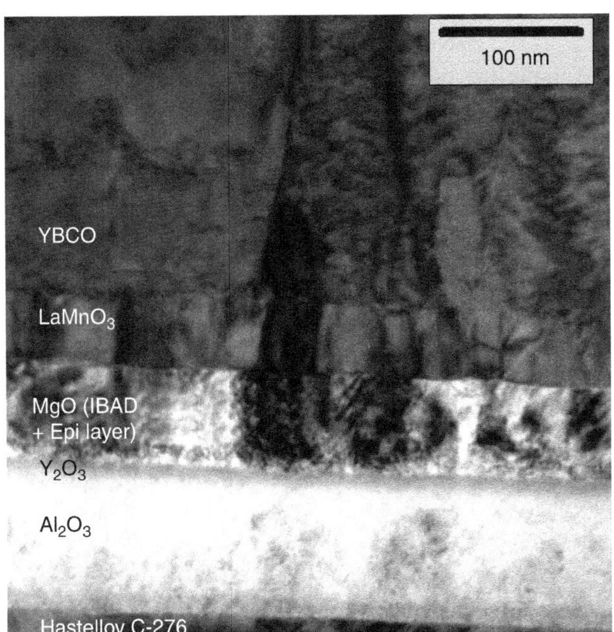

2.7 Cross-sectional microstructure of architecture of 2G HTS tapes based on IBAD MgO.

the state-of-the-art architecture based on IBAD films for biaxially-textured superconductors is shown. The architecture consists of a substrate that typically has a high degree of surface finish such as a surface roughness of about 1 nm. Such a high degree of surface finish is required since the IBAD MgO layer is only 10 nm, and biaxial texture develops in the nucleation stage itself. The architecture further consists of a barrier layer of about 80 nm-thick alumina, 7 nm-thick yttria, 10 nm-thick IBAD MgO, 30 nm-thick homo-epitaxial MgO and approximately 30 nm-thick $LaMnO_3$.

Rolling assisted biaxially textured substrates (RABiTS™) templates

In the case of IBAD templates, biaxial texture is achieved in an oxide layer through ion assist, and the flexible metal substrate itself does not impart any preferential texture. In 1995, an alternate type of textured template fabrication technique was developed: rolling assisted biaxially textured substrates (RABiTS) (Goyal *et al.* 1996). Here, metallurgical techniques previously developed for achieving cube texture in certain high-purity metals with high stacking-fault energy was effectively made use of for 2G HTS tapes. Certain FCC metals such as copper and Ni when deformed heavily (~99% cross-section reduction) develop a rolling texture. When the heavily deformed metal is recrystallized at a high-enough temperature (close to the melting point), the rolling texture is converted to a cube texture, i.e. (100) [001] where the grains within the substrate plane are oriented within 5–6° and also out-of-the-plane within a few degrees. Since even a small amount of impurities can lead to development of other orientations, high-purity metals are needed for RABiTS. Earlier, high-purity nickel was used as a substrate, but its softness required development of low-alloy-content binary alloys of Ni-W and Ni-Cr (de Boer *et al.* 2001). Typically 5–7.5% W is used to improve the yield strength from about 50 MPa for pure Ni to 180 MPa (Clickner *et al.* 2006, Eickemeyer 2008). A higher level of alloy content is preferred since these substrates are ferromagnetic (Ijaduola *et al.* 2004), leading to higher AC losses. However, use of even 9% W can cause problems with development of secondary textures, which can reduce the critical current density of the epitaxial superconductor film.

Also, the grain size of RABiTS templates is much larger than those prepared by IBAD because of the very high temperatures used in the processing of substrates. There has been concern about percolation of current through an epitaxial superconductor on RABiTS substrates when a few of the large grains are not properly oriented and thus block current flow (Hanisch *et al.* 2004). Indeed, it has been fortuitous that the most commonly used superconductor deposition technique with RABiTS substrates, namely metal organic deposition, yields a microstructure that overcomes such percolation issues.

As in the case of IBAD templates, buffer layers are required to transfer the biaxial texture epitaxially to the superconductor film as well as to provide a barrier to the diffusion of cations from the Ni alloy. Initially, there were concerns about nickel substrate oxidation when oxide buffer layers are deposited and metals such as Pd were used as an intermediate buffer. Later, it was found that processing of oxides in a forming gas (Ar-H_2) atmosphere enabled formation of the oxide buffer layer while maintaining a condition wherein NiO does not form (Norton *et al.* 1996). Alternatively, the surface of the Ni substrate was oxidized by a surface oxidation epitaxy where a biaxially-textured NiO is grown which is then used as a template for epitaxial growth of other buffers. In the first case, a 50 nm Y_2O_3 or CeO_2 is commonly employed as a seed layer to transfer the biaxial texture from the nickel substrate to an oxide. Then a 100–250 nm thick layer of YSZ is epitaxially grown and serves as a barrier to the diffusion of cations from the substrate. YSZ has a fluorite structure with a lattice parameter of 5.14 Å. YBCO has been found to grow epitaxially on YSZ after a 45° rotation in its basal plane but still, because of a nearly 6% lattice mismatch, misoriented grains can form in the superconductor. Hence, a lattice matching cap layer of CeO_2, about 50 nm thick, is epitaxially grown on the YSZ surface. The lattice parameter of CeO_2 is 5.41 Å and matches excellently with the YBCO after a 45° rotation. Again, it is fortuitous that CeO_2 is the chemically best-compatible candidate with the precursors used in the metal organic deposition (MOD) technique for the growth of YBCO.

Inclined substrate deposition (ISD) templates

A third, but not so-commonly used method for textured template fabrication, is the technique of inclined substrate deposition (ISD). In this case, a buffer layer such as YSZ or MgO is grown on a nickel alloy substrate at an angle of 55° to the incident flux. A shadowing mechanism appears to be responsible for biaxial texture, wherein grains oriented in a specific favourable direction shadow the growth of other grains. Pulsed laser deposition (PLD) (Ohmatsu *et al.* 2001) and evaporation techniques (Prusseit *et al.* 2005) have been used for ISD. Since texture development is based on preferential selection, a film thickness of more than 2 μm is required in ISD. However, since a rapid deposition rate also favours preferential orientation development, it compensates for the need for thick-film growth. ISD-based templates display a serrated morphology because of the inclined crystal growth (Koritala *et al.* 2005). This has become a problem in superconductor growth where the current has to traverse the serrated surfaces of the ISD template. An intermediate on-axis deposited buffer layer epitaxially grown over the ISD template has been employed to mitigate this issue (Prusseit *et al.* 2005).

2.2.4 Epitaxial superconductor deposition

Once a biaxially-textured template with a good lattice-matching cap layer is fabricated, superconducting films can be grown epitaxially by a number of chemical or physical deposition techniques. Metal organic precursors have been used in chemical deposition techniques such as metal organic chemical vapour deposition (MOCVD) and metal organic deposition (MOD). Oxide targets and elemental precursors are used in physical deposition techniques such as PLD and co-evaporation, respectively.

Metal organic chemical vapour deposition

MOCVD is a widely used technique for various electronic, optoelectronic, piezoelectric, ferroelectric, and multiferroic materials. In this process, precursors consisting of complex metal organic ligands of metal ions are used as a deposition source. A great benefit with MOCVD is that the precursors are maintained outside the deposition chamber, compared to the physical vapour deposition techniques, and hence can be refilled continuously during long deposition runs. Additionally, numerous mixtures of various precursors can be made with varying levels of dopants (to create flux-pinning sites, as described below) very quickly and effectively. Furthermore, high deposition rates of more than 0.5 μm/min have been achieved in superconductor film growth (Selvamanickam *et al.* 2004, Chou *et al.* 1995). Such high rates are unusual for MOCVD of other materials where rates of only about 1 μm/hour are used. Since the MOCVD precursors can be deposited over a large area, coupled with high deposition rates, high throughput can be achieved, which is very important for large-scale manufacturing.

Early work on MOCVD of superconducting films employed solid-state precursor delivery where solid sources of Y, Ba, and Cu with tetramethyl heptanedionate (thd) ligands were vaporized and the vapours were deposited on a hot substrate. Problems with unsteady vaporization of precursors as well as instability of Ba-thd prevented scalability of MOCVD (Becht 1996, Richards *et al.* 1995, Nagai *et al.* 1997, Burtman 1996, Busch 1993). Liquid precursor delivery wherein the -thd precursors of Y, Ba, and Cu are dissolved in a solvent such as tetrahydrofuran (THF) and then pumped into a vaporizer at a steady rate yielded a solution to this problem (Selvamanickam *et al.* 2001, Onabe *et al.* 2001, Stadel *et al.* 2000, Zhang *et al.* 1992). The vaporized precursors are then transported in a carrier gas of Ar, mixed with oxygen and then delivered onto a hot substrate through a showerhead. Substantial progress had to be made in achieving steady injection of precursors into the vaporizer, avoiding condensation of the vaporized precursors along the delivery lines as well as achieving uniform injection of precursors on the substrate over a large area. MOCVD is now the only technique that has been used to fabricate kilometre-long 2G HTS tapes, which by itself demonstrates the success of this technology.

Metal organic deposition

In a popular version of metal organic deposition, trifluoroacetate precursors of Y, Ba, and Cu are dissolved in trifluoroacetic acid and then deposited on a substrate

(McIntyre *et al.* 1990). Two steps have to be performed to convert the precursors into a superconducting film. The precursors have to be calcined at an intermediate temperature; there then follows their conversion in a water vapour atmosphere into a superconducting film. The advantage of this technique that nearly 100% of the precursors are used. One disadvantage is that the rate of superconductor formation is quite slow, of the order of 0.01 μm/min (Araki and Hirabayashi 2003). The rate limitation arises from diffusion of water vapour from the atmosphere to penetrate a rather thick film (~1 μm) to convert the precursor to film, and removal of hydrogen fluoride vapout through such a thickness. This rate limitation also makes it difficult to grow the thick films that are needed for high-current 2G HTS tapes.

Pulsed laser deposition

PLD is a well-used laboratory technique, and all the early work on 2G HTS tapes was done with PLD. In this process an excimer laser operating at 308 nm (XeCl) or 248 nm (KrF) is used to ablate a target made of a superconducting material and the plume is deposited on a hot substrate (Foltyn *et al.* 2000a, Iijima *et al.* 1998, Usoskin *et al.* 2001). A major advantage of PLD is the ability to form a stoichiometric film of nearly the same composition as the target. High deposition rates have also been demonstrated using industrial lasers (Foltyn *et al.* 2000b). Drawbacks have been with the very high cost of excimer lasers or laser tubes that need to be replaced often, and a limited deposition area.

Evaporation

Co-evaporation of Y, Ba and Cu metallic precursors has been explored for YBCO deposition (Prusseit *et al.* 2003). In this case, the metallic precursors are evaporated in individual sources using high-power electron beams or resistive heating and the vapour flux is deposited on a hot substrate. Low cost of precursors is an advantage. A challenge in making stoichiometric films is that the process requires advanced levels of control. This may be achieved using atomic absorption spectroscopy, controlled oxidation of metallic precursors to superconducting film, and techniques that involve an oxygen pocket or an oxygen shuttle.

2.2.5 Stabilizer

Once a superconducting film is deposited, it needs to be overcoated with metallic layers for electrical contacts. Nearly all metals react strongly with the superconductor, and so only silver or gold have been viable choices for an overlayer on top of the superconducting film. Sputtering has been the preferred choice to coat silver, but evaporation has also been employed by a few groups. In addition to a 2 μm-thick silver overlayer, a thicker copper layer also needs to be applied to the conductor to provide stabilization in case of overcurrents or a shunt as a parallel path around defects.

2.2.6 Properties of 2G HTS tapes

Critical current density (J_c) is the most important metric for 2G HTS tapes/wires. A four-probe technique is used to measure J_c where current is injected at the ends of the tape/wire and the voltage drop is measured at points over its length. From the current–voltage characteristic, the critical current (I_c) is defined as the current that results in a voltage drop of 1 μV/cm. The slope of the curve provides the n-value, which is also an important metric of the superconductor quality. Magnetic techniques using Hall probes have also been employed for 2G HTS tapes for rapid measurement of critical current density. The critical current density of 2G HTS tapes is dependent upon several microstructural features such as the degree of in-plane texture, out-of-plane texture, amount of misoriented (a-axis) grains, composition and oxygen content, to name a few. Additionally the thickness and roughness of buffer layers also significant affect J_c. A J_c of 5 MA/cm^2 can be achieved in 1 μm-thick films with a buffer in-plane texture of 5° or better.

Unlike superconducting films for electronic applications, high critical currents need to be achieved in 2G HTS tapes/wires in addition to a high J_c. An obvious approach to increase the I_c is to increase the superconductor film thickness. However, it has been found that the J_c of the superconducting film decreases with increasing thickness. As a result, the critical current increases with thickness only to a certain level, beyond which no additional gain is achieved. Figures 2.8a and 2.8b show the thickness dependence of critical current density and critical current respectively of 2G HTS tapes made by MOCVD on IBAD MgO templates. It is seen from Fig. 2.8b that critical currents increase up to a film thickness of about 3 μm, beyond which no improvement is achieved. Further, as seen in Fig. 2.8a, the J_c of a 2.5 μm thick film is only about 3 MA/cm^2 compared to 5.8 MA/cm^2 for a 0.5 μm-thick film. Two effects limit the critical current density of thick films. First, the microstructural quality is found to degrade with increasing thickness by the formation of misoriented a-axis grains as well as secondary phases. Figure 2.9 shows the top surface of 1 μm thick and 2 μm thick superconducting films grown by MOCVD. An abundant presence of a-axis grains can be observed in the thicker film. Cross-sectional transmission electron microscopy (TEM) analysis done on these films shows a-axis grains nucleating at sites of CuO in the superconducting film (Fig. 2.10). Misoriented grain growth has also been observed in thick films made by other techniques. Controlling the actual temperature of the superconducting film has been found to be a key to reducing the extent of a-axis grain growth and achieving higher J_c in thick films (Kakimoto *et al.* 2010).

A second possible cause for lower J_c in thick films could be the reduction in defect density away from the buffer–superconductor interface. Because of structural mismatch, an abundance of defects have been noticed at the buffer–

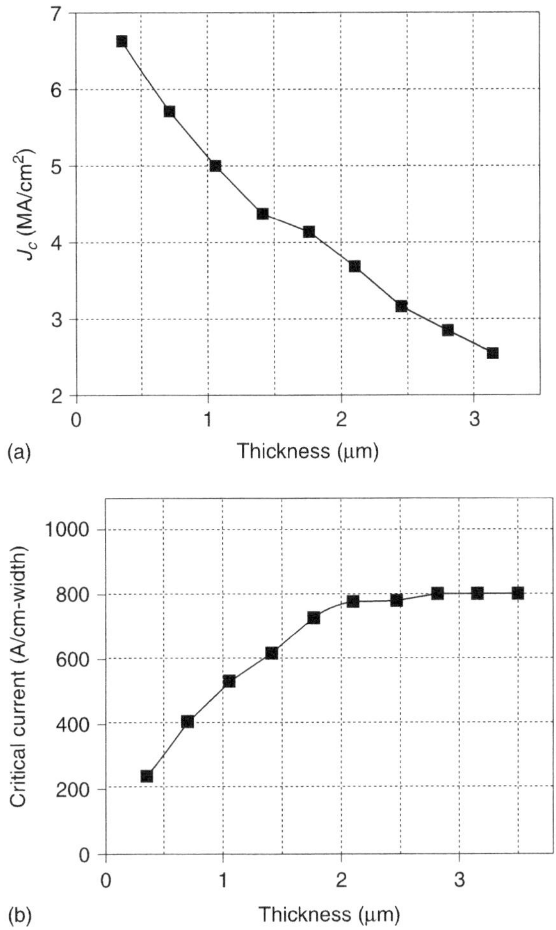

2.8 Film thickness dependence of critical current density (a) and critical current (b) of 2G HTS tapes.

superconductor interface, which is believed to contribute to improved pinning in thin films. It was shown that by reduction in the amount of these defects using an alternate substrate such as $NdGaO_3$, the improvement in critical current density near the interface observed in films on $SrTiO_3$ substrate can be eradicated (Foltyn *et al.* 2009). Work with PLD has also been done to create new interfaces within the superconducting film by alternate deposition of superconducting and dielectric layers. Critical currents as high as 1400 A/cm width were demonstrated in microbridged samples made by this multilayer technique (Foltyn *et al.* 2005).

In metre-long tapes, critical currents as high as 800 A/cm width were demonstrated by MOCVD. The critical current level in lengths of hundreds of metres is about 400 A/cm width, as described later in this chapter. There is plenty

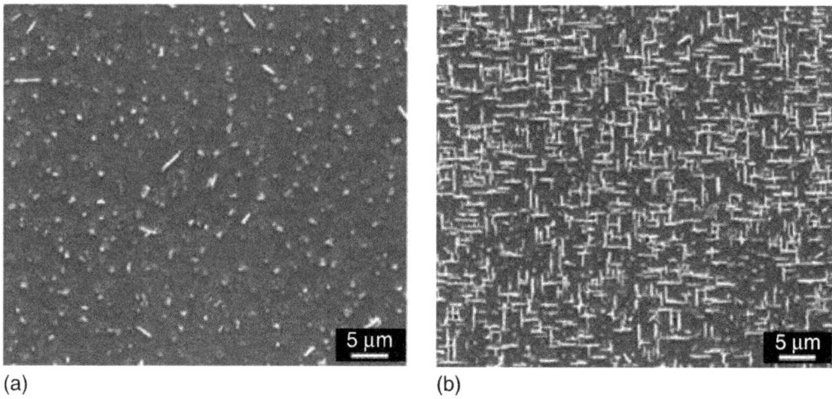

2.9 Microstructure of top surface of HTS film: (a) 1 µm thick, (b) 2 µm thick.

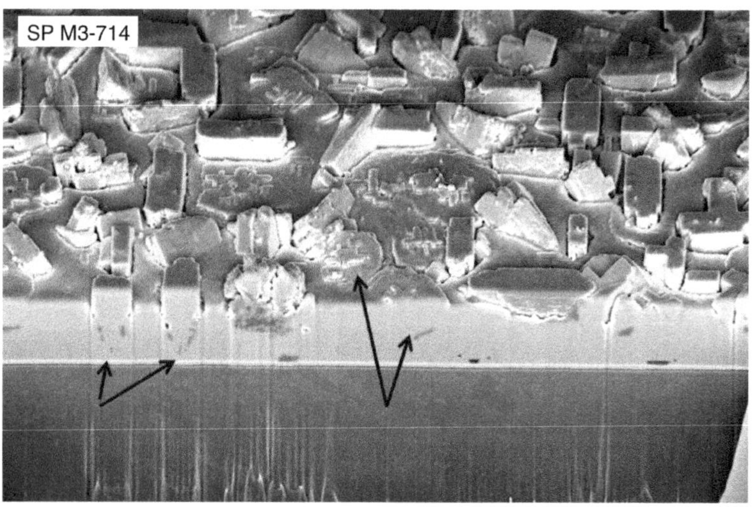

2.10 Cross-section of microstructure of HTS film showing growth of misoriented grains from CuO particles in the film.

of opportunity to raise the level of critical currents of long tapes to the levels achieved in short samples.

Since most applications of 2G HTS tapes/wires involve high magnetic fields, it is essential to maintain a high J_c in such magnetic fields over a range of temperatures. The problem has been that the critical current density of typical 2G HTS tapes/wires decreases by a factor of eight in a field of 1 T at 77 K when the field is applied perpendicular to the tape (Fig. 2.11). While this is still much better than with 1G tapes that have essentially zero J_c at that field, it requires that 2G HTS tapes need to be cooled down to temperatures of 30 K in order to achieve

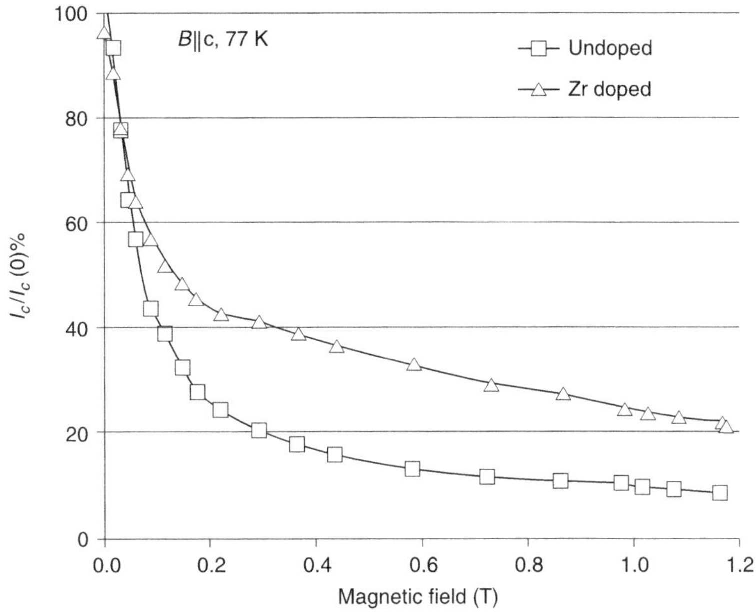

2.11 Magnetic field dependence of critical current of standard and Zr-doped 2G HTS tapes in the orientation of field perpendicular to tape at 77 K.

usable critical current values. An additional problem has been with the anisotropy in critical current, which in 2G HTS tapes is not as significant as with 1G HTS tapes, but still large enough to limit use at the higher temperatures. Figure 2.12 shows the angular dependence of critical current of a typical 2G HTS tape, where it can be seen that the I_c decreases rapidly as the magnetic field is moved away from the tape parallel direction (H || tape). Since the strongly conducting a–b plane is along the tape plane, a high J_c is achieved in the orientation of H || a–b. Intrinsic pinning due to the weakly superconducting layers between the Cu–O planes in the crystal lattice as well as pinning due to stacking faults in the a–b plane have been found to be responsible for pinning in the orientation of field parallel to the tape.

It was realized that defects need to be introduced perpendicular to the tape plane to enhance critical current in this orientation. Work on neutron, proton and heavy ion irradiation along the c-axis of single crystals of YBCO that resulted in columnar defect tracks in this orientation had shown significant improvements in J_c when the field is applied along the c-axis (Civale *et al.* 1990). This technique has been deemed to be impractical for commercial 2G HTS tapes. Work in the last five years on chemical substitution of the rare-earth element in YBCO and/or chemical doping of foreign elements have been shown to result in enhanced pinning. Pervoskites such as $BaZrO_3$ (BZO) or $BaSnO_3$ (BSO) have been

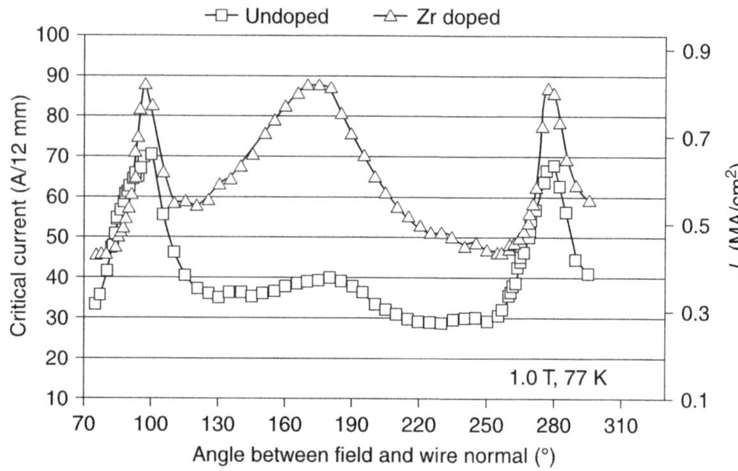

2.12 Angular dependence field dependence of critical current of standard and Zr-doped 2G HTS tapes at 1.0 T and 77 K.

demonstrated to be chemically compatible with YBCO and so have been effectively used as pinning centres in the form of self-assembled nanocolumns (Kang *et al.* 2006, Yamada *et al.* 2005b) or random particulate distribution (Macmanus-Driscoll *et al.* 2004). The lattice mismatch between the dopant phases and the superconductor results in a microstrain that is believed to be responsible for improved pinning (Puig *et al.* 2008, Rodriguez *et al.* 2008). In the case of *in situ* deposition techniques such as PLD and MOCVD, nanocolumnar structures of BZO about 5 nm in diameter have been observed primarily along the c-axis with a splay about this axis (Chen *et al.* 2009, Selvamanickam *et al.* 2009a). Figure 2.13 shows a cross-sectional TEM of a Zr-doped (Gd,Y)BCO conductor with abundant BZO nanocolumns along the c-axis. In addition, $(Gd,Y)_2O_3$ precipitates are found along the a–b plane. In MOCVD-based tapes, an optimum critical current is achieved with a Zr doping level of 7.5% (Selvamanickam *et al.* 2009b). A factor improvement of two has been achieved with Zr-doped HTS tapes in a magnetic field of 1 T applied perpendicular to the tape at 77 K, resulting in a zero-field critical current retention of 25–30% compared to about 10–15% in undoped tapes (Fig. 2.11). Such improved in-field performance is maintained even at lower temperatures. At 30 K and 2 T, a critical current of 2.5 times the critical current at 77 K, self field can be achieved. In comparison, the critical current of undoped films at 30 K and 2 T is 1.2 times the value at 77 K, self field.

Zr-doped HTS wires are now routinely made in production in lengths of hundreds of metres. Using a reel-to-reel system, the uniformity of critical current in a magnetic field in a 130 m-long Zr-doped tape was confirmed (Selvamanickam *et al.* 2011). Figure 2.14 shows the zero-field critical current, critical current in the orientations of $B \parallel c$, and at the angle corresponding to the minimum I_c in the

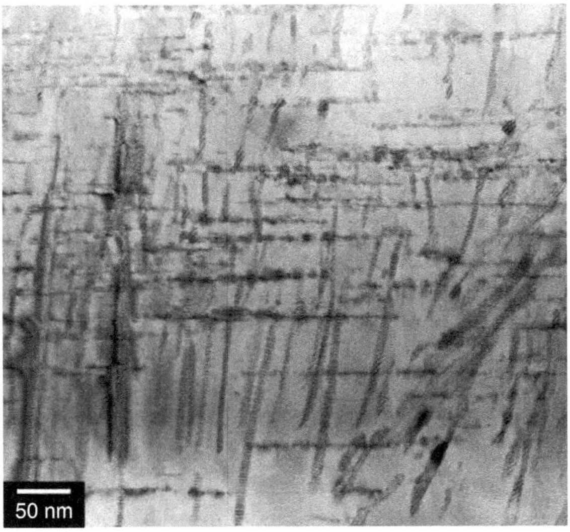

2.13 Cross sectional microstructure of HTS film made by MOCVD with 7.5% Zr showing self assembled nanocolumns of $BaZrO_3$.

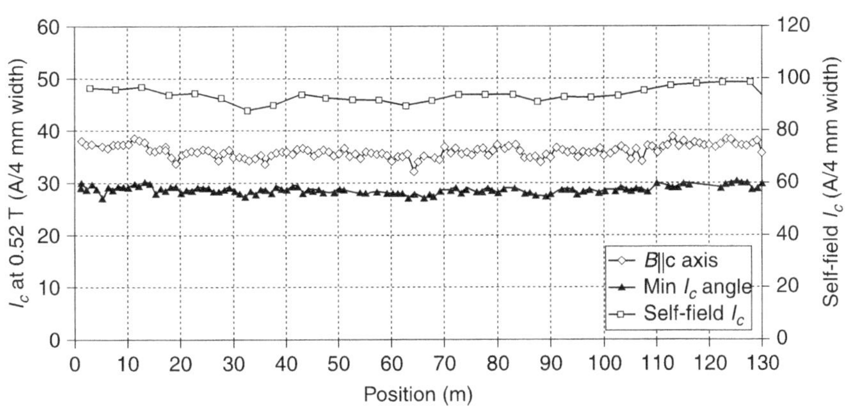

2.14 Critical current of 130 m Zr-doped 2G HTS tape made by MOCVD at zero applied field and at a field 0.52 T applied perpendicular to the tape and at an angle corresponding to minimum critical current value.

angular range. As shown in Fig. 2.14, the critical currents measured at 0.52 T applied along the c-axis and along the angle corresponding the minimum I_c are found to be very uniform over the entire length of 130 m. The average critical current at $B \parallel c$ is 36 A in the 4 mm-wide tape with an uniformity of 3.2% and the average critical current in the minimum I_c orientation is 28 A (over 4 mm) with a uniformity of 2.5%. The uniformity in in-field I_c values is comparable with the 3%

uniformity measured in the self-field I_c values over 130 m. Hence, it can be concluded that a uniform degree of pinning has been achieved in the Zr-doped tape over the entire length of 130 m. Also, the critical current retention of the average I_c values at $B \parallel c$ and at the minimum I_c angle in the applied field of 0.52 T are 38% and 30% respectively, of the average self-field I_c value of 94 A (4 mm width). These retention factors are consistent with the values measured in short samples cut from the ends of the long tape.

2.3 Manufacturing scale-up of 2G HTS tapes and prototype devices

2G HTS tapes are, at the time of writing, being manufactured by at least two manufacturers with an annual production capacity of a few hundred kilometres. Numerous process advances were made 2002–8 in the scale-up of 2G HTS tapes from just a metre length to a kilometre. Substrate electropolishing is conducted on multiple tapes simultaneously. Reactive sputtering with metal targets is now employed instead of radio frequency (rf) sputtering using oxide targets for high-rate deposition of alumina, yttria, MgO and $LaMnO_3$ buffers. Typical tape speeds are now about 600–750 m/h (4 mm wide equivalent) for alumina, yttria, and IBAD MgO layers and 360 m/h (4 mm wide equivalent) for homo-epi MgO and $LaMnO_3$ layers. Such high tape speeds are possible using helix-based tape handling, where a 12 mm-wide tape is wound around in a helical fashion multiple times over the deposition zone (typically 6–12 wraps). Such a helix tape configuration enables very high linear speeds, uniformity of deposition and long single-piece lengths. Additionally, simultaneous deposition of two layers (alumina and yttria, MgO and $LaMnO_3$) are conducted using two sequential deposition chambers with helix tape handling in both.

Several innovations have been done with superconductor processes too. Process stability in the MOCVD process has been increased to 40+ hours with uniform precursor delivery over such long time periods. Metre-long showerheads, helix tape handling, and high rates of precursor delivery all enable high linear tape speeds, even for deposition of thick films. On-line quality control is routinely used in superconductor and buffer deposition. On-line X-ray diffraction to monitor the relative intensities of (006) and (200) peaks of YBCO, on-line vision monitoring to examine surface inhomogeneities, buffer film thickness, on-line ellipsometry to examine film thickness in real time, and reflection high-energy electron diffraction (RHEED) to monitor texture developed during IBAD are a few of the *in situ* quality control techniques that have been developed in 2G HTS manufacturing.

Through process, equipment and materials innovations, 2G HTS wire manufacturing has been successfully scaled up to kilometre lengths. In addition, innovative test equipment has been developed to measure critical currents and n-values by continuous transport dc current using four-probe technique over 5 m intervals. A photograph of such a system is shown in Fig. 2.15.

2.15 Photograph of a transport critical current measurement for long tapes.

Figure 2.16 shows the critical current performance of a 1065 m long 2G HTS tape measured at 5 m intervals. A minimum critical current of 282 A/cm width was achieved over 1065 m corresponding to a performance level of 300–330 A m (Selvamanickam *et al.* 2009c). It can also be seen from the figure that the critical current levels are at 350–400 A/cm width over most of the length indicating that high performance can be achieved over long lengths, if not for a few drop outs. The reasons for the drop outs in I_c are typically mechanical damage (kinks), substrate defects or process fluctuations. Since helix tape handling and a large

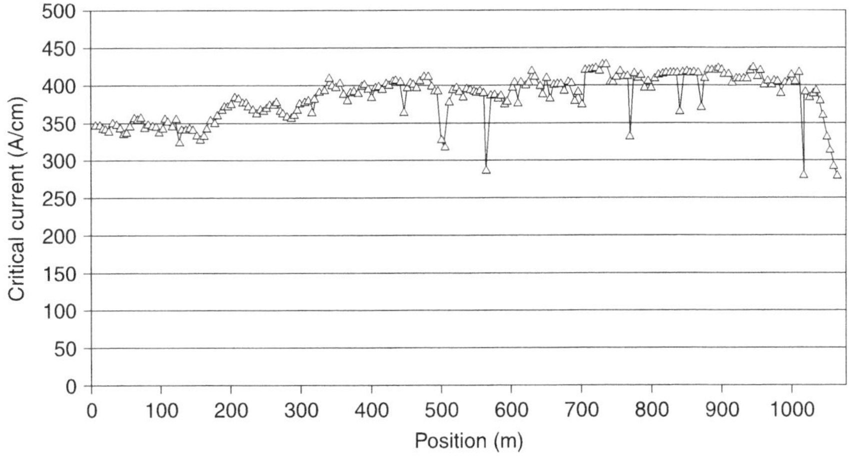

2.16 Critical current measurement of a 1065 m 2G HTS tape made by MOCVD.

deposition area are used, usually the tape is exposed over a length of about 5 m at any time in the deposition area. Therefore, any process fluctuations would impact a length of a few metres while substrate defects and mechanical damage cause I_c drops over a few millimetres.

In addition to transport current measurements, Hall probe-based techniques are used for determining local fluctuations in critical current down to a millimetre. Drops in critical current over a 5 m interval can thus be narrowed down to a millimetre using such measurements. Figure 2.17 shows critical current measured every millimetre by a Hall-probe technique over a 0.2 m segment of a kilometre long 2G HTS tape showing a localized drop in critical current. Other non-destructive techniques have been developed to monitor local variations in long tape critical current (Kiss *et al.* 2007).

An important area of processing of 2G HTS tapes includes silver overlayer deposition, slitting to required widths (typically 4 mm) and stabilizer application. Electroplating (Selvamanickam *et al.* 2005) and bonding (Schoop *et al.* 2005) are commonly used techniques for copper stabilizer application. Electroplating provides the benefits of rounded corners, which are desirable for high-voltage applications, absence of low melting point solders – thereby extending the usable temperature to 250 °C instead of 150 °C, and the ability to create a completely striated stabilizer for low AC loss multifilamentary architectures. Bonding

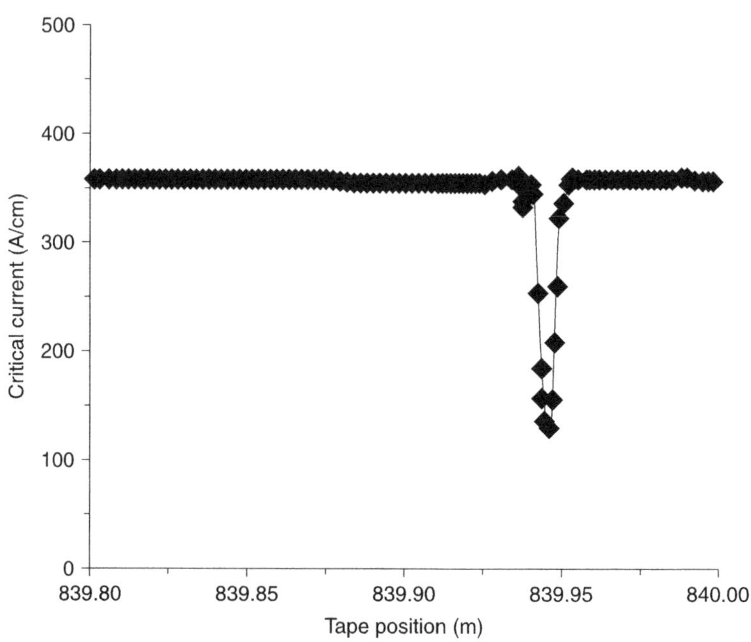

2.17 Critical current measurement obtained by Hall probe method from a 0.2 m segment of a 1 km 2G HTS tape.

provides the benefits of ease of use of various stabilizer materials such as stainless steel for high resistance, thicker stabilizers of the order of several hundred microns and lower cost.

Tremendous progress in this decade has resulted in 2G HTS wire performance levels being improved from just 90 A m in 2002 to over 300 000 A m in 2008. The progress during this time period is shown in Fig. 2.18. As shown in the figure, rapid progress occurred during 2006–8, which also marked the beginning of pilot manufacturing of 2G HTS wire. As of 2010, three other institutions have reported producing 500 m lengths of 2G HTS tapes (Rupich *et al.* 2010, Kutami *et al.* 2009, Izumi *et al.* 2009).

The first major delivery of 2G HTS wire was made for the Albany cable project, where nearly 10 000 m of wire was delivered with an average performance level of 200 A/cm width and meeting the required specification including 4 mm width and copper stabilizer (Selvamanickam *et al.* 2007). This wire was used to fabricate a 30 m-long, three-phase cable. The cable was tested to have a I_c of 2660–2820 A in the conductor layers and 2400–2500 A in the shield layers. The AC loss of the cable was found to be 0.34 W/m/phase at 800 A_{rms}. This cable was installed in the electric grid of National Grid in Albany, NY connecting two substations through a joint with a longer 1G HTS cable (Yumura *et al.* 2009). Photographs of the cross-section of the cable and installation in the underground duct are shown in Fig. 2.19 and Fig. 2.20 respectively.

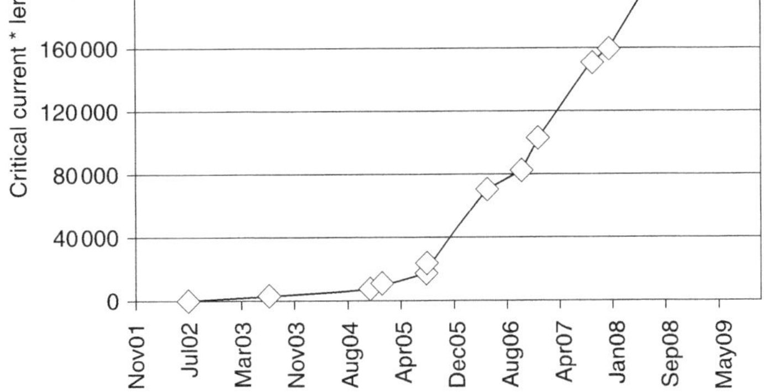

2.18 Critical current performance of long 2G HTS tapes fabricated by SuperPower between 2002 and 2008.

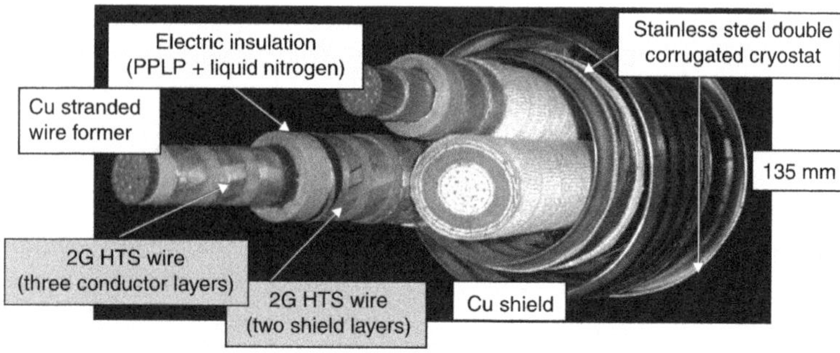

2.19 Internal construction of a 2G HTS cable.

2.20 Installation of the world's first device made with 2G HTS tape, in power grid.

Current–voltage characteristics obtained from the 30 m 2G HTS cable are shown in Fig. 2.21. As shown in the figure, an I_c of 2300 A was measured at 73 K whereas an I_c of 2750 A was measured at 69 K. No significant loss was found in the performance of the 2G cable after installation in the underground duct in the power grid. On 8 January 2008 the cable was energized in the power grid, marking this as the first in-grid device of any kind made with 2G HTS wire. The cable transmitted electricity in the grid powering 25 000 households in the Albany area without any problems.

Another application where prototypes of 2G HTS wire have been demonstrated is high-field insert coils. The high current density of 2G HTS wire at 4.2 K as well as its thin profile (total thickness of 0.1 mm) with the high-strength properties of the nickel alloy substrate provide excellent benefits for high-field insert coils. In 2007, an insert coil with 87 mm external diameter and 9.5 mm internal diameter was fabricated with 462 m of wire with an average critical current of 78 A/4 mm width. Details of the coil are shown in Table 2.1. As shown in Fig. 2.22, the

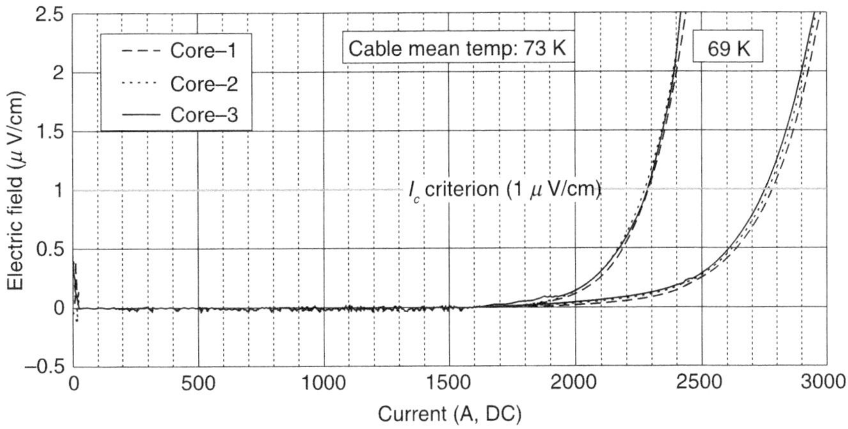

2.21 Current–voltage characteristic obtained from a 30 m 2G HTS cable installed in the National Grid.

Table 2.1 Details of coil made with 2G HTS tape and tested at 4.2 K

Coil ID (mm)	9.5
Winding inner diameter (mm)	19.1
Winding outer diameter (mm)	~87
Coil height (mm)	~51.6
Number of pancakes	12 (6 × double)
2G tape used (m)	~462
Average I_c of tapes in coil, 4 mm width (77 K, self field) (A)	78
Number of turns	~2 772
Coil J_e (A/mm^2 per A)	~1.569
Coil constant (mT/A)	~44.46
4.2 K coil I_c self field (A)	221
4.2 K amp turns at I_c self field	612 612
4.2 K central field, self field (T)	9.81
4.2 K coil I_c 19 T background (axial) (A)	175
4.2 K amp turns at I_c, 19 T background (axial)	485 100
4.2 K central field 19 T background (axial) (T)	36.8

coil was found to generate 9.81 T at 4.2 K in the absence of an external field which, to the author's knowledge, is the highest field generated by an HTS coil. In an applied field of 19 T, an additional field 7.8 T was generated resulting in a total field of 26.8 T (Fig. 2.22). Even higher fields of 27.4 T in a background field of 19 T and 33.8 T in a background field of 31 T have been generated with 2G HTS wire at 4.2 K.

The above coil was fabricated with a standard 2G HTS wire made without any dopants in the HTS film. Prototype coils have also been fabricated with Zr-doped HTS wires. A pancake coil with a clear bore of 21 mm was made

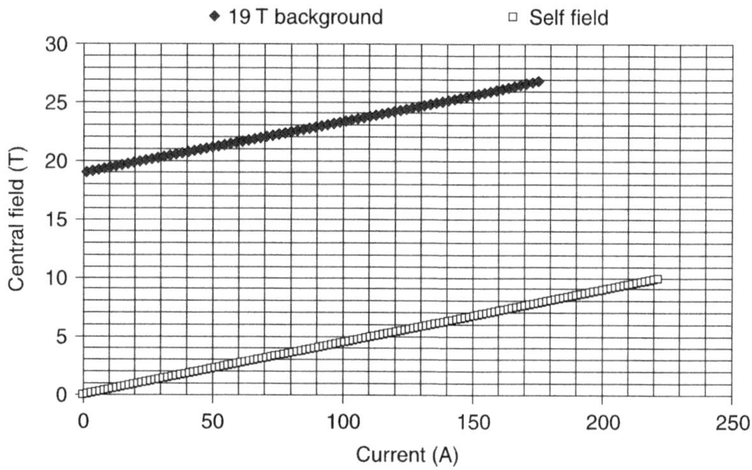

2.22 Load curves of coil made with 2G HTS tapes in zero background field and a background field of 19 T, at 4.2 K.

with six double pancakes. About 480 m of Zr-doped HTS wire with a critical current of 90–101 A in 4 mm widths was used. The engineering current density of the coil was measured to be nearly 168 A/mm^2 at 77 K. The coil generated a field of 1.09 T at 77 K, 2.5 T at 65 K and 3.0 T at 61 K. The field generated by the coil at 65 K is the same as that generated by the coil made with undoped wire. However, the coil with undoped wire had a much smaller bore diameter of 12.7 mm and used more wire (600 m) and much higher I_c wire (120–180 A). A comparison of the key parameters of the two coils is shown in Table 2.2. It is clearly seen that the same level of magnetic field is achieved in the coil made with Zr-doped wire even with less wire, lower critical current and larger bore. This outcome is a good demonstration of the benefit of Zr-doped wires for magnets and coils.

Table 2.2 Comparison of coils made with 2G HTS tapes with and without Zr doping

Coil property	With Zr-doped wire	With undoped wire
Coil inner diameter (mm)	22*	12.7*
Winding inner diameter (mm)	28.6	19.1
Number of turns (mm)	2688	3696
2G wire used (m)	~480	~600
Wire I_c (A)	90–101	120–180
Field generated at 65 K (T)	2.5	2.49
Wire I_c (A)	90–101	120–180

* Clear

2.4 Future of HTS wires and tapes and current research focus

Progress in the last ten years has brought 2G HTS wires to pilot-scale manufacturing and prototype demonstration. In order to reach full-scale commercialization and realize the vast market potential, the price/performance metrics of 2G HTS wires need to be improved. At present 2G HTS wires cost \$250–400/kA m based on their critical current performance at 77 K. The target is to achieve a 5–8 fold reduction in the wire cost to meet the requirements of various power applications. This reduction can be achieved by decreasing the unit cost (\$/m), increasing critical current (A) as well as improved retention of critical current at operating conditions of magnet field and temperature. In addition to improving cost/performance metrics, 2G HTS wires must be made available in large quantities of thousands of kilometres annually, up from the hundreds of kilometres available today. Given the expense of capital equipment, it is therefore important to improve throughput capabilities of wire fabrication technologies.

Issues with achieving high critical currents and approaches to obviate the challenges were discussed earlier in this chapter. Also, strategies to improve pinning in 2G HTS tapes for better in-field properties were also discussed. These are the primary areas of focus in 2G HTS wire research now.

Opportunities exist with reducing the number of thin-film layers in thin-film architecture using multifunctional layers. For example the alumina and yttria film in the IBAD-based architecture have been combined as a single $YAlO_3$ layer (Stan *et al.* 2009). Also, modified processes such as planarization are being developed as an alternate to electropolishing. In this case, two process steps, namely electropolishing and alumina barrier layer deposition, can be replaced with a single step. Techniques that enable improved process efficiency are very beneficial for both cost reduction as well as throughput increase. One such avenue that is being explored is to increase the conversion efficiency of MOCVD precursors to film. At present, the complex β-diketonate precursors of the HTS film are decomposed only by thermal activation, which is not very efficient resulting in a conversion efficiency of only about 15%. Plasma and UV activation of precursors are being explored to enhance the conversion efficiency, which will have a dramatic benefit on reducing the materials cost (which is a significant portion of the wire cost) as well as increasing throughput. Other process modifications such as electrodeposition of silver (Selvamanickam *et al.* 2010) instead of commonly used sputtering can lead to low-cost wires. Silver electrodeposition is also advantageous for multifilamentary 2G HTS wires with all striated silver and copper overlayers. Multfilamentary wires are being explored for reducing AC losses in HTS wires and thereby minimize the cryogenic burden in applications that involve significant AC fields.

2.5 Conclusions

Tremendous advances have been made in HTS wire technology development and scale up in the last decade. While first generation BSCCO-2223 superconductors were the workhorse for construction of various device prototypes in most of the decade, second-generation HTS tapes are rapidly becoming the choice for recent projects. Two techniques, namely ion beam assisted deposition (IBAD) and rolling assisted biaxially textured substrates (RABiTS) have been successfully developed to achieve excellent biaxial texture and have been scaled up to long lengths. Several processes such as metal organic chemical vapour deposition (MOCVD), metal organic deposition (MOD), pulsed laser deposition (PLD) and co-evaporation are being pursued for superconductor film deposition. High critical current densities have been achieved in lengths of hundreds of metres. The IBAD-MOCVD technique has been scaled up to over kilometre lengths with critical current performance of 300 000 A m. Prototype devices such as power transmission cable and high-field coils have been demonstrated with 2G HTS wires. The world's first 2G HTS device was inserted in the electric power grid in 2008. Ongoing research on 2G HTS tapes is focused on improving price/performance metrics. Significant advancements have been made in in-field performance of 2G HTS tapes by chemical doping. Further improvements in critical current densities of thick films and pinning characteristics in high fields are being actively addressed.

2.6 References

Araki T and Hirabayashi I, 'Review of a chemical approach to $YBa_2Cu_3O_{7-x}$ coated superconductors – metalorganic deposition using trifluoroacetates' (2003) *Supercond. Sci. Technol.* **16**, R71–R94.

Arendt P, and Foltyn S, 'Biaxially Textured IBAD-MgO Templates for YBCO-Coated Conductors' (2004), *MRS Bulletin*, 543–550.

Becht M, 'Metal-organic chemical vapor deposition as an emerging technique for the fabrication of $YBa_2Cu_3O_x$ tapes' (1996) *Appl. Supercond.* **4**, 465–474.

Bransky I, Maartense I, and Peterson T L, 'Dependence of transport critical current density on sample size for ceramic RBa2Cu3O7-$_x$ (R≡Y, Ho, Gd)' (1989) *J. Appl. Phys.* **66**, 5510–5514.

Burtman V, Schieber M, Brodsky I, Hermon H, and Yaroslavsky Y, 'Raman and infrared spectroscopy and transport properties of aged organometallic precursors used for the deposition of high-Tc superconducting films' (1996) *J. Cryst. Growth* **166**, 832–835.

Busch H, Fink A, Müller A, and Samwer K, 'Beta-diketonate precursors for the chemical vapour deposition of YBaCuO' (1993) *Supercond. Sci. Technol.* **6**, 42.

Chen I W, Wu X, Keating S J, Johnson P A, and Lien T Y, 'Texture development in YBa_2Cu_3Ox by hot extrusion and hot-pressing' (1987) *J. Am. Ceram. Soc.* **70**, C388–C390.

Chen K C, Maheswaran B, Liu Y P, Giessen B C, Chan C *et al.*, 'Critical current enhancement in field-oriented $YBa_2Cu_3O_{7-}\delta$' (1989) *Appl. Phys. Lett.* **55**, 289–291.

Chen Y, Selvamanickam V, Zhang Y, Zuev Y, Cantoni C *et al.*, 'Enhanced flux pinning by $BaZrO_3$ and $(Gd,Y)_2O_3$ nanostructures in metal organic chemical vapor deposited GdYBCO high temperature superconductor tapes' (2009) *Appl. Phys. Lett.* **94**, 062513.

Chou P C, Zhong Q, Li Q L, Abazajian K, Ignatiev A *et al.*, 'Optimization of J_c of YBCO thin films prepared by photo-assisted MOCVD through statistical robust design' (1995) *Physica C.* **254**, 93–112.

Civale L, Marwick A D, McElfresh M W, Worthington T K, Malozemoff A P *et al.*, 'Defect independence of the irreversibility line in proton-irradiated Y-Ba-Cu-O crystals' (1990) *Phys. Rev. Lett.* **65**, 1164–1167.

Clickner C C, Ekin J W, Cheggour N, Thieme C L, Qiao Y *et al.*, 'Mechanical properties of pure Ni and Ni-alloy substrate materials for Y–Ba–Cu–O coated superconductors' (2006) *Cryogenics*, **46**, 432–438.

D'Ovidio C A, Fiscina J E, and Esparaza D A, 'Influence of the sample geometry on critical current density of high T_c granular superconductors' (1991) *Appl. Phys. Lett.* **69**, 8265–8267.

De Boer B, Reger N, Fernandez G R, Eickemeyer J, Holzapfel B *et al.*, 'Biaxially textured Ni-alloy tapes as substrates for buffer and YBCO film growth' (2001) *Physica C.* **351**, 38–41.

Dimos D, Chaudhari P, and Mannhart J, 'Superconducting transport properties of grain boundaries in $YBa_2Cu_3O_7$ bicrystals' (1990) *Phys. Rev. B.* **41**, 4038–4049.

Eickemeyer J, Huhne R, Guth A, Rodig C, Klauß H *et al.*, 'Textured Ni–7.5 at.% W substrate tapes for YBCO-coated conductors' (2008) *Supercond. Sci. Technol.* **21**, 105012.

Ekin J W, Braginski A I, Panson A J, Janocko M A, Capone II D W *et al.* 'Polycrystalline $Y_1Ba_2Cu_3O_x$' (1987) *J. Appl. Phys.* **62**, 4821–4828.

Ekin J W, Hart, Jr H R, and Gaddipati A R, 'Transport critical current of aligned polycrystalline $Y_1Ba_2Cu_3O_{7-}\delta$ and evidence for a nonweak linked component of intergranular current conduction' (1990) *J. Appl. Phys.* **68**, 2285–2295.

Evetts J E, and Glowacki B A, 'Relation of critical current irreversibility to trapped flux and microstructure in polycrystalline $YBa_2Cu_3O_7$' (1988) *Cryogenics* **28**, 641–649.

Foltyn S R, Arendt P N, Depaula R F, Dowden P C, Coulter J Y *et al.*, 'Development of meter-long YBCO coated conductors produced by ion beam assisted deposition and pulsed laser deposition' (2000a) *Physica C.* **341–348**, 2305–2308.

Foltyn S R, Dowden P C, Arendt P N, and Coulter J Y, 'Proc. Annual Peer Review of the Department of Energy's Superconductivity Program for Electric Systems', Washington DC, July 17–19 (2000b).

Foltyn S R, Wang H, Civale L, Jia Q X, Arendt P N *et al.*, 'Overcoming the barrier to 1000 A/cm width superconducting coatings' (2005) *Appl. Phys. Lett.* **87**, 162505–162507.

Foltyn S R, Wang H, Civale L, Maiorov B and Jia Q X, 'The role of interfacial defects in enhancing the critical current density of YBa2Cu3O7-δ coatings' (2009) *Supercond. Sci. and Technol.* **22**, 125002.

Goldschmidt D, 'Multiple superconducting transition in ceramic $YBa_2Cu_3O_{7-}$' (1989) *Phys. Rev. B.* **39**, 2372–2376.

Goyal A, Norton D P, Budai J D, Paranthaman M, Specht E D *et al.*, 'High critical current density superconducting tapes by epitaxial deposition of $YBa_2Cu_3O_x$ thick films on biaxially textured metals' (1996) *Appl. Phys. Lett.* **69**, 1795–1797.

Groves J R, Arendt P N, Foltyn S R, Jia Q X, Holesinger T G *et al.*, 'Improvement of IBAD MgO template layers on metallic substrates for YBCO HTS deposition' (2003) *IEEE Trans. Appl. Supercond.* **13**, 2651–2654.

Gyrogy E M, van Dover R B, Jackson K A, Schneemeyer L F and Waszczack J V, 'Anisotropic critical currents in $Ba_2YCu_3O_7$ analyzed using an extended Bean model' (1989) *Appl. Phys. Lett.* **55**, 283–285.

Hanisch J, Sarma V S, Zeimetz B, Schindler F, Eickemeyer J *et al.*, 'Simulation of the critical current density and its dependence on geometrical factors in RABiTS based coated conductors' (2004) *Supercond. Sci. Technol.* **17**, 1003.

Heinig N F, Redwing R D, Tsu F, Gurevich A, Nordman J E *et al.*, 'Evidence for channel conduction in low misorientation angle [001] tilt $YBa_2Cu_3O_{7-x}$ bicrystal films' (1996) *Appl. Phys. Lett.* **69**, 577–579.

Holesinger T G, Kennison J A, Liao S, Yuan Y, Jiang J *et al.*, 'A shell model for the filament structure of Bi-2223 conductors' (2005) *IEEE Trans. Appl. Supercond.* **15**, 2514–2517.

Iijima Y, Hosaka M, Tanabe N, Sadakata N, Saitoh T *et al.*, 'Growth structure of yttria-stabilized-zirconia films during off-normal ion-beam-assisted deposition' (1998) *J. Mater. Res.* **13**, 3106–3113.

Iijima Y, Kakimoto K, Yamada Y, Izumi T, Saitoh T *et al.*, 'Research and development of biaxially textured IBAD-GZO templates for coated superconductors' *MRS Bulletin* (2004) **29**, 564–571.

Iijima Y, Kaneko N, Sutoh Y, Kakimoto K, Saitoh T *et al.*, 'The influence of assisting ion elements on texture improvement of IBAD-$Gd_2Zr_2O_7$ films' (2005) *Physica C.* **426–431**, 899–903.

Iijima Y, Tanabe N, Ikeno Y and Kohno O, 'Biaxially aligned $YBa_2Cu_3O_{7-x}$ thin film tapes' (1991) *Physica C* **185–189**, 1959–1960.

Ijaduola A O, Thompson J R, Goyal A, Thieme C L H and Marken K, 'Magnetism and ferromagnetic loss in Ni–W textured substrates for coated conductors' (2004) *Physica C.* **403**, 163–171.

Imayev M F, Kazakova D B, Gavro A N and Trukhan A P, 'Grain growth in a $YBa_2Cu_3O_{7-x}$ superconductive ceramics', (2000) *Physica C* **329**, 75–87.

Izumi T, Yoshizumi M, Miura M, Nakaoka K, Sutoh Y *et al.*, 'Development of TFA-MOD Process for Coated Conductors in Japan' (2009) *IEEE Trans. Appl. Supercond.* **19**, 3119–3122.

Jia Q X, Foltyn S R, Arendt P N, Groves J R, Holesinger T G *et al.*, 'Role of $SrRuO_3$ buffer layers on the superconducting properties of $YBa_2Cu_3O_7$ films grown on polycrystalline metal alloy using a biaxially oriented MgO template' (2002) *Appl. Phys. Lett.* **81**, 4571–4573.

Jin S and Graebner E, 'Processing and fabrication techniques for bulk high-Tc superconductors. A critical review' (1991) *Mat. Sci. and Eng.* **B7**, 243–260.

Jin S, Sherwood R C, Tiefel T H, Van Dover R B and Johnson Jr D W, 'High *TC* superconductors—composite wire fabrication' (1987) *Appl. Phys. Lett.* **51**, 203–204.

Jin S, Tiefel T H, Sherwood R C, Davis M E, van Dover R B *et al.*, 'High critical currents in YBaCuO superconductors' (1988) *Appl. Phys. Lett.* **52**, 2074–2076.

Johnson D W Jr., Gyorgy E M, Rhodes W W, Feldman L C and van Dover R B, 'Fabrication of ceramic articles from high Tc superconducting oxides' (1987) *Adv. Ceram. Mater.* **2**, 364–371.

Kakimoto K, Igarashi M, Hanada Y, Hayashida T, Tashita C *et al.* 'High-speed deposition of high-quality RE123 films by a PLD system with hot-wall heating' (2010) *Supercond. Sci. & Technol.* **23**, 014016.

Kang S, Goyal A, Li J, Gapud A A, Martin P M *et al.*, 'High-performance high-T_c superconducting wires' (2006) *Science* **311**, 1911–1914.

Kawahara N, Kawabata S, Enami H, Shinohara T, Hoshizaki H *et al.*, 'Preparation of bulky Bi(Pb)-Sr-Ca-Cu-O superconductor by magnetized twin-roll' (1990) *Jpn. J. Appl. Phys.* **29**, L284–L286.

Kiss T, Inoue M, Shoyama T, Koyanagi S, Mitsui D *et al.*, 'Visualizing Transport Properties in IBAD Based YBCO Coated Conductors by Multiple Analysis Techniques' (2007) *IEEE Trans. Appl. Supercond.* **17**, 3211–3214.

Koritala R E, Ma B, Miller D J, Uprety K K, Fisher B L *et al.*, 'Surface roughness of magnesium oxide buffer layers grown by inclined substrate deposition' (2005) *IEEE Trans. Appl. Supercond.* **15**, 3031–3033.

Kumakura H, Uehara M and Togano K, 'Localized large superconducting critical currents in $Ba_2Y_1Cu_3O_{9-y}$' (1987) *Appl. Phys. Lett.* **51**, 1557–1560.

Kupfer H, Apfelstedt I, Flukiger R, Keller C, Meier-Hirmer R *et al.*, 'Investigation of inter- and intragrain critical currents in high T_c ceramic superconductors' (1988) *Cryogenics* **28**, 650–660.

Kutami H, Hayashida T, Hanyu S, Tashita C, Igarashi M *et al.*, 'Progress in research and development on long length coated conductors in Fujikura' (2009) *Physica C.* **469**, 1290–1293.

Lo W, and Glowacki B, 'Effect of cold rolling on microstructure, grain alignment, and transport critical current anisotropy of silver clad $(Bi, Pb)_2Sr_2Ca_2Cu_3O_x$ superconducting tapes' (1991) *Supercond. Sci. Technol.* **4**, 361–363.

Macmanus-Driscoll J L, Foltyn S R, Jia Q X, Wang H, Serquis A *et al.*, 'Strongly enhanced current densities in superconducting coated conductors of $YBa_2Cu_3O_{7-x}$ + $BaZrO_3$' (2004) *Nature Materials* **3**, 439–443.

Mannhart J and Tsuei C C, 'Limits of the critical current density of polycrystalline high-temperature superconductors based on the current transport properties of single grain boundaries' (1989) *Z. Physik B.* **77**, 53–59.

McHenry M E, Maley M P and Willis J O, 'Systematics of Critical Current Density Hysteresis in Polycrystalline Y-Ba-Cu-O' (1989) *Phys. Rev. B.* **40**, 2666–2669.

McIntyre P C, Cima M J and Ng M F, 'Metalorganic deposition of high Jc $Ba_2YCu_3O_{7-x}$ thin films from trifluoroacetate precursors onto (100) $SrTiO_3$' (1990) *J. Appl. Phys.* **68**, 4183–4187.

Murakami M, Morita M, Doi K and Miyamoto K, 'A new process with the promise of high J_c in oxide superconductors' (1989) *Jpn. J. Appl. Phys.* **28**, 1189–1194.

Nagai H, Yoshida Y, Ito Y, Taniguchi S, Hirabayashi I *et al.*, 'Preparation of films by metal organic chemical vapour deposition using liquid-state nonfluorinated sources' (1997) *Supercond. Sci. Technol.* **10**, 213.

Narumi E, Song L W, Yang F, Patel S, Kao Y H and Shaw D T, 'Critical current density enhancement in $YBa_2Cu_3O_{6.8}$ films on buffered metallic substrates' (1991) *Appl. Phys. Lett.* **58**, 1202–1204.

Norton D P, Goyal A, Budai J D, Christen D K, Kroeger D M *et al.*, 'Epitaxial $YBa_2Cu_3O_7$ on Biaxially Textured Nickel (001): An approach to superconducting tapes with high critical current density' (1996) *Science*, **274**, 755–757.

Ohmatsu K, Muranaka K, Hahakura S, Taneda T, Fujino K *et al.*, 'Development of in-plane aligned YBCO tapes fabricated by inclined substrate deposition' (2001) *Physica C.* **357–360**, 946–951.

Onabe K, Akata H, Higashiyama K, Nagaya S, and Saitoh T, 'Superconducting property of $Y_1Ba_2Cu_3O_x$ films formed on silver substrates by continuous chemical vapor deposition technique' (2001) *IEEE Trans. Appl. Supercond.* **11**, 3150–3153.

Osabe G, Ayai N, Kikuchi M, Tatamidani K, Nakashima T *et al.*, 'Recent progress of high performance Ag-sheathed Bi2223 wire' (2010) *Physica C*, **470**, 1365–1368.

Paranthaman M, Aytug T, Christen D K, Arendt P N, Foltyn S R *et al.*, 'Growth of thick $YBa_2Cu_3O_{7-}$ films carrying a critical current of over 230 A/cm on single $LaMnO_3$-buffered ion-beam assisted deposition MgO substrates' (2003) *J. Mater. Res.* **18**, 2055–2059.

Peterson R L and Ekin J W, 'Airy pattern, weak-link modeling of critical currents in high-Tc superconductors' (1989) *Physica C.* **157**, 325–333.

Poeppel R B, Flandermeyer B K, Dusek J T, and Bloom I D, Chemistry of high temperature Superconductors, Fabrication of $YBa_2Cu_3O_7$ superconducting ceramics' (1987) *ACS series 351*, American Ceramic Soc. Washington D.C., **351**, 261–265.

Prusseit W, Nemetschek R, Hoffmann C, Sigl G, Lümkemann A *et al.*, 'ISD process development for coated conductors' (2005) *Physica C.* **426–431**, 866–871.

Prusseit W, Nemetschek R, Semerad R, Numssen K, Metzger R *et al.*, 'Evaporation—the way to commercial coated conductor fabrication' (2003) *Physica C.* **392–396**, 801–805.

Prusseit W, Sigl G, Nemetschek R, Hoffmann C, Handke J *et al.*, 'Commercial coated conductor fabrication based on inclined substrate deposition' (2005) *IEEE Trans. Appl. Supercond.* **15**, 2608–2610.

Puig T, Gutierrez J, Pomar A, Llordes A, Gazquez J *et al.*, 'Vortex pinning in chemical solution nanostructured YBCO films' (2008) *Supercond. Sci. Technol.* **21**, 034008.

Richards B C, Cook S L, Pinch D L, Andrews G W, Lengeling G *et al.*, 'MOCVD of high quality $YBa_2Cu_3O_{7-}\delta$ thin films using a fluorinated barium precursor' (1995) *Physica C.* **252** 229–236.

Roas B, Schultz L and Saemann-Ischenko G, 'Anisotropy of the critical current density in epitaxial $YBa_2Cu_3O_x$ film' (1990) *Phys. Rev. Lett.* **64**, 479–482.

Rodriguez J P, Barnes P N and Varanasi C V, 'In-field critical current of type-II superconductors caused by strain from nanoscale columnar inclusions' (2008) *Phys. Rev. B.* **78**, 052505–052508.

Rupich M W, Li X, Thieme C, Sathyamurthy S, Fleshler S, Tucker D *et al.*, 'Advances in second generation high temperature superconducting wire manufacturing and R&D at American Superconductor Corporation' (2010) *Supercond. Sci. Technol.* **23,** 014015.

Salama K, Selvamanickam V, Gao L and Sun K, 'High current density in bulk YBa_2Cu_3Ox superconductor' (1989) *Appl. Phys. Lett*, **54**, 2352–2354.

Schoop U, Verebelyi D T, Rupich M W, Thieme C, Li X *et al.*, 'Performance and mechanical properties of copper stabilized coated conductor wire' (2005) Epitaxial Growth of Functional Oxides; *Proc. Electrochem. Soc.* PV **2003–29**, 254–262.

Selvamanickam V, Carota, Funk M, Vo N, Haldar P *et al.*, 'High-current Y-Ba-Cu-O coated conductor using metal organic chemical-vapor deposition and ion-beam-assisted deposition' (2001) *IEEE Trans. Appl. Supercond.* **11**, 3379–3381.

Selvamanickam V, Chen Y, Kesgin I, Guevara A, Shi T *et al.*, (2011) 'Progress in performance improvement and new research areas for cost reduction of 2G HTS wires' *IEEE Trans. Appl. Supercond* (in print). *IEEE Trans. Appl. Supercond.* **21,** 3049 (2011).

Selvamanickam V, Chen Y, Xiong X, Xie Y Y, Martchevski M *et al.*, 'High performance 2G wires: from R&D to pilot-scale manufacturing' (2009a) *IEEE Trans. Appl. Supercond.* **19**, 3225–3220.

Selvamanickam V, Chen Y, Xiong X, Xie Y, Zhang X *et al.*, 'Progress in scale-up of second-generation HTS conductor' (2007) *Physica C*. **463–465**, 482–487.

Selvamanickam V, Dackow J and Xie Y, 'U.S. Department of Energy Annual Peer Review' (2009b).

Selvamanickam V, Guevara A, Zhang Y, Kesgin I, Xie Y *et al.*, 'Enhanced and uniform in-field performance in long (Gd, Y)–Ba–Cu–O tapes with zirconium doping fabricated by metal–organic chemical vapor deposition' (2010) *Supercond. Sci. Technol.* **23**, 014014.

Selvamanickam V, Knoll A, Xie Y, Li Y, Chen Y *et al.*, 'Scale up of applications-ready practical Y-Ba-Cu-O coated conductors' (2005) *IEEE Trans. Appl. Supercond.* **15**, 2596–2599.

Selvamanickam V, Xie Y, Reeves J and Chen Y, 'MOCVD-based YBCO-coated conductors' (2004) *MRS Bulletin*, **29**, 579–582.

Selvamanickam V, Chen Y, Xie J, Zhang Y, Guevara A *et al.*, 'Influence of Zr and Ce doping on electromagnetic properties of (Gd,Y)–Ba–Cu–O superconducting tapes fabricated by metal organic chemical vapor deposition' (2009c) *Physica C*. **469**, 2037–2043.

Shimizu E and Ito D, 'Critical current density obtained from particle-size dependence of magnetization in yttrium barium copper oxide ($YBa_2Cu_3O_{7-}\delta$) powders' (1989) *Phys. Rev. B*, **39**, 2921–2923.

Stadel O, Schmidt J, Wahl G, Jimenez C, Weiss F *et al.*, 'Continuous YBCO deposition onto moved tapes in liquid single source MOCVD systems' (2000) *Physica C*. **341–348**, 2477–2478.

Stan L, Feldmann D M, Usov I O, Holesinger T G, Maiorov B *et al.*, 'Composite as diffusion barrier/nucleation layer for HTS coated conductors based on IBAD MgO' (2009) *IEEE Trans. Appl. Supercond.* **19**, 3459–3462.

Ueyama M, Hikata T, Kato T and Sato K, 'Microstructures and J_c-B characteristics of Ag-sheathed bi-based superconducting wires' (1991) *Jpn. J. Appl. Phys.* **30**, L1384–L1386.

Usoskin A, Dzick J, Issaev A, Knoke J, García-Moreno F *et al.*, 'Critical currents in long-length YBCO-coated conductors' (2001) *Supercond. Sci. and Technol.* **14**, 676.

Usoskin A and Freyhardt H C, 'YBCO-coated conductors manufactured by high-rate pulsed laser deposition' (2004) *MRS Bulletin*, 583–589.

Usov I, Arendt P N, Stan L, Depaula R, Wang H *et al.* 'Characteristic of alumina diffusion barrier on Hastelloy' (2004) *J. Mat. Res.* **19**, 1175–1180.

Verebelyi D T, Christen D K, Feenstra R, Cantoni C, Goyal A *et al.*, 'Low angle grain boundary transport in $YBa_2Cu_3O_{7-}\delta$ coated conductors' (2000) *Appl. Phys. Lett.* **76**, 1755–1757.

Wang C P, Do K B, Beasley M R, Geballe T H and Hammond R H, 'Deposition of in-plane textured MgO on amorphous Si_3N_4 substrates by ion-beam-assisted deposition and comparisons with ion-beam-assisted deposited yttria-stabilized-zirconia' (1997) *Appl. Phys. Lett.* **71,** 2955–2957.

Wang H, Foltyn S R, Arendt P N, Jia Q X, Macmanus-Driscoll J L *et al.*, 'Microstructure of $SrTiO_3$ buffer layers and its effects on superconducting properties of $YBa_2Cu_3O_{7-}\delta$ coated conductors' (2004) *J. Mater. Res.* **19**, 1869–1875.

Xiong X (2000) U.S. Patent Application 20070238619.

Yamada Y, Takahashi K, Kobayashi H, Konishi M, Watanabe T *et al.*, 'Epitaxial nanostructure and defects effective for pinning in $Y(RE)Ba_2Cu_3O_{7-x}$ coated conductors' (2005a) *Appl. Phys. Lett.* **87**, 132502–4.

Yamada Y, Watanabe T, Muroga T, Miyata S, Iwai H *et al.*, 'Rapid production of buffered substrates and long length coated conductor development using IBAD, PLD methods and "Self-Epitaxial" ceria buffer' (2005b) *IEEE Trans. Appl. Supercond.* **15**, 2600–2603.
Yu L S, Harper J M E, Cuomo J J and Smith D, 'Alignment of thin films by glancing angle ion bombardment during deposition' (1985) *Appl. Phys. Lett.* **47,** 932–933.
Yumura H, Ashibe Y, Itoh H, Ohya M, Watanabe M *et al.*, 'Phase II of the Albany HTS Cable Project' (2009) *IEEE Trans. Appl. Supercond.* **19**, 1698–1701.
Zhang J, Gardiner R A, Kirlin P S, Boerstler R W and Steinbeck J, 'Single liquid source plasma-enhanced metalorganic chemical vapor deposition of high quality $YBa_2Cu_3O_{7-x}$ thin films' (1992) *Appl. Phys. Lett.* **61**, 2884–2886.

3
High temperature superconductor (HTS) cables

S. I. SCHLACHTER and W. GOLDACKER,
Karlsruhe Institute of Technology (KIT), Germany

Abstract: Many superconductor applications such as rotating machinery, transformers and magnets with low inductance require high current cables with low AC losses. This chapter gives an overview on cabling techniques for the high temperature superconductors (HTS) BSCCO (2212), BSCCO (2223) and (RE)BCO. A short review is given of the basic properties of HTS wires and tapes and the basic requirements of HTS cables for different applications. Cabling concepts for the different HTS materials are presented, and current performance and AC loss behaviour are discussed. After a short description of remaining challenges and future trends, cabling techniques are summarized.

Key words: high temperature superconductor, Roebel cable, Rutherford cable, AC losses.

3.1 Introduction

Several applications as superconducting magnets for particle accelerators, fusion magnets, large rotating machinery and transformers need a substantial current-carrying capacity for an optimization of the device, which may reach several thousand A. In fusion magnets, for example, high currents are needed to reduce the number of winding turns to limit the inductance. Further requirements for many high-current applications are sufficiently low operational losses and residual magnetisation. Low-loss cable designs limit the energy losses in AC applications and therefore investment and running costs for refrigeration and cooling power. Cabling techniques with NbTi strands have been developed for many years and now are well established. As an example, different NbTi cable layouts have been developed for five of the six coils tested in the Large Coil Task (LCT), e.g. the LCT-Roebel-cable of the EURATOM coil (Beard *et al.*, 1988). However, for NbTi cables, high-field applications are restricted due to the low upper-critical field. Extending the application to higher fields, Nb_3Sn Rutherford cables and cable-in-conduit (CIC) assemblies were prepared and tested in particular for fusion magnet purposes.

Meanwhile, interest in cables from high temperature superconductors (HTS) was emphasized due to the very low degradation of the transport currents in high magnetic fields when operated at low temperatures. However, HTS wires or tapes have very strict non-trivial preparation schemes such as the powder-in-tube technique (PIT) for wires and tapes in the case of the BSCCO ($22n(n+1)$)/Ag systems (BSCCO ($22n(n+1)$) = $Bi_2Sr_2Ca_nCu_{n+1}O_{2n+6+x}$, where n = 1 or 2 for technical superconductors) or the coated conductor technology for the (RE)BCO group of materials (*RE*)BCO = $RE\mathrm{Ba_2Cu_3O_{7-\delta}}$ where *RE* = Rare Earth element), a flat substrate-thin film composite. The complex preparation schemes of these conductors are a consequence of the specific material properties such as phase formation conditions and chemical reactivity. The anisotropic crystal structure and superconducting properties require an aligned crystal growth (texture), good grain connectivity and effective flux pinning to ensure high transport supercurrents. As a consequence, the flexibility for application of specific conductor modifications is quite limited, e.g. filamentary structure or low AC loss structures, and the required superconducting performance, especially the transport current level. Technical applications, however, need more sophisticated conductor features as the fundamental conductors can provide. Examples are a filamentary percolation path of the supercurrents with an applied transposition along the conductor for AC currents with low losses, improved mechanical strength to withstand the hoop stress from Lorentz forces in magnets, or simply a much higher transport current for devices as transformers or power generators. Most of these demands can only be satisfied by assembling cable composites from the fundamental conductors. From low temperature superconductors (LTS) as NbTi, different cable structures are already known and applied, e.g. the Roebel cable (RC), the Rutherford cable (RFC) and the cable-in-conduit (CIC) technique. LTS strands have the advantage of being ductile, providing much flexibility for the cabling process and the cable design. In the case of the different HTS materials, individual limitations fix the form and deformability of the strands and determine therefore the potential for a cabling technique. Required features of the cables are filaments or strands, mechanical strength, transposition of filaments and strands, stabilisation of the superconductor, ability to bend the cable into coil shapes, and finally a high cable filling factor, providing high engineering current densities. This chapter gives an overview about the stage of cable research and development for the three HTS systems, which are of technical interest, addressing the aspects mentioned.

3.2 Short survey on properties and availability of technical long length high temperature superconductors (HTS)

3.2.1 BSCCO (2212) tapes and wires

Favoured by the very flat field dependence of the critical currents due to the extremely high irreversibility field, BSCCO (2212) tapes and wires have been in focus for application at temperatures of 4–20 K for a long time. BSCCO (2212) shows in principle a similar sensitive grain-boundary angle dependency for the currents as (RE)BCO and requires therefore a grain alignment for optimized currents. BSCCO (2212) wire and tape development took different routes for optimized current carrying capability of the conductors; thin films and thick-film-coated tapes for optimized texture and excellent grain connectivity have been developed; a third very successful route was multifilamentary powder-in-tube (PIT) Ag-sheathed wires and tapes. The optimized thick-film tapes prepared by the pre-annealing and intermediate rolling method (PAIR) of Showa achieved a current carrying capacity unreached by other preparation methods due to the optimized texture and phase compaction, but the conductor length achievable with reasonable efforts and costs was limited to about 60–100 m lengths (Hasegawa *et al.* 2000). The round PIT wires have been optimized with regard to the best transport currents, resulting in multifilamentary wires with high numbers of filaments (up to 889) and small filament diameters, down to 15 μm (Hasegawa *et al.*, 2004). Wires mainly from four different suppliers (IGC-AS, Oxford (OST), Nexans-Korea and Showa) were used for cables and were available in lengths of a few hundred metres. Superconducting intergrowths between the filaments in the Ag sheath often help to improve the transport current performance (Miao *et al.*, 2005), but are also responsible for limiting the AC loss reduction due to the coupling currents between bridging filaments. The OST wire showed strong filament coupling, whereas the other producers claim an absence of intergrowths. So far, BSCCO (2212) wires are not handled on the commercial market due to limited demand and high material costs, in particular of the Ag sheath.

3.2.2 BSCCO (2223) tapes

BSCCO (2223) tapes found much more interest compared to the BSCCO (2212) system, since remarkable transport currents could be reached at 77 K due to the high critical temperature of 110 K. The low irreversibility field, however, limits applications at 77 K to fields below 1 T. Conductors were developed with several options, with an Ag/AgMg composite sheath for improved mechanical strength and with AgAu alloy sheath for drastically reduced thermal conductivity and application in current leads. Meanwhile, commercial conductor lengths exceeding 2000 m are offered by three companies: Bruker HTS, Innost and Sumitomo. A typical tape cross-section dimension is 4 × 0.25 mm. Innost (Bejing, China) is able to deliver 37 filamentary BSCCO (2223)/Ag/AgMg tapes in up to 1000 m lengths and critical currents of >110 A (77 K, self field). Bruker HTS is able to produce up to 2500 m lengths of 121 filamentary tapes with Ag/AgMg sheathing and a critical current of about 120 A (77 K, self field). BSCCO (2223) tapes from Sumitomo can be ordered with up to 180 A critical current (77 K, self field) in lengths of up to 1500 m, with Ag or reinforced Ag/AgMg sheaths. All companies offer the option of insulated wires, made by coextrusion (Bruker HTS), laquer impregnation or tin clad (Innost) or wrapped with polyimide tape (Sumitomo).

3.2.3 (*RE*)BCO tapes (CC)

(*RE*)BCO coated conductors (CCs) are prepared by a broad variety of methods and conductor architectures. Commercially available long length conductors are offered by two manufacturers, AMSC and SuperPower, made by using two completely different approaches. The method adopted by AMSC is based on the use of textured cube recrystallized Ni-W-substrates transferring the texture to the $CeO_2/YSZ/Y_2O_3$-buffer/superconductor architecture. The (RE)BCO layer is applied by slot-dye chemical solution deposition (metal-organic deposition with the TFA-method) with Dy-doping of the superconducting phase. The typical length of a sample piece is 300 m, with the potential to produce 1000 m lengths. The sample width is typically 40 mm for the coating procedure (wide-strip technology) and final splicing leads to 10 samples of 4 mm-wide tapes. An option is to slit customer-specific other widths of the conductor and apply various lamination techniques with stabilising copper or reinforcing stainless steel tapes. The current carrying capacity on a 100 m length of 4.4 mm wide (344 superconductor) was given as at minimum 262.5 A/cm width. The installed capacity for this conductor performance at AMSC is claimed to be 500 km per year.

Coated conductors from SuperPower are prepared on Hastelloy substrates and the texture is introduced via the MgO/LMO based buffer architecture. This approach is using metal-organic chemical vapour deposition (MOCVD) for (RE) BCO and the application of additional Zr flux pinning centres. Tape thickness can be customized and is usually 100 μm including a 20 μm surrounding Cu-stabilizer. Commercial tape performance is well exceeding 300 A/cm width depending on piece length. This company's CC material is available in lengths greater than 1000 m with minimum 282 A/cm width and can be ordered in 4 and 12 mm widths. The mechanical performance of SuperPower CC is outstanding with >800 MPa tensile stress tolerance and bending tolerance <11 mm bending radius.

Other potential industrial producers of CCs, e.g. Zenergy, THEVA and Bruker HTS have intensive research on long conductor lengths and may join the market in future. A couple of CC developers exist in Japan and Korea, however their CCs are not commercially available on the market.

3.2.4 Summary

A common feature of all HTSs is the dependence of the intergrain current on the grain boundary angle but with differently strong consequences for the different systems. The (RE)BCO CC is the most sensible case and requires a 2D orientation of the grains with tolerable grain-boundary angles of only a few degrees. As a consequence, very sophisticated preparation methods have to be applied for the production of wires and tapes from the different materials with poor flexibility to change the composite structure.

BSCCO (2223) conductors require a c-axis texture for high intergrain currents achieved by alternating tape rolling and recrystallization steps and an Ag matrix due to the required permeability for oxygen during the annealing treatment. The consequence is a tape geometry with filaments having a high aspect ratio and relatively high conductor costs.

In BSCCO (2212) conductors a *c*-axis texture leads to high inter-grain currents as demonstrated for multilayer tapes prepared by the PAIR technique (Hasegawa *et al.*, 2000) but also allows still quite high currents (approximately reduced to 1/3) with a kind of fibre texture in round wires with round filaments, prepared by the PIT- technique and conventional deformation. The limitation to Ag or Ag-alloy matrix is caused by the same chemical reasons as for BSCCO (2223). The mechanism of texture formation in filaments requires a large filament/Ag interface which requires preparation of multifilamentary wires with small filament diameters. Round wires are most suitable for cabling techniques as will be shown later.

As consequence, for HTS we have the fixed geometry of flat tapes for the (RE) BCO and BSCCO (2223) material and the possibility of round wires for BSCCO (2212) with moderate degradation in current-carrying performance.

BSCCO (2212) as well as BSCCO (2223) conductors require an Ag matrix that is oxygen-permeable, which gives limited scope for modification of the sheath properties, in particular with respect to improved mechanical strength. Tensile axial stress tolerance can be improved to approximately 250 MPa using AgMg dispersion-hardened alloy sheaths. In contrast to the BSCCO systems, the strength of (RE) BCO CC composites depends crucially on the preparation routes, the nature of the sheath material, and the involved heat treatment temperature levels. Use of strong substrate materials such as cold-worked Hastelloy provides excellent mechanical strength with tolerable axial stresses up to 800–1000 MPa, meeting all technical requirements of the applications. Cube-textured substrates are usually of Ni-W showing some magnetism. The shape of the CC is strictly limited to the substrate-buffer-HTS-Ag architecture; thermal stabilisation requires application-specific coating with soldered copper or copper deposited by electroplating. The field dependence of the critical currents of CCs with sharp T_c transition at 77 K is superior to BSCCO materials. All HTS tapes show anisotropy of the transport critical current in external magnetic fields up to a typical factor of three. Anisotropy can be reduced or even tailored in the case of CCs by doping artificial pinning centres.

3.3 Basic requirements for cables for different applications

3.3.1 Current carrying capacity

There are many parameters that determine the required properties of an HTS cable: the size or power level and the correlated operation current of the device, the background field at the cable, the operation temperature of the device and the nature of the current, DC or AC and, in particular, the AC frequency. For transformers and big magnets the inductance of the winding determines the required cable current. In order to minimize losses due to AC currents and fields a suitable cable design is required. An essential requirement for applications is the engineering current density of the cable, which determines the energy density in the device and, therefore, for example, the size reduction compared to conventional solutions. In Chapter 2 characteristic values of transport currents of the different fundamental wires are given which strongly depend on operation temperature and field.

Fault current limiters (FCL) are considered to be used in the mid-voltage level grid (10–20 kV) and require typically up to about 2–4 kA (ac) operating current when being implemented in a medium-voltage 10 kV grid. This can usually be realized by a few parallel-stacked conductors of a few metres length (CC), with negligible influence of the self field degradation of the currents, when applied in a quite open winding and at a low AC loss level.

For transformers the rated current I is determined by the rated power of the device S and also the grid voltage level U:

$$I = S/(\sqrt{3}\, U) \qquad [3.1]$$

A 100 MVA transformer, for example, requires nearly 3 kA current when operated in a 20 kV grid. In this case the self field degradation is not negligible at field levels of up to a few tenths of a Tesla.

Magnets and windings for rotating machinery require a certain number of ampère-turns $F = N{\cdot}I$ for field generation, which gives some freedom for the selected conductor current. The magnetic field scales proportionally to the number of ampère-turns, and the winding induction L scales proportionally to N^2. If the design of large devices requires low operating currents, the required high number of turns therefore leads to a high inductance of the winding. If the device should be optimized for ramped or AC currents, inductive voltages are generated in the windings, which need to be limited and scaled proportionally to $L{\cdot}\Delta I/\Delta t$. As a consequence the windings need to be protected from partial discharge and high-voltage breakdown events that might damage the winding pack. Necessary high-voltage insulation of the conductors reduces the engineering current density. Therefore, the conductors for very large devices as fusion or accelerator magnets are usually designed for high operating currents leading to a low inductance, with the consequence that the protection requirements against high voltages are reduced significantly. Operating currents can reach the order of several 10 kA, e.g. 13 kA in the CERN LHC accelerator dipole magnets, which are actually made from LTS, or 68 kA in the case of the ITER toroidal field coil cable-in-conduit conductors (CIC) assembled from Nb_3Sn LTS wires.

Rotating machinery such as motors or generators can typically require up to about 5 kA operating current depending on the size of the device (the largest conventional hydropower generators have up to 25 kA operating current), a performance which is also well above the current carrying capacity of single HTS and which requires cabled conductors.

3.3.2 Mechanical strength and load

Mechanical stress effects have different origins that have to be considered for different situations. First, the mechanical performance of the fundamental conductor gives the boundary conditions for the robustness of the cabling process, in particular the bending ability. In coil windings, thermal stresses are caused by the different material thermal behaviours during cool down. In rotating machinery, conductors will experience centrifugal forces; whereas in magnetic windings Lorentz forces will lead to mechanical stress in the conductor. All high temperature superconducting materials are brittle, and irreversible damage by crack formation occurs if they are strained beyond the tensile regime. All wire and tape composites, and in particular the cable designs, have to provide compressive stress in the superconducting layer or filament causing a prestrain which can balance externally applied strains. The range of tolerable axial stresses of all high temperature superconducting materials is more than one order of magnitude, for BSCCO conductors 80–350 MPa (from AgAu-sheath to steel laminated tape) and for (RE) BCO-CC about 150–800 MPa (NiW to Hastelloy substrate).

3.3.3 AC losses

AC losses occur in AC-operated conductors, in AC background fields and in rapidly ramped fields. The orientation of the conductor in the external field plays an important role. AC loss measurements for fields perpendicular and parallel to the cable are therefore mandatory. Cable features to limit AC losses are a filament and strand structure to limit magnetisation losses. A resistive matrix between the filaments, high inter-strand resistance and transposition of the superconducting paths are required to limit coupling losses. Finally, a design of the cable and material choice to limit eddy current losses in the matrix or in the structure materials is necessary.

3.4 HTS cable configurations

HTS cable configurations for windings with good bending properties focus on Rutherford Cable (RFC) designs and Roebel cables (RC), also known as continuously transposed cable (CTC). Common to both cable types is the transposition of the strands (HTS wires, tapes or subcables). Rutherford cables consist of strands assembled side-to-side on top and below a substrate tape with a twist angle that allows each strand to approach the edge of the substrate and cross over to the other side (transposition effect). Such a cable has a large aspect ratio in width/height. The Roebel or CTC arrangement, however, consists of two parallel stacks of conductors with periodic step over for the strand to the opposite stack, located on upper and lower stack side.

3.4.1 BSCCO (2212) Rutherford cables (RFC)

BSCCO (2212) cables found special interest for the application in accelerator magnets with elevated fields (Collings *et al.*, 1999), for the application in SMES systems (Ohsemochi *et al.*, 2006) and transformers (Ha *et al.*, 2008), for open MRI systems and also for insert coils for high-field magnets (Weijers *et al.*, 2003). Beside the high critical current density at low temperatures (>10^5 A cm^{-2} at 4–20 K) the weak degradation of the currents in magnetic background fields is very attractive. The competing low temperature superconducting material Nb_3Sn, also a brittle superconductor, but with potential to be used in higher fields than NbTi, was already cabled to a Rutherford structure with success (Sumption *et al.*, 1999), so it was a straightforward challenge to investigate a RFC prepared from round BSCCO (2212) wires.

Preparation method

A Rutherford cable structure is very suitable for the assembly of superconducting strands with a round cross-section, in particular round multifilamentary wires, and to apply a transposition in the cable. For round wires the transposition of the filaments is a standard option for industrial fabrication and it is in particular welcome for the strands of a Rutherford Cable. As already mentioned, BSCCO (2212) superconductors are not as strongly fixed to the tape geometry as BSCCO (2223). Wires can achieve about 10–30% of the current values of tapes, have sufficient grain connectivity and the necessary texture for high transport currents (perpendicular to the filament axis) can be achieved for thin filaments (15–35 μm diameter). In multifilamentary wires up to 889 filaments could be produced by the different producers. The phase formation of the superconductor needs a heat treatment (HT) that peaks in the range 870–890°C for partial melt-assisted BSCCO phase formation followed by different temperature ramps for crystal growth and oxygen load. This requires chemical compatibility of sheath materials and wire components for the processing. Only Ag or Zr-doped Ag is chemically compatible with BSCCO with this heat treatment. At such temperatures Ag sheaths are sensitive to poisoning from other materials with which they are in contact.

Since the RFC requires a step over of the strands around the side edge at each half transposition length, crack formation in the superconducting filaments has to be avoided. The only way to handle this difficulty is applying an assemble-and-react method in the cabling process with the reaction heat treatment applied after the cable assembling procedure. All cable components need to be compatible with this HT, i.e. temperatures up to about 890 °C and oxygen-containing atmosphere. Collings *et al.* (1999) applied a central core to the cable for better mechanical performance of the cable which met this requirement. To avoid poisoning of the strands from the substrate tape during HT, systematic investigation led to NiCr80 alloy as a suitable material for the central core. Two different core thicknesses were investigated and pitch lengths of >20–30 mm were found to have no negative influence on the transport currents. Kim *et al.* (2007) prepared their cable in a similar way but without applying a central core (Fig. 3.1). Cabling of the strands

Type	Cross-section	Appearance
8 strand Rutherford cable	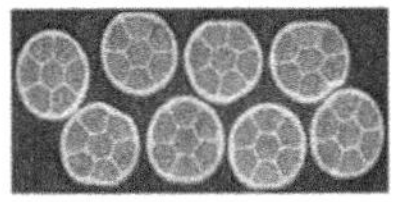	
20 strand Rutherford cable	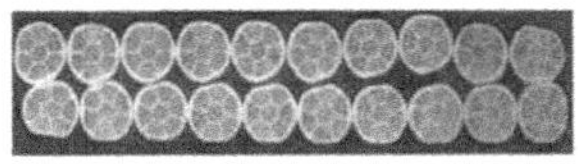	
30 strand Rutherford cable	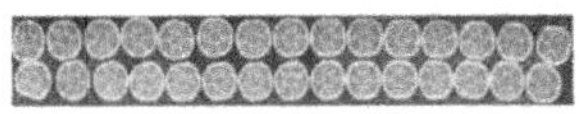	

3.1 Micrographs and pictures of Rutherford cables. (Kim *et al.*, 2009, copyright 2009 IEEE.)

was always performed via a cabling machine equipped with a turk head where the strands were fed together and moderately pressed into the RFC cross-section. Table 3.1 shows an overview of selected BSCCO (2212) Rutherford cables from several groups. BSCCO (2212) wires from IGC-AS, OST (Fig. 3.2), Nexans-Korea and Showa with quite different numbers of strands, cross-sections and wire lengths were used.

Cross-section of wire	Specification	
	Filament number Diameter Ag/SC ratio I_c (4.2 K, 0 T)	61×7 0.82 mm 2.8 >5000 A
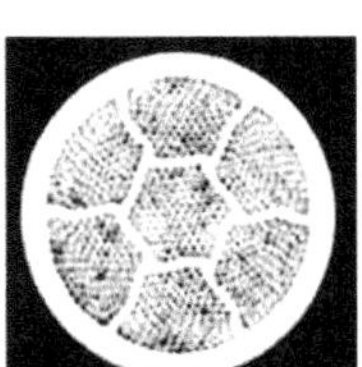	Filament number Diameter Ag/SC ratio I_c (4.2 K, 0 T)	127×7 1.02 mm 2.8 >900 A

3.2 Specification of double bundled strands for 12 kA Rutherford cable. (Hasegawa *et al.*, 2004, copyright 2004 IEEE.)

Table 3.1 Selected high current Roebel and Rutherford Cables from BSCCO (2212) and BSCCO (2223) superconductors

Reference	HTS material	I_c (4.2K, s.f.) (A)	No. of strands	Transposition pitch (mm)	Filaments per strand	I_c (4.2K, s.f.) (A) per strand	Cable dimensions Width × height (mm) (length (m))
Leghissa *et al.*, 2006	BSCCO (2223)	1180†	13	2457	55	135†	9 × 2.5 (50)
Hasegawa *et al.*, 2004	BSCCO (2212)	12 000*	20	73	889	> 5000	11.35 × 1.9 (147)
	BSCCO (2212)	12 000*	30	87	427	> 900	13.28 × 1.55 (300)
Kim *et al.*, 2009	BSCCO (2212)	3146*	20	55	385	225*	8.26 × 1.46 (110)
	BSCCO (2212)	6960*	30	92	385	225*	12.39 × 1.46 (10)
Collings *et al.*, 1999	BSCCO (2212)	3495	19	27.5	305	170	7.77 × 2.73 (5–10)

*Extrapolated; †at 77 K.

Transport currents

The transport currents achieved in the RFCs depend strongly on the strand performance and the number of strands applied in the cable. In Collings *et al.* (1999) the RFC with central NiCr core and 19 strands (Fig. 3.3) required protection from the core material to avoid poisoning of the HTS strands (Fig. 3.4). The cable carried the expected transport current (19-fold the strand values) with 3495 A at 4.2 K and self field and a field dependence equivalent to that of the strands. A chemical protection of the strands seems a mandatory feature for BSCCO (2212) RFCs. Other examples for RFCs with 20 and 30 strands were presented in Hasegawa *et al.* (2004) and Ha *et al.* (2008). The transport critical current could only be measured in 1 T background field, being approximately 9 kA at 4.2 K. An extrapolation to self field conditions yields a value of about 12 kA. The bending performance of cable and strands was investigated. The strands allowed a bending strain of 0.5% whereas the RC showed a much better performance with nearly 1% tolerable bending strain. Kim *et al.* (2009) presented RFCs with 8, 20 and 30 strands without a central core (Fig. 3.1). Transport currents at 4.2 K could not be measured; however, an estimation

3.3 Micrograph of Rutherford cable strand before heat treatment. (Collings *et al.*, 1999.)

3.4 Micrograph of cable cross section of Rutherford cables with central core. (Collings *et al.*, 1999.)

of the current potential was developed on the basis of the strands' properties and taking into account self field effects. The prospected transport currents of the cables were claimed to be 4785–6960 A at 4.2 K, self field.

AC losses

A detailed analysis of the AC loss contributions of three different RFCs was performed by Collings *et al.* (1999). Measurements were performed on 8-fold stacks of short (20–24 cm) RFC samples, insulated against each other. Field was applied in both directions perpendicular and parallel to the plane of the cable (Fig. 3.5). AC fields were applied up to 400 mT with 10–450 Hz. AC losses were expressed in terms of inter-strand resistances to describe the inter-strand coupling losses as the most important source of losses. The different cable structures with 19 strands addressed this aspect with different preparation schemes. One cable had a naturally oxidized NiCr core, one an additional plasma-sprayed MgO clad and the third one had each second strand replaced by a NiCr-strand. The first two cables showed low losses in perpendicular field orientation and very low losses in parallel field orientation, whereas the double insulated strands, separated by NiCr strands showed only the frequency independent hysteretic loss component. The strands themselves turned out to be fully bridged and had only a hysteretic component and no eddy current contributions. Detailed considerations on the contact resistances are also presented by Collings *et al.* (1999) showing the positive effect of the central core. The difference comparing the cables with oxidized NiCr and with additionally MgO covered core was small, indicating the effectiveness of the surface oxidation. A comparison of the results to the LHC inner specifications gave a rather good situation with a moderately satisfying balance between limited strand coupling and necessary strand stabilisation.

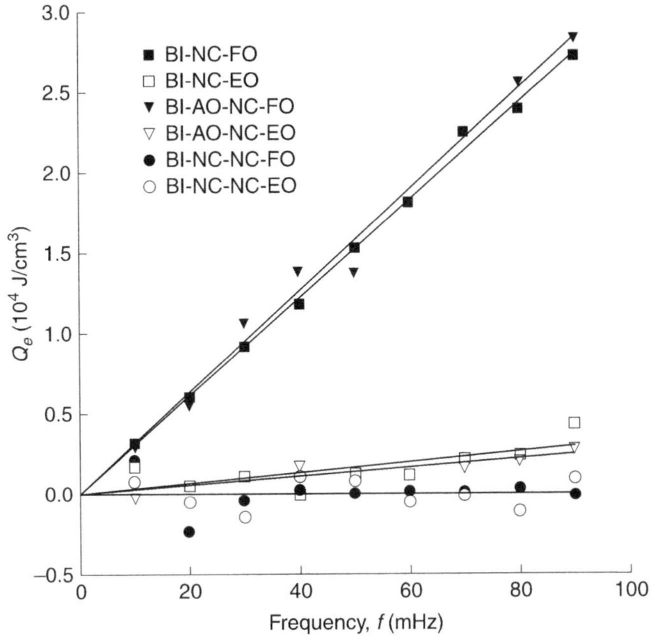

3.5 Coupling losses (at B_m = 400 mT) for cables BI-NC (core of NiCr), BI-AO-NC (NiCr-core and plasma-sprayed Al_2O_3 coating), BI-NC-NC (alternating 9 HTS, 9 NiCr strands). FO: face orientation, EO: edge orientation of field. (Collings *et al.*, 1999.)

3.4.2 BSCCO (2223) Roebel cables

BSCCO (2223) is fixed to the tape geometry and for cables the annealing-assembling route is mandatory since the heat treatment of the tape is usually split into three annealing steps with intermediate rolling steps for core densification and grain orientation. This boundary condition limits the possible procedures for assembling a cable from BSCCO (2223) tapes. A quite successful route was developed by Siemens AG for transformer applications following the Roebel bar approach (Leghissa *et al.*, 2002).

Preparation technique

Assembling BSCCO (2223) tapes as strands to a Roebel bar structure requires in-plane bending of the tapes to realize the step-over of the strands between the two strand stacks of the cable. In-plane bending is limited by the tolerable tensile strain range of the order of 0.25% for reinforced Ag/AgMg sheaths. When cabling 13 strands by means of a semi-automatic tool, a full transposition length of 2457 mm or longer was achievable (Leghissa *et al.*, 2006). The cable was finally wrapped with glass-fibre tape for mechanical stabilisation of the structure. The cable cross-section corresponds roughly to double the strand width (7.35–9 mm) and the cable thickness to 7–8 times the strand thickness, leading to 2.3–2.54 mm cable thickness. The resulting filling factor was 80–85%. Cable lengths between 50 and 160 m length were fabricated for performance tests of BSCCO (2223) Roebel cable windings.

Current carrying behaviour

Strands of different current performance between 48–135 A (77 K, self field) were used to fabricate the different RCs. Resulting DC cable currents ranged between 345–1180 A (77 K, self field) depending on current performance of the strands. This is significantly lower than expected from calculating the design values of the cable current, which is simply the sum of the strand currents (Fig. 3.6). The explanation is given by the strong field dependence of the critical current which is in addition strongly anisotropic with respect to the field orientation (Fig. 3.7). Since the cable current is roughly ten times the strand current, the field generated by the cable current (self field) leads already to a significant degradation of the transport current, reaching about 35% for a transport current of 1180 A. A linear relationship between design current and self field-effected current was found (Fig. 3.6). Measurements in additional external fields show strong contributions of the perpendicular self field component up to about 20 mT, whereas in parallel field the self field effects contribute significantly up to about 300–400 mT

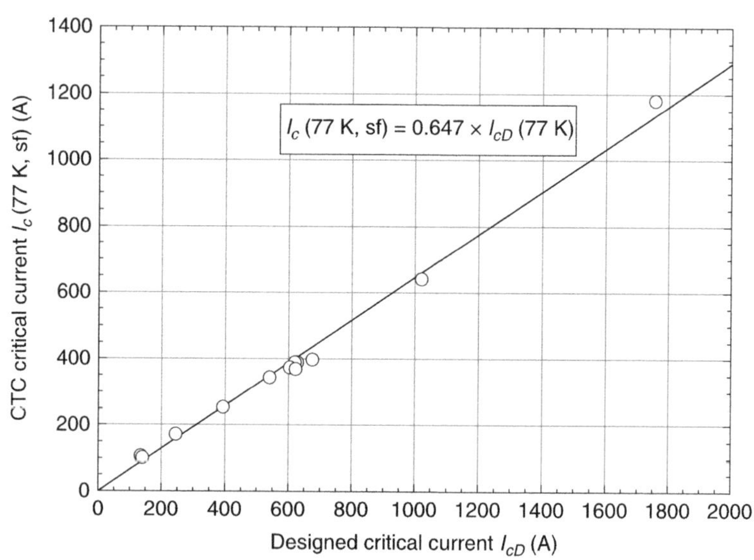

3.6 Critical current of CTCs at 77 K, self-field, versus the designed critical current. (Leghissa *et al.*, 2006.)

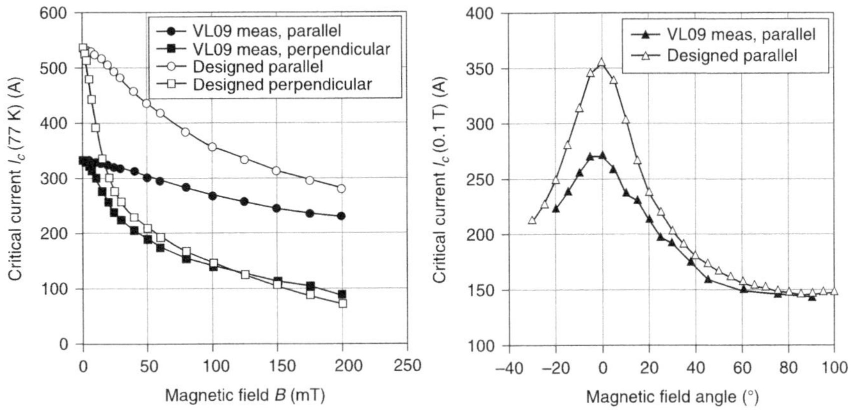

3.7 Critical current of VL09 (BSCCO (2223) CTC) as a function of magnetic field and field orientation at 77 K. (Leghissa *et al.*, 2006.)

(Leghissa *et al.*, 2006). The measured self field influence indicates a very anisotropic field distribution in the cable. The mechanical performance of the cable under tensile stress was reduced by about 25% of tolerable stress compared to single strands which can be explained by the pre-strain load from the bending of the strands in the cable structure.

AC losses

BSCCO (2223) tapes with different transposition lengths of the filaments down to 5 mm were produced and investigated by several authors. AC loss reduction of 50% was found in parallel fields for a twist pitch of 10 mm at 20% reduced currents (Hussennether *et al.*, 2004). Further reductions were hindered by the filament coupling from the high conductive Ag matrix. Interfilamentary barriers introduced to the tapes showed the expected effects on the losses; however, too large current degradations had to be taken into account compensating the progress in loss reduction. In the RC no inter-strand insulation was applied. For the cables, AC transport losses were measured and modelled with a modified Norris model for elliptical cross-sections (Leghissa *et al.*, 2006) (Fig. 3.8). Three different modifications of the loss formula were considered:

1. assuming N independent tapes with losses being N-times that of the single strand,
2. a monoblock model with cable design current,
3. a monoblock model with a cable current reduced by the self field effect.

Version (3) gave the most satisfactory description. The losses are solely determined by the cable current, not by the presence of a filament twist in the strands, which

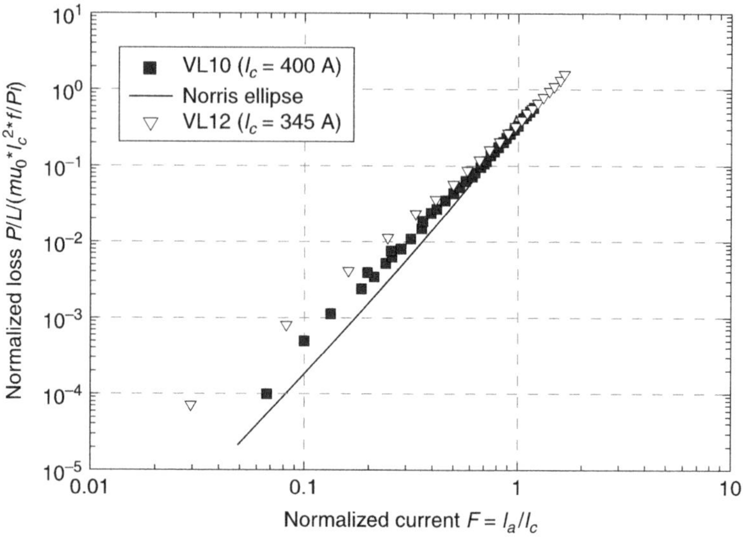

3.8 Scaled transport current AC loss of CTCs VL10 and VL12. (Leghissa *et al.*, 2006.)

was also checked, and the main contributions are associated with the perpendicular field component. The cable behaves as a monoblock conductor.

3.4.3 CC Roebel cables

The DC performance of coated conductor tapes has been enhanced significantly in the past years. Important steps on the way to better current performance, especially in high magnetic fields, were the improvement of the homogeneity of the superconducting layer and the improvement of flux pinning properties by doping. The $I_c(B)$ anisotropy could be drastically reduced by the introduction of oriented pinning sites, which leads to an enhancement of I_c in the 'bad' field direction perpendicular to the applied field. Meanwhile, pinning sites can be applied in a tailored orientation to modify the anisotropy in a large range.

The commercial coated conductor tapes currently available, however, are not really suitable for AC applications due to their poor AC loss behaviour. The high aspect ratio of tape and superconducting layer leads to high hysteresis losses in perpendicular fields. To reduce hysteresis losses reduction of the aspect ratio can either be achieved by reduction of the tape width or by filamentation of the superconducting layer, e.g. by laser scribing. Magnetic AC loss measurements on short striated and unstriated samples showed a drastic reduction of the losses for the striated samples, especially at low frequencies, where the filaments are completely decoupled. At high frequencies, and in samples with large transposition length or with filaments which are connected at the end of the conductor, however, the conductor behaves like a single superconducting block. A full transposition of the filaments with short transposition lengths would help to reduce coupling losses effectively; however, this could not be realized so far for full superconducting percolation paths.

Abraimov *et al.* (2008) presented a technique to achieve a transposition of filaments in a Rutherford geometry by diffusion bonding of two silver-clad coated conductor tapes with zigzag patterned filaments that were partially separated by a dielectric layer. In AC loss measurements they observed a tenfold decrease of losses at frequencies between 20–400 Hz. However, for long length cables the diffusion bonding technique is a complex procedure because the zigzag filaments have to be connected with very high precision to ensure a continuous current path from end to end. The connection of filaments between the two tapes must have very low resistances to avoid additional ohmic losses. Abraimov *et al.* (2008) achieved contact resistances of the order of 10^{-8} Ω cm^2. The availability of superconducting joints would reduce the losses considerably; however no such joining technique is known at the time of writing.

Another method to fully transpose coated conductor tapes is the Roebel technique, which has already been applied for LTS and BSCCO cables. The design, preparation and properties of coated conductor Roebel cables, which are also called continuously transposed conductors (CTC), are described below. Table 3.2 gives an overview of the properties of Roebel cables from different publications.

Preparation technique

The idea of preparing Roebel conductors from HTS tapes had already been published in 1999 (Wilson, 1999). The first Roebel assembled coated conductor (RACC) cable with a current carrying capability of 500 A at 77 K, self field, was prepared by Forschungszentrum Karlsruhe (now Karlsruhe Institute of Technology, KIT), Germany (Goldacker *et al.*, 2006). Commercial DyBCO coated conductor tapes from THEVA Dünnschichttechnik GmbH, Germany, were punched into meander-shaped geometry with a specially designed punching tool. Sixteen punched tapes were then assembled into the final RACC cable. Figure 3.9 shows the different preparation steps for RACC cables from a commercial coated conductor tape via the punched strands to the assembly of the final cable. On the right side a schematic figure of an incomplete cable with five strands illustrates the transposition of the strands.

Automated devices for the production of longer length RACC cables have been developed. At KIT an automatic device for long length punching of up to 12 mm-wide tapes is available. Industrial Research Limited (IRL), New Zealand, has developed an automated device for the punching of multiple strands from 40 mm-wide tapes (Long *et al.*, 2008). A critical point for punching of long length Roebel strands is the absolutely precise value of the transposition length. For two 100 m long strands with transposition length L_T = 10 cm, for example, a continuous

Table 3.2 Overview of properties of REBCO coated conductor Roebel cables from different groups

Reference	I_c (77 K, s.f.) (A)	No. of tapes (stack number × tape number)	Trans-position pitch (mm)	Filaments per strand	I_c (77 K, s.f.) (A) of punched tape (of original tape)	Width (mm) of punched strand (of original tape)	Cable dimensions: width × height (mm) (cross section (mm^2))	Cable length (m)
Goldacker *et al.*, 2006	> 500 A	16* (16 × 1)	180	1	~63 (~232)	Not given (10)	~10 × not given (not given)	0.38
Goldacker *et al.*, 2007	1020	16 (16 × 1)	190	1	89.7–94.2 (~227)	Not given (12)	~12 × ~0.75 (9)	0.45
Goldacker *et al.*, 2009a	2628	45 (15 × 3)	188	1	~149.5 (~'359)	Not given (12)	~12 × ~2.3 (27.6)	1.1
Terzieva *et al.*, 2010	465	14 (14 × 1)	109	1	Not given (~125)	2 (4)	4.3 × 1 (4.3)	0.5
	1060	39 (13 × 3)	109	1	Not given (~125)	2 (4)	4.4 × 2.7 (11.88)	0.5
	1195	50 (10 × 5)	109	1	Not given (~125)	2 (4)	4.4 × 3.8 (16.72)	0.5
Lakshmi *et al.*, 2009	435	11	90	1		2 (40)		1
Lakshmi *et al.*, 2010a	203	5	90	1	50.2 ± 0.8 A (not given)	2 (12)		0.72
Lakshmi *et al.*, 2010b	407	10	90	1		2 (40)		0.09
Badcock *et al.*, 2009	407.2	10	90	1		2		0.18
			300	1	~98.4 (not given)	5 (12)		
Long *et al.*, 2008	~400	10	90	1		2 (40)		
Long *et al.*, 2010	203	5		1		2 (12)		0.72
	309	9		1		2 (12)		0.54
	1100	15		1		2 (12)		5
Jiang *et al.*, 2009	203	5	90	1	49.4–50.7 (not given)	2 (12)		0.72
Jiang *et al.*, 2010	196.6‡	5	90	1	46.5–46.7 (not given)	2 (12)		0.72
	309.4†/318.8‡	9	90	1	46.8–48.7 (not given)	2 (12)		0.54
Lee *et al.*, 2009		10	80	1		4.5 (12)		0.12
		10	80	4		4.5 (12)		0.12
		10	80	6		4.5 (12)		0.12

* +1 Cu strand, †parallel connection of strands, ‡serial connection of strands.

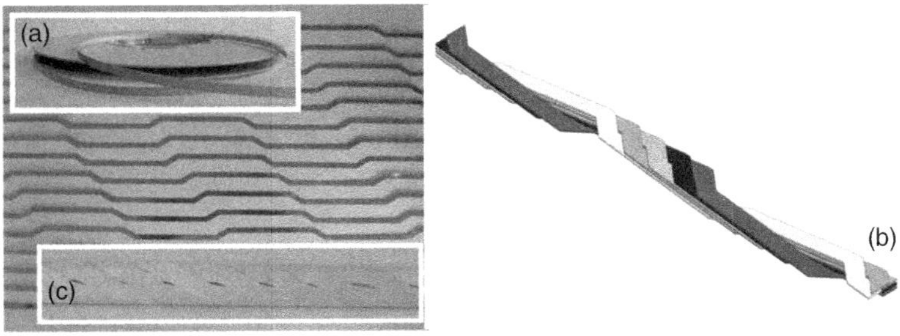

3.9 Fabrication of Roebel cables: (a) coated conductor tapes are cut in a meander shaped geometry, e.g. by punching or laser cutting. The meander shaped tapes (b) are subsequently assembled to the final Roebel cable (c).

error in the transposition length of ΔL_T=100 µm in one strand would sum up to a phase shift of a full transposition length at the end of the meandering strand and cabling would therefore not be possible due to the mismatch. Parallel punching of all strands of a cable from wider tapes (40 mm for AMSC CC) helps to avoid this problem because then the phase shift is the same for all strands assembled in the cable.

For assembling long lengths of RACC cables, cabling machines based on planetary winding methods for different cable geometries have been developed at IRL (Beck and Mulholland, 2008; Badcock *et al.*, 2009).

For many applications strand insulation is an important issue to control current sharing and losses in the cable. An insulation system with typically 20 µm thickness withstanding 250 V applied potential and repeated cryogenic cycling has been developed by IRL. The process uses an adaption of laser micro-machined print roller technology to apply an epoxy acrylate polymer formulation to the strands (Badcock *et al.*, 2009). Another insulation technique tested at IRL is the extrusion of a coating of a copolymer of tetrafluoroethylene-hexafluoropropylene (PTFE). A tape to ground breakdown voltage (BDV) of 2 kV could be realized for a 100 µm thick insulation layer. The advantage of this technique is the better edge coverage compared to the first method; however, a reduction of the thickness of the insulation layer below 100 µm is difficult. A third method for insulation of strands is application of a layer of polyimide tape (Kapton™) to the strands. The polyimide tape can be punched and assembled together with the strand. The disadvantage of this technique is the missing edge coverage; however, the insulation layer thickness can be far below 100 µm. With a 25 µm thin foil a resistance >499 Ω could be achieved at 250 V (Long *et al.*, 2010).

DC currents in CC Roebel cables

The maximum current that can be transferred through a RACC cable is determined by the cable width, the current-carrying capability of the strands and the number of assembled strands. The current carrying capability of the single strands strongly depends on the absence of defects and on homogeneity of the original tapes that are punched into a meander-shaped geometry. Punching of long lengths of inhomogeneous tapes often leads to a strong reduction of I_c/cm width because in some areas the good sections with high-current-carrying capability are punched out of the tape and the sections with low I_c/cm width remain in the meander shaped strand. It has been shown that with high-precision punching techniques and selected homogeneous coated conductor tapes the average reduction of I_c/cm width can be as low as 3% (Goldacker *et al.*, 2007), which is in the order of the punching accuracy.

The current-carrying capability of a Roebel cable can easily be enhanced by insertion of additional strands. However, for a fixed twist length L_T the maximum number of strands N_{max} is limited by the step-over angle α and the strand width in the stepover region $w_{step\text{-}over}$:

$$N_{max} \leq (L_T/w_{step\text{-}over}) \sin\alpha \qquad [3.2]$$

A method to enhance the number of tapes over N_{max} is to assemble stacks of punched tapes instead of single tapes as strands of a cable (Goldacker *et al.*, 2009a; Terzieva *et al.*, 2010). Figure 3.10 shows such RACC cables with the same width (~4.4 mm) and transposition length (109 mm) but with different numbers of assembled tapes. The cable with 14 single-tape strands has a critical current of 465 A at 77 K, self field. With 13 strands consisting of three-fold stacks of tapes (i.e. 39 tapes in total) and with ten strands consisting of five-fold stacks of tapes (i.e. 50 tapes in total) the current carrying capability of the tapes could be increased to 1060 A and 1195 A, respectively (Terzieva *et al.*, 2010). For a 12 mm-wide RC and three-fold stacking of 15 strands a transport current of 2.6 kA at 77 K, self field was achieved. In principle, the number of tapes in a stack can be further

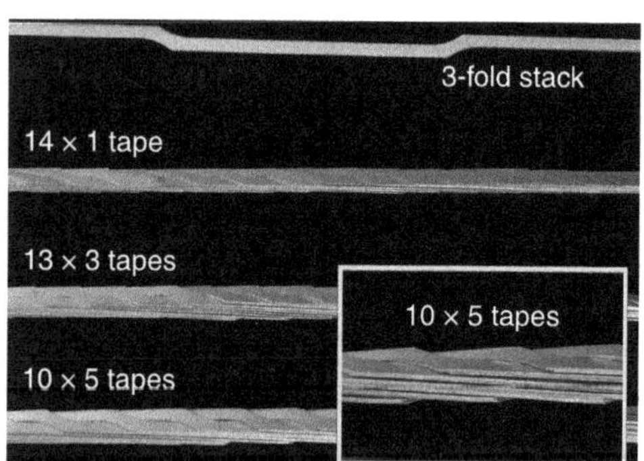

3.10 Multi-stacking of tapes in strands for Roebel cables with enhanced current carrying capability and reduced AC losses. Three- or five-fold stacks were used in the cables with 13 × 3 and 10 × 5 tapes, respectively, to enhance the current-carrying capability of the cable without changing the width or transposition length. (Terzieva *et al.*, 2010.)

increased to enhance the current-carrying capability. However, for strands with thicker stacks the cabling process becomes more and more difficult and the stresses applied during the cabling process become larger.

The critical current of RACC cables has been found to be up to 30–60% lower than the sum of critical currents of the punched tapes, the design current of the cable (Goldacker *et al.*, 2007; Goldacker *et al.*, 2009a). The strong reduction is a consequence of the higher self field in the cable. This effect is very pronounced for measurements at elevated temperatures without background field, because the change of I_c with field ($\Delta I_c/\Delta B$) is very strong under these conditions. In high magnetic background fields and at low temperatures, the self-field effect can almost be neglected because $\Delta I_c/\Delta B$ is much smaller.

A severe problem for testing the $E(I)$ properties of RACC cables is the inhomogeneous transfer of currents into the strands. Due to the insulating buffer layers in the coated conductor tapes the current flow from the substrate side into the superconducting layer is only possible via a conducting layer that electrically connects the substrate side to the superconducting layer (e.g. Ag cap layer and a surrounding Cu stabilizer). However, in punched strands the electrical connection from substrate to tape is punched out at the cross-over sections and the sections that are later in the centre of the cable. When the whole cable end is soldered as a block to the current leads, the current has to be transferred from the outermost tape to the tapes in the core of the cable. The partially missing connections from substrate to superconducting layer result in varying contact resistances from tape to tape, and lead to complicated distribution of currents in the different tapes. In short cables with strands that are not well connected to each other between the current leads a redistribution of currents is only possible at the end of the cable. At low currents this can lead to a very inhomogeneous current distribution in the cable with the current flowing only in those single strands with the lowest contact resistances. In cables with strands consisting of single tapes the current transfer to the strands can be improved by soldering at least one transposition length to the current leads so that the superconducting layer of each tape is at least once on the outside of the cable with good contact to the current leads. A method for obtaining equal currents in the single strands of short cables has been demonstrated at IRL by connecting all strands in series (Jiang *et al.*, 2010). For future longer cables a well-defined moderate strand connection to allow current redistribution without too strong enhancement of coupling losses has to be considered (Schuller *et al.*, 2007).

AC losses in CC Roebel cables

The magnetic AC loss in superconducting tapes and cables has three major contributors. Hysteresis losses in the superconducting layer are the main loss component in coated conductor tapes with a high aspect ratio. A loss component that is strongly dependent on the interfilament or interstrand resistance in striated conductors or cables is the coupling losses due to currents circulating between different filaments or strands across the metal matrix. The resistance of matrix and structure materials of tapes and cables determines the magnitude of eddy current losses. In coated conductors with ferromagnetic substrate (e.g. rolling assisted biaxially textured substrates (RABiTS™) tapes with Ni-alloys) the ferromagnetic losses also contribute to the total losses.

Magnetic AC loss measurements have been performed by several groups. The AC loss of Roebel cables with different strand numbers, magnetic or non-magnetic substrate material and with insulated or non-insulated strands was investigated by Lakshmi *et al.* (2009, 2010a, 2010b). For the magnetic field applied parallel to the tape surface they found no measurable losses for a sample with non-magnetic substrate or very small losses for samples with magnetic substrate. In this case the main loss contribution comes from hysteresis losses in the ferromagnetic Ni-W substrate.

In the case of a perpendicular applied field the losses are much higher than in parallel field. In Roebel samples with non-magnetic substrate the losses are mainly determined by hysteresis losses in the superconducting layer. While for low field amplitudes the losses can be reduced by Roebel assembly of strands, at higher field amplitudes the losses of assembled cable and of single strands converge, showing that coupling of strands does not play a role in this regime. A slight frequency dependence of the losses which is different in different field ranges was interpreted to arise from eddy currents in the Cu stabilizer and from the intrinsic frequency dependence of hysteresis losses. In Roebel cables with ferromagnetic substrate hysteresis losses in the ferromagnetic substrate are the main loss contribution at low and medium fields. The superconducting layer is partly shielded by the ferromagnetic substrate. In higher magnetic fields with full penetration the loss contribution from the substrate is negligible and the main losses are due to hysteresis losses in the superconducting layer. Loss contributions due to coupling currents between the strands were not observable, a slight frequency dependence of the losses was caused by eddy currents.

In Roebel cables with stacked tapes a reduction of the losses per unit volume could be observed compared to losses from a single strand (Fig. 3.11, Terzieva *et al.*, 2010). With an increasing number of tapes in a stack the losses decreased. The reduction of losses could be explained by the shielding effect of neighbouring tapes in the stack.

A paper by Lee *et al.* (2009) addresses the effect of striations, transposition and coupling of strands at the end of the conductor on AC losses in magnetic measurements. The authors show that width reduction of strands is effective in reducing hysteresis losses from coated conductors. However, connection of the ends of two narrow edge-to-edge conductors (Y6_2C in Fig. 3.12) increases the losses to essentially the same value as for a conductor with twice the width of the narrow conductors (Y12 in Fig. 3.12).

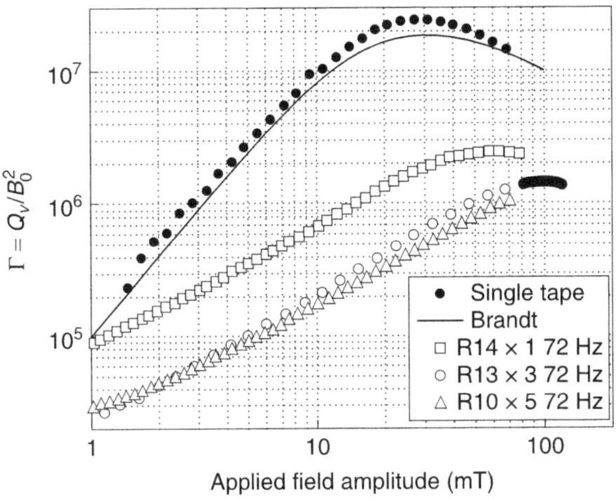

3.11 Loss factor, $Q_V/(B_0^2)$, for the magnetization AC losses per unit cycle and cable volume, Q_V, as a function of the applied field amplitude B_0. The losses per unit volume are much smaller than in the original tape due to the stacking effect. (Terzieva *et al.*, 2010.)

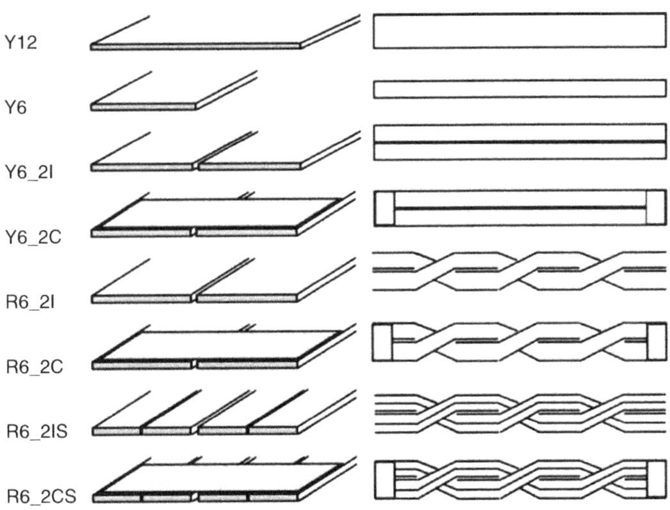

3.12 Shapes of samples for measuring the magnetization losses to estimate the effects of striation and transposition. (Lee *et al.*, 2008.)

In Fig. 3.13 the losses of two parallel strands with free (Y6_2I) and connected (Y6_2C) ends and of transposed strands with free (R6_2I) and connected (R6_2C) ends are shown. For the parallel strands the losses are much lower when the ends are not connected. In the case of transposed strands the losses are lower than for the parallel strands and nearly no difference is visible between connected and free ends.

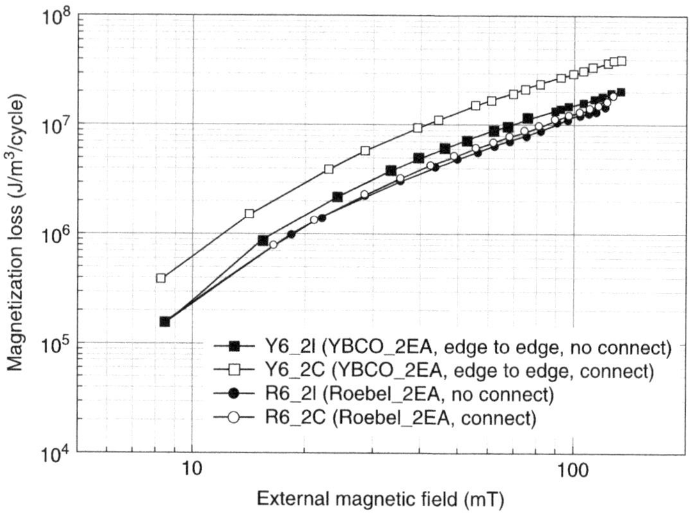

3.13 Measured magnetization losses of non-transposed and transposed samples with and without the end-connections. (Lee *et al.*, 2008.)

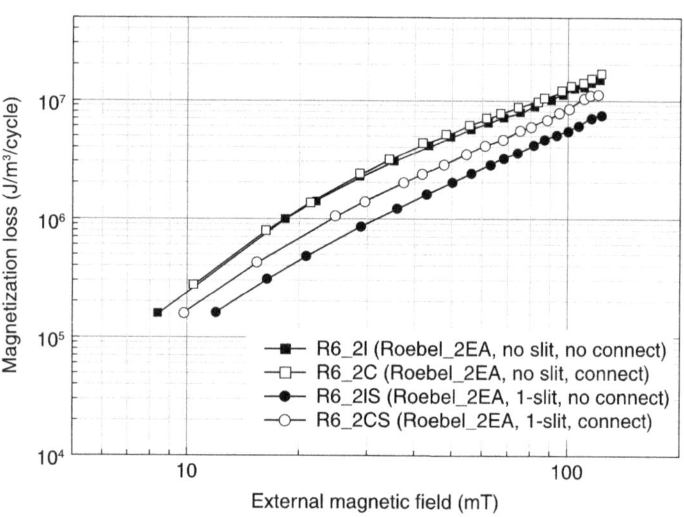

3.14 Measured magnetization losses of transposed samples with and without the end-connections and striations. (Lee *et al.*, 2008.)

Figure 3.14 shows the losses of the transposed strands with free (R6_2I) and connected (R6_2C) ends and of transposed and striated strands with free (R6_2IS) and connected (R6_2CS) ends. The striation of strands further reduces the losses in transposed conductors with free and connected ends. However, with connected ends the loss reduction is less effective.

This experiment shows very nicely that loss reduction by transposition and width reduction by striation is very effective in Roebel cables.

Measurements of transport AC losses are still rare due to the non-trivial boundary condition in the complex Roebel cable structure. Jiang *et al.* (2009, 2010) made investigations on five- and nine-strand cables with 2 mm-wide insulated strands with different AC current frequencies. Transport AC loss measurements were performed monitoring the single strands via individual voltage loops following the established method of Gömöry *et al.* (2005). A strong scattering of the results between the strands was found in particular for low frequencies. The observation of reduced scattering at high frequencies was explained by the influence of the differing resistances (mainly the contact resistance) for the different strands. Since the reactance increases with increased frequency, the influence of the different contact resistances of the strands and the scattering between the AC loss values becomes smaller.

Evaluating the effect of the transposition on the Roebel cable, the loss data for a Roebel cable and for a single strand multiplied by the number of strands in a cable were compared. The estimations were done for a cable current of $0.85 \cdot I_c$. The transport AC loss per conductor of the nine-strand Roebel cable are 6.2–6.7 times those in the single strand, compared to the nine-times predicted for the closed bundles of nine conductors, whereas for the five-strand Roebel cable the losses are 2.2–3.5 times those in the single strand compared to five times predicted for the single strand. The authors concluded that transport AC loss increases with increasing strand number. It is roughly equivalent to that of two parallel stacks where each conductor in the stack carries the same current.

3.4.4 Status of CC Rutherford cable concepts

Some applications, e.g. fusion magnets or busbars for accelerator magnets, require current carrying capabilities exceeding 10 kA. In contrast to busbars, which are normally not exposed to high magnetic fields, the magnetic fields in fusion magnets can exceed 10 T. For future fusion reactors the use of HTS magnets with an operation temperature around 65 K could save cooling costs and increase the efficiency of the reactor. Furthermore, the complex thermal shield that is required for LTS fusion magnets operating at 4.5 K (e.g. ITER magnets) could be omitted resulting in a reduction of complexity and size of the reactor.

However, transport currents of the order of >10 kA in magnetic fields >10 T and at temperatures around 65 K are difficult to realize with a simple scale-up of the Roebel technique. Instead, the Roebel cables themselves could be used as strands to form a larger cable. Depending on the application, the Roebel cables could simply be wound around a circular former. This concept is similar to the concept of power cables that consist of layers of tapes that are wound around a circular former and placed in a cryostat for LN_2 cooling. However, this concept is not suitable for the winding of big magnets, because the engineering current density is quite small

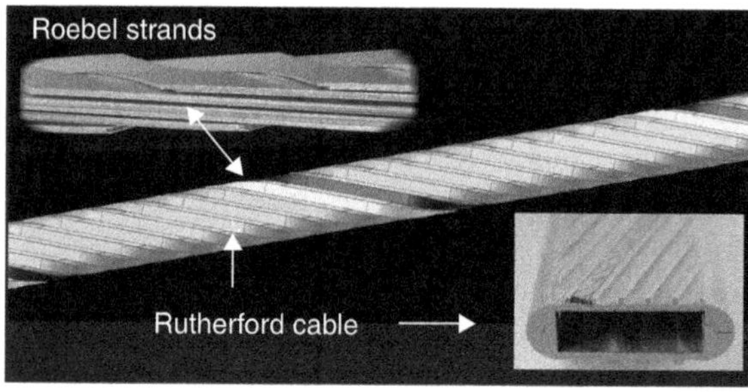

3.15 For cables with current-carrying capabilities of several kA, Roebel cables can be used as strands to build a larger Rutherford cable.

due to the small superconductor fraction that is achievable with a rather big 'central hole'. The superconductor fraction as well as the bending properties for magnet winding can be improved by winding Roebel cables around a flat former with round edges (Fig. 3.15). In such coated conductor Rutherford cables (CCRCs) the strands are fully transposed and experience similar fields and forces in a magnet winding. The transposition of strands helps to reduce coupling losses. The minimum twist pitch and the number of strands that can be assembled in a CCRC are determined by the winding angle and the width and thickness of the CCRC former. Maximum winding angle and minimum thickness of the former have to be adjusted according to the bending properties of the strands. At the time of writing such concepts for CCRCs are under development at KIT (Schlachter *et al.*, 2010).

3.4.5 Alternative CC cable concepts

One concept for the realization of transposed filaments in a Rutherford-type geometry by diffusion bonding of two silver-clad coated conductor tapes with zigzag patterned filaments that were partially separated by a dielectric layer (Abraimov *et al.* 2008) was discussed in Section 3.4.3. The AC losses could be successfully reduced; however, for long length cables the diffusion bonding technique is a complex procedure because the zigzag filaments have to be connected with a very high precision to ensure a continuous current path from end to end.

Sumption *et al.* (2005) investigated AC losses for various concepts of strand twisting. For a single twisted flat tape the engineering critical current density J_e becomes rather low. One possibility to realize strand twisting retaining a high J_e is to twist a round stack of tapes (Fig. 3.16a). The minimum twist length of such a round stack of wires is determined by the maximum allowable strain at the edges of the widest tape in the centre of the cable. For a 1 cm-wide central tape in the cable, the minimum twist length is around 30 cm. For a strong reduction of twisting induced strain, Sumption *et al.* (2005) suggest cabling with the superconducting tapes arranged on a hexagonal or polygonal former (Fig. 3.16b).

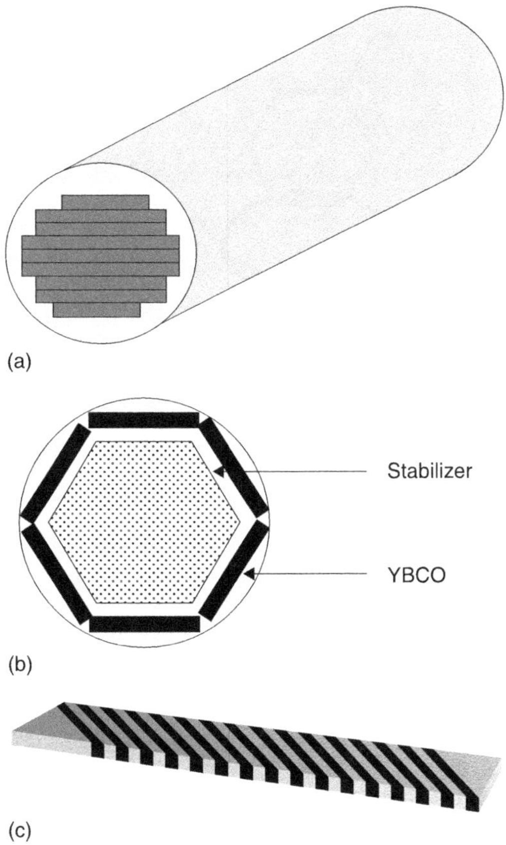

3.16 (a) A stacked, round, twisted conductor assembled from several coated conductor tapes of different widths, (b) hexagonal geometry conductor with internal Cu stabilizer and hexagonal YBCO elements, (c) Rutherford geometry conductors. (Copyright 2009 IEEE, adapted from Sumption *et al.* 2005.)

As a good compromise for performance and manufacturability the authors suggest a Rutherford-type winding process for single tapes (Fig. 3.16c). A modification and extension of this concept was presented by Goldacker *et al.* (2009b) proposing six stacks of CC applied around a hexagonal former from copper or copper alloy which served as a cooling channel, surrounded by a tube: a coated conductor in conduit cable (CCIC) (Fig. 3.17).

A new concept of cabling (RE)BCO coated conductors for high-field magnet and low AC loss applications has been introduced by van der Laan (2009), and a number of prototypes have been constructed to demonstrate the feasibility of the concept. This idea is to wind several layers of coated conductor tape around a thin former (Fig. 3.18). By placing the tapes with the superconductor to the inner side of the cable, compressive strains are applied to the tape depending on former

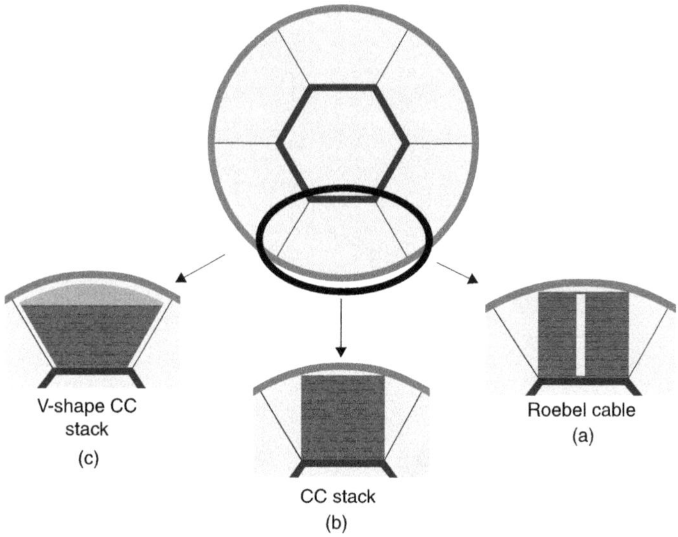

3.17 CCIC concept applying strands from CC. (a) Roebel strand, (b) simple stack, (c) 'V' shaped stack for optimized engineering current. The hexagonal central former is twisted around the longitudinal direction. (Goldacker *et al.*, 2009b.)

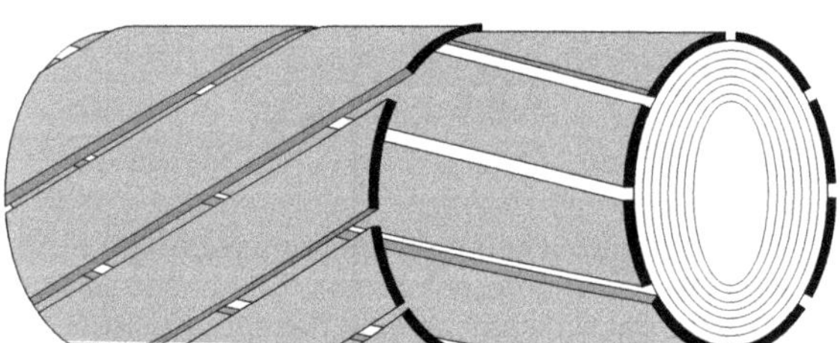

3.18 Schematic of a multi-layer cable, showing how each layer is configured out of multiple conductors. (van der Laan, 2009.)

radius and twist length. The very high elastic limit of coated conductor tapes for compressive strain allows the use of thin formers. At a compressive strain of −2% a reduction of I_c of 95% was observed, however, this degradation is fully reversible indicating that the superconductor is not damaged. Van der Laan was able to wind tapes around a former of 3.19 mm diameter without any irreversible degradation of the current-carrying capacity, because the strain was still below −2%.

3.5 Future trends and challenges

The development of HTS cables for magnet and energy applications has made a lot of progress in the last few years. Cabling methods have been developed for round BSCCO (2221) wires and flat BSCCO (2223) and (RE)BCO coated conductor tapes. Due to the superior superconducting properties of (RE)BCO coated conductors, especially their better performance in high magnetic fields at elevated temperatures, this material seems to be very promising for applications in rotating machinery, fault current limiters, transformers and magnets. However, for broad market penetration, the availability of this material and especially the costs have to be reduced. Today the cabling concepts for (RE)BCO coated conductors mainly concentrate on reduction of AC losses and increase of current performance; however, research and development is also necessary for jointing techniques, mechanical reinforcement and thermal stabilization of the tapes and cables. Especially for (RE)BCO coated conductors, long cables that allow fabrication of larger windings are only available in limited quantity, mainly due to the high cost of the coated conductor material. BSCCO (2223) CTCs have been used for high-current coils, including traction transformers, reactors and racetrack-shaped coils for rotating machinery (Leghissa *et al.*, 2006). For magnet applications, the development of quench detection and quench damping techniques is an important issue, because of the slow quench propagation velocities. The development of round HTS cables with a high filling factor could be beneficial for magnet windings.

3.6 Conclusions

HTS cables are required for many applications in which high current-carrying capacity and low AC losses are mandatory, e.g. high-field magnets with low inductance, rotating machinery and transformers. The brittle nature of HTS materials and the necessity of textured grain growth lead to conductor concepts with strong limitations regarding sheath material (e.g. Ag or Ag-alloys for BSCCO (2212) and BSCCO (2223) material) or conductor geometry (e.g. tape geometry for BSCCO (2223) or (RE)BCO conductors). Round BSCCO (2212) wires could be cabled successfully in a Rutherford geometry. Due to the brittleness of the filaments in reacted BSCCO (2212) conductors a cable-and-react approach had to be applied to ensure a good current performance. As the heat treatment is performed after the cable assembling procedure, all components of the cable have to be compatible with the high heat-treatment temperature of approximately 870–890 °C in an oxygen atmosphere. Resistive cores or barriers had to be introduced in the cable cross-section in order to reduce coupling currents, which lead to high AC losses due to the high conductance of the Ag-alloy sheath.

For tape geometries cabling is difficult due to bad bending performance in the tape plane. The Roebel concept could be successfully applied for BSCCO (2223) tapes by Siemens AG. The achieved transposition length, however, was very large due to the limited bending properties in the tape plane.

For (RE)BCO tapes the Roebel technique can be applied as well. The possibility to punch or cut (RE)BCO tapes in a meander-shaped geometry before the assembly process is a great advantage with respect to achieving a short transposition length. This punching process is not possible for BSCCO (2223) tapes because of the multifilamentary structure of these conductors. By assembling stacks of meander-shaped (RE)BCO tapes, the current-carrying capacity of the Roebel cables could be enhanced without changing the transposition length or width of the cables, matching the requirements for the different applications. A Rutherford cabling concept with (RE)BCO Roebel cables as strands has been suggested to realize (RE)BCO cables with current-carrying capacities exceeding 10 kA in magnetic fields >10 T, e.g. for application in high-field magnets for future fusion machines. Other cabling concepts for (RE)BCO conductors include twisting of round stacks of tapes, Rutherford winding of (RE)BCO tapes and arranging the tapes on a twisted hexagonal former.

The cabling concepts for HTS with high current capacity and low AC losses are promising; however, long-term research and development efforts are still necessary to optimize HTS cables for the variety of applications.

3.7 References

Abraimov D, Gurevich A, Polyanskii A, Cai X Y, Xu A *et al.* (2008), 'Significant reduction of AC losses in YBCO patterned coated conductors with transposed filaments', *Supercond. Sci. Technol.*, **21**, 082004 (4pp).

Badcock R A, Long N J, Mulholland M, Hellmann S, Wright A *et al.* (2009), 'Progress in the manufacture of long length YBCO Roebel cables', *IEEE Trans. on Appl. Supercond.*, **19**, 3244–3247.

Beck P J and Mulholland M G (2008), *Apparatus and method for producing composite cable*, European Patent Application EP 1936706A2.

Beard D S, Klose W, Shimamoto S and Vescey G (1988), 'The IEA Large Coil Task', *Fusion Engineering and Design*, **7**, 23–53.

Collings E W, Sumption M D, Scanlan R M, Dieterich D R, Motowidlo L R *et al.* (1999), 'Bi:2212/Ag-based Rutherford cables: production, processing and properties', *Supercond. Sci. Technol.*, **12**, 87–96.

Goldacker W, Nast R, Kotzyba G, Schlachter S I, Frank A *et al.* (2006), 'High current DyBCO-ROEBEL assembled coated conductor (RACC)', *Journal of Physics: Conference Series*, **43**, 901–904.

Goldacker W, Frank A, Heller R, Schlachter S I, Ringsdorf B *et al.* (2007), 'ROEBEL assembled coated conductors (RACC): preparation, properties and progress', *IEEE Trans. on Appl. Supercond.*, **17**, 3398–3401.

Goldacker W, Frank A, Kudymow A, Heller R, Kling A *et al.* (2009a), 'Status of high transport current ROEBEL assembled coated conductor cables', *Supercond. Sci. Technol.*, **22**, 034003 (10pp).

Goldacker W, Schlachter S I, Kudymow A, Drechsler A, Terzieva S *et al.* (2009b) 'Promising new concepts for AC CC cables with very high currents', *Proceedings of CCA-Barcelona.*

Gömöry F, Frolek L and Šouc J (2005), 'Non-uniform current distribution as the cause of false voltage signals in the ac loss measurement on a superconducting cable', *Supercond. Sci. Technol.*, **18**, 780–90.

Ha D W, Kim S C, Han I Y, Oh J G, Lee J H *et al.* (2008), 'Study on Bi–2212 Rutherford Cabling Process for SMES', *IEEE Trans. on Appl. Supercond.*, **18**, 1192–1195.

Hasegawa T, Koizumi T, Ohtani N, Kitaguchi H, Kumakura H *et al.* (2000), 'Development of multilayer $Bi_2Sr_2CaCu_2O_x$ superconducting tapes', *Supercond. Sci. Technol.*, **13**, 23–28.

Hasegawa T, Ohtani N, Koizumi T, Aoki Y, Nagaya S *et al.* (2001), 'Improvement of superconducting properties of Bi–2212 round wire and primary test results of large capacity Rutherford cable', *IEEE Trans. on Appl. Supercond.*, **11**, 3034–3037.

Hasegawa T, Nishioka J, Ohtani N, Hikichi Y, Scanlan R *et al.* (2004), '12 kA HTS Rutherford Cable', *IEEE Trans. on Appl. Supercond.*, **14**, 1066–1069.

Hussennether V, Oomen M, Leghissa M and Neumüller H W (2004), 'DC and AC properties of Bi–2223 cabled conductors designed for high-current applications', *Physica C*. **401**, 135–139.

Jiang Z, Staines M, Badcock R A, Long N J and Amemiya N (2009), 'Transport AC loss measurement of a five strand YBCO Roebel cable', *Supercond. Sci. Technol.*, **22**, 095002 (7pp).

Jiang Z, Badcock R A, Long N J, Staines M, Thakur K P *et al.* (2010), 'Transport AC loss characteristics of a nine strand YBCO Roebel cable', *Supercond. Sci. Technol.*, **23**, 025028 (8pp).

Kim S C, Ha D W, Oh S S, Han I Y, Bruzek C E *et al.* (2007), 'Study on thermo-mechanical treatment in fabrication of Bi2212/Ag HTS wire', *Cryogenics*, **47**, 437–441.

Kim S C, Ha D W, Oh S S, Song K J, Kim S J *et al.* (2009), 'Fabrication and properties of Bi2212 Rutherford superconducting cables', *IEEE Trans. on Appl. Supercond.*, **19**, 3076–3079.

Lakshmi L S, Thakur K P, Staines M P, Badcock R A and Long N J (2009), 'Magnetic AC loss characteristics of 2G Roebel cable', *IEEE Trans. on Appl. Supercond.*, **19**, 3361–3364.

Lakshmi L S, Staines M P, Thakur K P, Badcock R A and Long N J (2010a), 'Frequency dependence of magnetic ac loss in a five strand YBCO Roebel cable', *Supercond. Sci. Technol.*, **23**, 065008 (6pp).

Lakshmi L S, Staines M P, Badcock R A, Long N J, Majoros M *et al.* (2010b), 'Frequency dependence of magnetic ac loss in a Roebel cable made of YBCO on a Ni–W substrate', *Supercond. Sci. Technol.*, **23**, 085009 (10pp).

Lee J K, Park M J, Lee S W, Kim W S, Park C *et al.* (2008), 'Effects of striations on magnetization loss of transposed coated conductors', *Physica C*. **468**, 1727–1730.

Lee J K, Kim W S, Park C, Byun S, Han B W *et al.* (2009), 'Estimation of magnetization loss at angular magnetic field using perpendicular magnetization loss in striated and transposed YBCO coated conductor', *Physica C*. **469** 1432–1435.

Leghissa M, Gromoll B, Rieger J, Oomen M, Neumüller H W *et al.* (2002), 'Development and application of superconducting transformers', *Physica C*. **372–376**, 1688–1693.

Leghissa M, Hussennether V and Neumüller H W (2006), 'kA-class high-current HTS conductors and windings for large scale applications', *Advances in Science and Technology*, **47**, 212–219.

Long N J, Badcock R, Beck P, Mulholland M, Ross N *et al.* (2008), 'Narrow strand YBCO Roebel cable for lowered AC loss', *Journal of Physics: Conference Series*, **97**, 012280 (7pp).

Long N J, Badcock R A, Hamilton K, Wright A, Jiang Z *et al.* (2010), 'Development of YBCO Roebel cables for high current transport and low AC loss applications', *Journal of Physics: Conference Series*, **234**, 022021 (6pp).

Miao H, Marken K R, Meinesz M, Czabaj B and Hong S (2005), 'Development of round multifilamernt Bi–2212/Ag wires for high field magnet applications', *IEEE Trans. on Appl. Supercond.*, **15**, 2554–2557.

Ohsemochi K, Koyanagi K, Kurusu T, Tasaki K, Ono M *et al.* (2006), 'Test results of an experimental coil with Bi–2212 Rutherford cable for high energy-density HTS-SMES', *Journal. of Phys. Conf. Series*, **43**, 825–828.

Schlachter S I, Goldacker W, Grilli F, Heller R and Kudymow A (2010), 'Coated conductor Rutherford cables (CCRC) for high-current applications: concept and properties', submitted to *Proceedings of the Applied Superconductivity Conference, Aug. 1–6, 2010*, Washington DC.

Schuller S, Goldacker W, Kling A, Krempasky L and Schmidt C (2007), 'AC-loss measurement of a DyBCO-Roebel assembled coated conductor cable (RACC)', *Physica C*. **463–465**, 761–765.

Sumption M D, Barnes P N and Collings E W (2005), 'AC losses of coated conductors in perpendicular fields and concepts for twisting', *IEEE Trans. on Appl. Super.*, **15**, 2815–2818.

Sumption M D, Collings E W, Scanlan R M, Nijhuis A and Ten Kate H H J (1999) 'AC loss and contact resistance in Nb3Sn Rutherford cables with and without a stainless steel core', *Adv. Cryogen. Eng. Mater.*, **44**, 1077–84.

Terzieva S, Vojenciak M, Pardo E, Grilli F, Drechsler A *et al.* (2010), 'Transport and magnetization ac losses of ROEBEL assembled coated conductor cables: measurements and calculations', *Supercond. Sci. Technol.*, **23**, 014023 (8pp).

van der Laan D C (2009), '$YBa_2Cu_3O_{7-}$ coated conductor cabling for low ac-loss and high-field magnet applications', *Supercond. Sci. Technol.*, **22**, 065013 (5pp).

Weijers H, Trokiewiz U, Mbaruku A and Schwartz J (2003) 'Properties of Bi–2212PIT conductor and reinforced Bi–2212 coils under axial tension and compression', MT18-Morioka-Japan.

Wilson M (1999), 'Superconductivity and accelerators: the good companions', *IEEE Trans. on Appl. Super.*, **9**, 111–121.

4

Bulk high temperature superconductor (HTS) materials

T. COOMBS, University of Cambridge, UK

Abstract: This chapter concentrates on bulk materials. A bulk superconductor is one in which the superconductor has been formed into a lump, usually cylindrically shaped, but can also be hexagonal, rectangular or even square. Bulk superconductors are typically 3–5 cm across and 1 cm thick. They have many uses but the principal one is as extremely compact high-field permanent magnets in superconducting machines. A 2.6 cm (RE)BCO puck has been magnetised to 17.24 T: this is an order of magnitude greater than the flux density available from a conventional permanent magnet. This chapter describes the materials, manufacturing process, magnetisation process and some examples of machines.

Key words: bulk superconductor, permanent magnets, high fields, electrical machines.

4.1 Introduction

High temperature superconductivity was born in 1986 when Bednorz and Muller succeeded in inducing superconductivity in lanthanum barium copper oxide (LaBaCuO, also known as LBCO); the oxide's critical temperature (T_c) was 35 K, a full 12 K higher than the previous record. This discovery stimulated a great deal of additional research in high temperature superconductivity on cuprate materials with structures similar to LBCO, soon leading to the discovery of compounds such as bismuth strontium calcium copper oxide ($Bi_2Sr_2Ca_{n-1}Cu_nO_{2n+4+x}$, where n can be 1, 2 or 3 with 2 being the most commonly studied) commonly known as BSCCO (T_c 107 K) and yttrium barium copper oxide ($YBa_2Cu_3O_{7-x}$) commonly known as YBCO (T_c 92 K). Amongst many others the cuprates BSCCO and YBCO have for many years, in engineering terms, remained the most important. BSCCO and YBCO were joined relatively recently by magnesium diboride (MgB_2). Although the T_c of MgB_2 is much lower than either BSCCO or YBCO it is much cheaper and easier to fabricate. Still more recently, a whole new family of superconductors, the pnictides, have been discovered. Superconductors have many useful properties and come in a multitude of different forms:

- single crystal,
- thin and thick films,
- melt-processed bulks,
- wires both single and multi-cored,

- tapes either rolled using powder-in-tube (PIT) or similar techniques or deposited using techniques such as ion beam assisted deposition (IBAD) or rolling assisted bi-axially textured substrate (RABiTS).

High temperature superconducting quantum interference devices (SQUIDS) and electronic devices switching many times faster than silicon and with lower losses have been proposed and demonstrated. Machines using superconducting wires and tapes have been built and tested along with a multitude of other devices such as fault current limiters.

This chapter concentrates on bulk material. A bulk superconductor is one in which the superconductor has been formed into a lump, commonly cylindrically shaped, but can also be hexagonal, rectangular or even square. Bulk superconductors differ from other forms in that they have thicknesses that are measured in millimetres as opposed to microns or nanometres. In addition, they typically do not rely on a substrate and (in the case of YBCO) are textured using a family of processes based on the melt powder melt growth (MPMG) process first identified by Murakami *et al.*[1] They are black in appearance and typically have diameters of 3–5 cm although smaller and larger samples are made. A typical thickness would be 1 cm. They are manufactured in a two-stage process: the MPMG stage first, and then an oxygenation phase.

Many of the high temperature superconductors (HTS) are suitable for making into bulk materials but the two with the most promise are the so-called (RE)BCO cuprates (the most famous of which is YBCO) and MgB_2. There is a considerable body of knowledge behind the (RE)BCO cuprates in bulk form and they have been made since the first discovery of YBCO in 1987. Originally the materials were sintered, but this leads to small grains with poor connectivity and, although the intra-grain currents are high, the inter-grain currents were very poor so that the materials were of low quality. Relatively early on though, Murukami, Morita, Doi and Miyamoto working in Japan at the Nippon Steel laboratories reported that a process known as melt powder melt growth (MPMG) could be used to produce high-quality YBCO. This was first reported in Murakami *et al.*[1] and it is worth reproducing the abstract from that paper:

> We report a new process which promises high critical current density in oxide superconductors. The process consists of three stages. Firstly a YBa_2Cu_3Ox sample is rapidly heated and quenched from the Y_2O_3 plus liquid region. Subsequently the quenched sample is reheated to the Y_2BaCuO_5 plus liquid region, and then slowly cooled with a temperature gradient in flowing oxygen. The process enables us to grow a superconducting phase unidirectionally and to suppress the second phase intrusion, leading to the production of well textured YBa_2Cu_3Ox which yields a high J_c value in the presence of magnetic fields. It is also found that Bean's critical state is realised in such high J_c samples.

More than twenty years later (although they have evolved considerably) the methods used for growing (RE)BCO bulks of high quality can still be said to have their roots in this process.

Bulk superconductors are particularly important because they can provide very high magnetic fields in a very compact package. To understand their significance it is important to understand the origins of magnetism.

4.1.1 Origins of magnetism

Magnetism arises from two sources:

- Electric currents or, more generally, moving electric charges, which create magnetic fields.
- Many particles have non-zero 'intrinsic' (or 'spin') magnetic moments. Just as each particle, by its nature, has a certain mass and charge, each has a certain magnetic moment, possibly zero.

A material is magnetised when the individual magnetic moments are aligned. How they align and the degree of the alignment dictates the type of magnetism. The main types of magnetism are diamagnetism, ferromagnetism, ferrimagnetism/ antiferromagnetism and paramagnetism.

Diamagnetism

Diamagnetism is a fundamental property of all matter, although it is usually very weak. Diamagnetic substances are usually composed of atoms that have no net magnetic moments (i.e., all the orbital shells are filled and there are no unpaired electrons). However, when exposed to a field, a negative magnetisation is produced and thus the susceptibility is negative. Water is weakly diamagnetic and the resulting magnetohydrodynamic (MHD) effect has been used to power ships in Japan and China.

In addition, superconductors are themselves diamagnetic. When exposed to a magnetic field they produce a field in opposition to the magnetising field. All of the superconductors described in this chapter are Type II superconductors. Type II superconductors have two critical fields: H_{c1} and H_{c2} (where H is the applied magnetic field in A/m). Below H_{c1}, which is typically a few millitesla, superconductors are perfectly diamagnetic and they precisely cancel the magnetic field. The upper critical field, H_{c2}, is the field at which the superconductor stops being superconducting. This field is temperature dependent but can be as high as tens of tesla. It is this high H_{c2} that makes high temperature superconductors so useful.

Ferromagnetism

Ferromagnetism is what people usually mean when they talk about magnets. Common ferromagnets are iron, cobalt and nickel. Ferromagnetic materials exhibit parallel alignment of moments resulting in large net magnetisation, even in the absence of a magnetic field. At high temperatures thermal activation destroys this alignment, leading to the existence of a magnetic ordering temperature that is known as the Curie point.

Ferrimagnetism/antiferromagnetism

Ferrimagnetism differs from ferromagnetism in that there are two distinct magnetic sublattices. These, when ordered, have antiparallel spins but the spins are not equal. This results in a net magnetic moment. Ferrimagnets, like ferromagnets, have a Curie point. In an antiferromagnet the spins in the two sublattices are equal, resulting in no magnetic moment.

Paramagnetism

In this class of materials, some of the atoms or ions in the material have a net magnetic moment due to unpaired electrons in partially filled orbitals. However, the individual magnetic moments do not interact magnetically and the magnetisation is zero when the field is removed. In the presence of a field, there is now a partial alignment of the atomic magnetic moments in the direction of the field, resulting in a net positive magnetisation and positive susceptibility.

Common to all forms of magnetism that arise out of alignment of spins is the fact that the magnetism is limited by the length scale of the orbitals. In addition, if we draw a diagram we can see that adjacent spins can effectively cancel each other out, leading to a permanent magnet being the equivalent of a single-turn solenoid where the magnetism comes from a surface current. This means that, no matter how physically big the magnet is, as long as we keep the geometry consistent we will not change the magnetisation (Fig. 4.1).

Since the field is produced by a surface current, the field profile is similar to that produced by a single-layer solenoid or coil of similar overall dimensions. In a

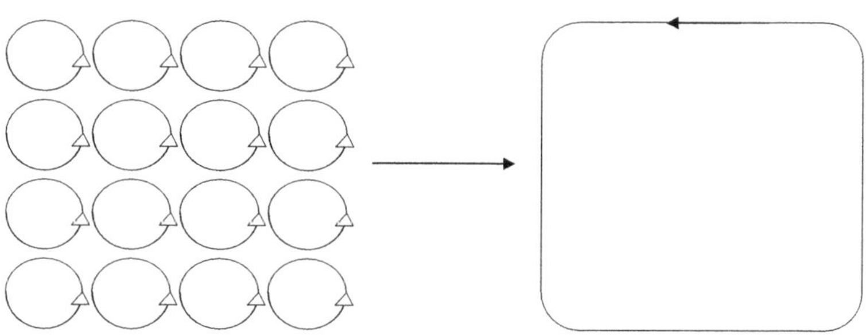

4.1 The sketch on the left represents the individual spins as circulating currents. Adjacent spins have anti-parallel currents where they abut and hence internally the currents cancel out, leading to one single surface current as is shown on the right.

long, thin magnet magnetised parallel to its long access, the field across its end surfaces is essentially uniform. In a short, fat magnet the field will be a maximum at the edge and will dip in the middle.

A superconducting bulk, however, performs much like a multi-layer coil. Currents, once induced, flow on the length scale of the superconductor and, instead of being equivalent to a single surface current, are equivalent to multiple concentric currents. Once the bulk is fully magnetised these currents flow over the whole volume of the superconductor and produce a characteristic peak in the B field at the centre of the superconductor, regardless of the aspect ratio of the sample.

The magnetic field available from a bulk superconductor is therefore much higher than that which is available from a permanent magnet. In addition, a bulk superconducting magnet is much more compact than a solenoid wound from superconducting wire or tape. The reason for this is that the whole of the superconductor carries current in a bulk, whereas in a wound solenoid allowance has to be made for the substrate and for insulation between layers. This means that the engineering current density for a bulk is one or two orders of magnitude higher than that of a wire or tape.

4.1.2 Maximum obtainable trapped fields

Type II superconductors have two critical fields H_{c1} and H_{c2} (Fig. 4.2). H_{c1} is the field at which flux starts to penetrate the superconductor and H_{c2} is the field at which superconductivity is lost entirely. H_{c2} depends on the temperature but generally is very high in high temperature superconductors such as YBCO or its

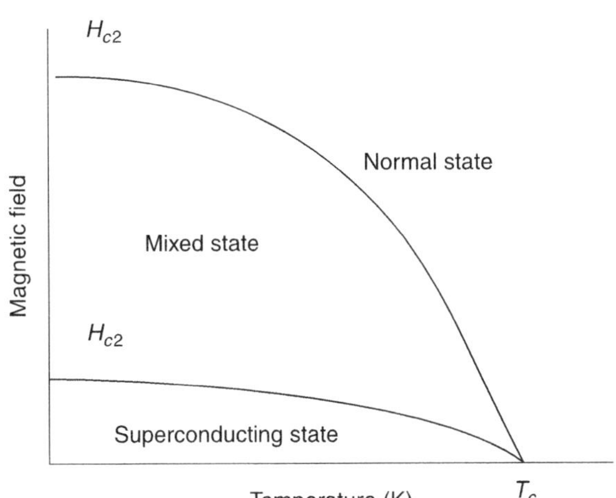

4.2 Superconducting states at different applied magnetic fields. H_1 (also known as H_{c1}) is the lower critical field below which no magnetic field enters the superconductor; H_{c2} is the upper critical field at which the material loses superconductivity.

(RE)BCO analogues, up to 100 T by some reports. In bulk superconductors it is not H_{c2} that limits the maximum obtainable magnetisation or trapped field, rather it is the Lorentz force that leads to structural breakdown at high fields and has limited the maximum trapped field in a 2.6 cm puck to just over 17 T[2].

Apart from the effect of the Lorentz forces, three major factors combine to limit the maximum obtainable trapped field.

Critical current

The critical current is temperature-dependent but is also strongly influenced by pinning centres. Without pinning centres, flux would be able to move freely within the body of the superconductor and the critical current density J_c would effectively be zero. Pinning centres, which are non-superconducting regions on the micron or nanometre scale, can be formed in many different ways. The principle method is to make sure that there are second-phase materials in there that do not themselves superconduct classically; in YBCO this has involved the use of 211 or yttria particles.

Grain boundaries

In bulks the ideal is to eliminate grain boundaries entirely. Typical inter-grain J_{cs} are an order of magnitude less than intra-grain J_{cs} and a bulk with no grain boundaries will allow the currents to flow on the length scale of the bulk itself and will hence produce the maximum possible magnetic moment. This is entirely different from coated conductors. In coated conductors the technique is to texture the grains so that they are themselves very small; this is done to maximise the areas of the common boundaries between grains.

Diameter

The maximum B field is a product of the critical current and the diameter of the bulk. Hence, a larger bulk will produce a larger moment. This works only up to a point, as very large bulks will have more grain boundaries and this in itself will limit the available moment. In addition, a stoichiometric mix of YBCO is not a good superconductor; in order to make it superconduct it must be oxygenated and it is very difficult to get oxygen to penetrate all the way into a large sample. This limits the average J_c. Typically, the maximum obtainable trapped field will occur in samples of about 3 cm diameter.

4.2 Design of bulk high temperature superconductor (HTS) materials

The commonly accepted method of preparing bulk superconductors is the melt powder melt growth (MPMG) method. In this technique precursor powders are pressed into a puck and a seed placed on top of the puck. The temperature of the puck is then raised to the melting point of the powders and from there is slowly lowered. The seed, which does not melt, acts as a nucleation site and when correctly managed the material solidifies and grows from this single point.

Although the original material was YBCO this has expanded to include a whole family of cuprates known as (RE)BCO where Re can be a range of rare earth elements such as neodymium, samarium, gadolinium or indeed yttrium itself. All of these variants have different lattice parameters and melting points and, accordingly, a range of different seeds have been tried. Until recently the seeds were primarily crystals but recent developments have included thin film seeds.

Pinning centres are vital if we are going to obtain a high J_c and a high magnetic moment. By far the most common method of including pinning centres is to have so-called 211 inclusions. The superconducting phase is $(Re)Ba_2Cu_3O_y$ and an additional $(Re)_2BaCuO_5$ 211 phase is included, which is non-superconducting. Finely distributed, these non-superconducting particles act as barriers to the movement of flux lines. In this section we present a number of different processing routes for some of the most important bulks superconductors.

4.2.1 Materials and methods of preparation

This section will look at the three main materials that are available in bulk form. These are (RE)BCO, BSCCO (in both 2223 and 2212 forms) and MgB_2.

(RE)BCO

Of the materials that are generally available in bulk form by far the most studied is YBCO and by extension (RE)BCO. The major steps that took place in the development of YBCO/(RE)BCO were:

- solid state sintering (1987)
- melt textured-growth (MTG) processing (1989)
- top-seeded melt textured growth (TS-MTG) (1993)
- solute rich liquid crystal pulling (SRLCP) (1996)
- (RE)BCO – substitution of various rare earths such as Nd, Sm Gd, Eu can lead to enhanced pinning and changes in T_c.

The main preparation methods for (RE)BCO bulk materials are summarised as follows:

- Melt texture growth (MTG)[3,4]. The melt-textured growth relates to slow cooling in a temperature gradient, which results in directional solidification at a growth front.
- Quench melt growth (QMG)[5]. The quench melt growth process fabricates the bulk by calcining, melting and quenching, grinding into powders and then the MTG process is used. A feature of the QMG method is that the second-phase

Y211 is very fine and has homogeneous distribution in the matrix of the materials, so that the flux pinning and J_c are increased.

- Liquid-phase process (LPP)[6]. In the liquid-phase process method, slow cooling through the peritectic transformation (1030–980 °C) has been shown to control the microstructure of superconductors.
- Powder melt process (PMP)[7]. The powder melt process of PMP is the same as MTG but uses precursor powders of YBaCuO (211) phase and Ba-Cu-O as starting materials instead of Y123 materials in the PMP process.
- Melt powder melt growth (MPMG)[8]. The attractive feature of the melt powder melt growth (MPMG) process is that other components such as fine Ag powders can be added during solid-state mixing.
- Top-seeded melt-textured growth (TS-MTG)[9–11]. The standard composition of the top-seeded melt-textured growth method starting material is $YBa_2Cu_3O_{7-x}$ (Y-123) with an Y-211 excess of 25 mol%.
- Solute rich liquid crystal pulling (SRLCP)[12–16]. This method uses high-quality (3N) Y_2O_3, $BaCO_3$ and CuO powders as the solute and solvent. Natural flow and forced flow dissolve the 211 particles, and Y-123 phase is precipitated to Y-123 crystal.
- Preparations for LRE-123 bulks[17–19]. The oxygen-controlled-melt-growth (OCMG) process has been developed to prepare the LRE-123 bulks. MgO seeds prove to be efficient in the growth orientation control of large-grain LRE-123 pellets when a small quantity of ZnO is added, accompanied by a substantial reduction of liquid phase loss.

BSCCO

There are a number of preparations for BSCCO bulk[20–29]. In addition to this, Bi-2223 rod is available, which is usually prepared by normal sintering, hot forging process, and cold isostatic pressing methods. One of the more successful techniques for fabricating Bi-2212 tubes is the melt cast process (MCP)[27]. Another method that has been used is the fabrication of cylinders strengthened using MgO whiskers.

MgB_2

MgB_2, although recently discovered, has attracted much less attention than (RE) BCO, and its material properties are poorer at equivalent temperatures. Nevertheless it is important chiefly because it is cheap to fabricate and can be made in much bigger sizes than the rare earth bulks. MgB_2 is mainly produced using a sintering route and the challenge lies in densifying it. Various preparations for MgB_2 are available, the main ones of which are hot pressing (HP)[30], reactive liquid Mg infiltration (RLI)[31], solid phase reaction[31] and the powder-in closed-tube (PICT) diffusion method[32].

4.2.2 Properties

Levitation

A superconductor will resist changes in magnetic flux. According to Maxwell's equations a changing flux produces an electric field that in turn leads to a current. The magnetic field produced by the current acts to resist the original change in magnetic field. According to the Bean model this whole process occurs with no loss occurring in the superconductor, hence the change in magnetic field induced exactly cancels the field that produced it, and this results in a strongly diamagnetic effect. The only thing that prevents it being perfectly diamagnetic is that the currents induced flow within the body of the superconductor. Perfect diamagnetism requires the currents to flow on the surface, as occurs in the Meissner effect.

The net result is that if a magnet is brought up to a superconductor the superconductor behaves in a similar manner to an image viewed in a mirror, producing a mirror image of the magnet that is being brought up to the superconductor. These two magnets repel each other and thus a levitation force is produced. The closer the magnet to the superconductor, the closer the real and image magnet are and the greater the levitation forces.

This system is thus stable in the direction of motion of the magnet since the rate of change of the force developed is always of the opposite sign of the change in the gap (a reducing gap produces an increasing gap and increasing produces an increasing force).

Less obvious is the fact that this system is also stable in the lateral direction. As the magnet moves across the superconductor its image moves with it and so do the flux lines. If the magnet moves far enough laterally the flux lines move from their original pinning centres to new ones. The pinning centres resist the movement of the flux lines and hence resist the movement of the magnet. Thus an additional stabilising force, involving trapped flux, begins to act, stabilising the system laterally.

For a Type II superconductor, the equilibrium position of a levitated permanent magnet (PM) depends on its position when the superconductor transitions below the critical temperature. One may distinguish two cases: zero-field cooling (ZFC), where the superconductor is cooled with the PM far away, and field cooling (FC), where the superconductor is cooled with the PM in close proximity. In the ZFC case, as the PM is brought close to the superconductor, the flux tends to be excluded or trapped near the superconductor's surface, and the repulsive force between the PM and the superconductor is strong. In the FC case, a substantial amount of flux is trapped in the superconductor, and very little force is exerted on the PM when it is at the FC position.

Trapped field characteristics

The strong pinning in (Re)BCO means that high magnetic fields can be trapped. The amount of field that can be trapped is a function of the pinning force, the diameter and the temperature. A typical puck is 30 mm in diameter. Typical trapped fields for YBCO are 1.3 T at 77 K[33], 16 T at 24 K[34] and 17.24 T at 29 K between two 26.5 mm samples[2]. It is possible that these records may be far exceeded in (RE)BCO; compounding the RE-123[35,36] with several rare-earth ions was found to dramatically enhance flux pinning. For example (Nd,Eu,Gd)-Ba-Cu-O exhibits an extremely high-irreversibility parallel magnetic field of 15 T at 77 K with a critical current value exceeding 104 A/cm^2 at 10 T[37,38].

Magnetic levitation characteristics

A strong diamagnetic effect leads to useful levitation forces. Greater than 80 N has been achieved for zero field cooled 3 cm-diameter YBCO pellets at 77 K at a distance of 0.5 mm from a 0.4 T 2.5 cm-diameter SmCo magnet[39]. This represents magnetic pressures of >10 N/cm^2 and is limited by the magnetic field strength of the permanent magnet[40–42], not the properties of the (RE)BCO.

Magnetic shielding characteristics

HTS bulks have a frequency-independent magnetic shielding ability. Shielding occurs because of the Meissner/diamagnetism effects in which the induction in the superconductor is proportional to the applied magnetic field and not, as is more usual, the rate of change of magnetic flux[43].

In superconductors the shielding factor is defined as the ratio of the magnetic field inside the shield at its centre to the applied field. A high-degree shielding is achieved up to a critical external field that is determined by the critical current density of the material[44].

Typically the critical field is 23 oersted (Oe) at 77 K extending up to 105 Oe at 4 K. Tesla and gauss are units of magnetic flux density (B) with 1 T equal to 10^4 gauss, oersted is used for magnetic field intensity (H). In SI units this is commonly described using A/m. 1 A/m equals $4\pi \times 10^{-3}$ Oe. B and H are related by the local permeability and the presence or absence of any magnetic material by the relationship $B = \mu(H + M)$ where M is the magnetisation of the magnetic material. Effective shielding using high T_c superconducting shields can be achieved up to critical magnetic fields of 10–100 Oe. Higher current densities will shield higher fields. This can be achieved either by lowering the operating temperature or improving the quality of the HTS bulks.

4.3 Development and application of bulk HTS materials

As was seen in the previous section a great deal of effort has gone into the development of HTS bulk materials and a large number of process routes have been tried and perfected. In parallel with this a number of applications have been investigated and these are examined below. Although a lot of these applications are relatively mature there has as yet been relatively little take up and there are still challenges to be solved.

In no particular order the applications that have received the most attention are:

- Bearings/levitation (diamagnetic effect).
- The fact that a superconductor will react to external changes in magnetic field enables us to produce an entirely passive non-contact bearing. This has been demonstrated both as a thrust bearing and a journal bearing.
- Bearings for flywheels (diamagnetic effect).
- In principle the non-contact bearings could be used in a range of different machines. An energy storage flywheel is the most suitable environment for two reasons. The first is that in a very real sense a superconducting bearing provides a quantifiable advantage over a rolling element bearing. The losses are extremely low and a very efficient flywheel can be built as a result. The second reason is that an energy storage flywheel is a static environment there are no external forces and consequently design of the system is relatively simple.
- Motors/generators (diamagnetic effect/trapped field magnet).
- Many different arrangements have been proposed and considered. These fall into two classes. The first is the ones that use the same property as the bearings. A moving rotor field drags the superconducting rotor around. These are loosely termed reluctance machines. They are of interest but do not offer substantial advantages over conventional machines. Of more importance are designs that use the superconductors as trapped field magnets. Potentially at least a superconducting trapped field magnet is an order of magnitude stronger than a conventional permanent magnet. Consequently magnets of this type would lead to considerable improvement over conventional machines.
- Magnetic separation (trapped field magnet).
- Magnetic separators can and have been built that use trapped field superconducting magnets.
- Magnetrons (trapped field magnet).
- Magnetron sputtering: superconductors could be used as trapped field magnets in a magnetron sputtering device to deposit films. This would yield considerable advantages over conventional machines.
- Current leads (current carrying capability at high temperatures).
- A very practical application of HTS bulks is as current leads. Compared to tapes, bulks have a very low thermal conductivity and this makes them ideal for use as current leads feeding an LTS magnet system. Their use cuts down expensive thermal losses and (potentially) reduces loss of helium. Both BSCCOs and (RE)BCOs have been used as current leads.
- Fault current limiters (current carrying/normal state resistivity).
- A lot of work has gone into the development of fault current limiters, both resistive and inductive. Both BSCCO and (RE)BCO bulks have been used to build demonstrator fault current limiter projects.

4.3.1 Bearings and levitation

One of the well known properties of HTS is the ability to levitate magnetically. This ability extends to large masses. Early demonstrations involved people[45]. Famously a Sumo wrestler plus a platform containing magnets with a total levitated mass (wrestler plus magnet) of 200 kg were suspended in mid-air. Even larger masses have been suspended, e.g. 410 kg[46] and 1050 kg[47].

Magnetic levitation could be used to suspend cryogenic tanks or superconducting transmission lines. This reduces heat leak significantly by removing the physical conduction path and, of course, since a cryogenic environment is already present this can be used to cool the HTS. One of the earliest studies was the use of HTS levitation in hydrogen storage tanks for automobiles[48].

Other applications include levitated conveyor systems in clean rooms, where high purity requirements mandate no mechanical contact[49]. Alternatively we could consider the use of trapped-field HTSs to replace the superconducting coils aboard maglev vehicles[50]. Much higher levitation heights could be achieved than with conventional permanent magnets. Diamagnetic HTSs could be placed over a PM guideway[47]. In China, the Qingchengshan maglev vehicle test line was proposed using high T_c SCPMs to replace the normal electromagnets as the suspension magnets for the maglev vehicles[51]. A demonstration HTS maglev test vehicle was built to study the integrated performance of an HTS maglev system, such as shown in Fig. 4.3[51,52]. The YBCO bulks are cylinders of 30 mm diameter and 18 mm in thickness.

The combination of diamagnetic behaviour as the superconductor attempts to negate any change in its internal magnetic field, together with the ability to lock magnetic field lines in pinning centres within the HTS, allows the stable levitation

4.3 The man-carrying high-temperature superconducting maglev test vehicle.

of a PM above or below an HTS[53]. If the PM is cylindrically symmetric, it freely rotates about its axis of symmetry and forms a low-loss magnetic bearing that is passively stable[54]. HTS bearings have seen considerable development and have been suggested for many applications[54,55].

HTS bearings have very low coefficients of friction, as has been demonstrated for several experimental HTS bearings[54], and the limit to obtaining lower values seems to be the homogeneity of the permanent magnet rather than the superconductor itself. Mechanical roller bearings typically have a coefficient of friction (COF) of the order of 10. Damping time constants in directions other than the rotational one are typically of the order of seconds for HTS bearings, so even though they exhibit low rotational drag, they have a remarkably high damping.

Many bearings have been built and demonstrated. A recent example for loads up to 10 kN was developed by Nexans SuperConductors (Huerth, Germany) together with TU Braunschweig and Siemens AG (Erlangen, Germany)[56].

A novel feature of this design is the subdivision of the cylindrical stator, which consists of two half-shells. The lower half-shell can be moved perpendicular to the rotation axis of the bearing, thus allowing an advanced force activation (Fig. 4.4). The stator shells in Fig. 4.5(a) consist of 270 single-domain YBaCuO monoliths produced by the top-seeded-melt growth process. Each of the trapezoidal tiles of about 33 × 35 × 10 mm exhibits a trapped magnetic field of 1.4 T and a critical current density of 1.3 × 10^5 A/cm^2 at self-field and 77 K. The cryogen-free cooled stator with an inner diameter of 325 mm has a length of

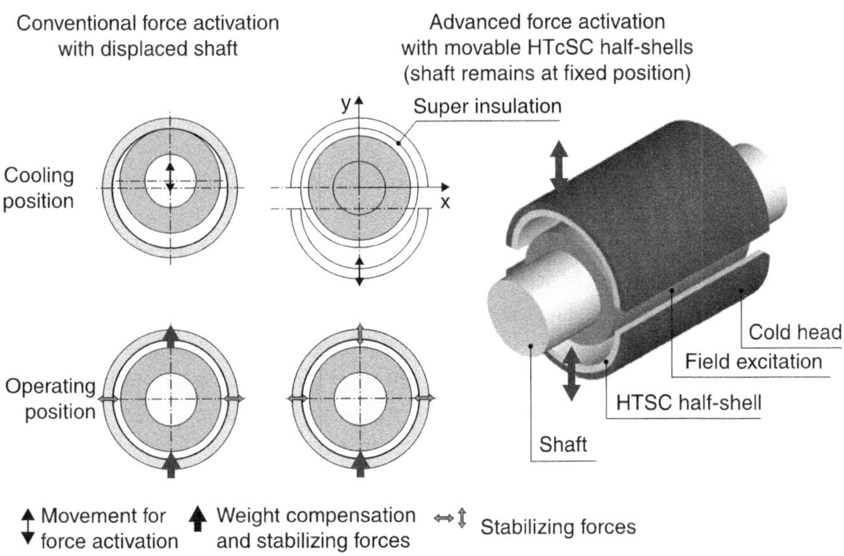

4.4 Superconducting magnetic bearing with conventional and advanced force activation with movable HTS half-shells.

4.5 (a) Superconducting magnetic bearing: Cu half-shell stator with HTS tiles. (b) Superconducting magnetic bearing rotor, assembled from NdFeB permanent magnet ring segments.

305 mm. The rotor in Fig. 4.5(b) is assembled from NdFeB PMs. With its outer diameter of 319 mm, this allows a working gap of 3 mm. The radial bearing capacity of 5 kN at 26 K was measured both under static and rotating (typically 3600 rpm) conditions. The bearing exhibits a linear force-displacement curve over a ±0.3 mm displacement range and shows a vertical-bearing stiffness of 5.1 kN/mm. At present, this appears to represent the world's largest high-capacity superconducting magnetic bearing (SMB).

4.3.2 Flywheel energy storage

Flywheels are an obvious and relatively simple application of HTS bearings. Flywheels have relatively low energy density but very high power density in comparison to batteries. The energy stored is proportional to $(1/2)J\omega^2$, Thus the energy stored increases with the square of the speed of rotation. An HTS bearing with its low coefficient of friction does not only reduce losses, it permits higher speed operation. With modern graphite fibre/epoxy materials, the inertial section of a flywheel rotates with rim speeds well in excess of 1000 m/s and achieves energy densities greater than those of even the most advanced batteries. The kinetic energy in a 10 × 30 cm diameter flywheel with this rim speed is approximately 1 kWh, and a person-sized flywheel could store 20–40 kWh. Experimental versions of flywheels that employ HTS bearings, such as that shown in Fig. 4.10 below, have already stored 2 kWh.

The availability of HTS bearings that are so nearly friction-free naturally leads to their consideration for flywheel energy storage. Several major projects have investigated this application[57–60]. Flywheels with conventional bearings typically experience high-speed idling (that is, no power input or output) losses on the order of 1% per hour. With HTS bearings, rotational losses as little as 0.001% (of the energy in a full-speed rotor) per hour have been achieved in laboratory experiments[61]. When conduction losses to the cryostat are included, and energy is taken from the flywheel to power the refrigerator that cools the HTSs, total parasitic energy losses for the bearing of 0.1% per hour are believed to be achievable. When coupled with efficient motors/generators and power electronics (capable of losses as low as 4% on input and output), the potential exists for constructing flywheels with a 90% diurnal storage efficiency. Probably only one other technology is capable of achieving such high diurnal storage efficiencies: superconducting coils hundreds of meters in diameter.

There are many different flywheel systems under development in various parts of the world. In Germany an example is the 10 kW class DYNASTORE (Fig. 4.6, 4.7). The characteristics of the HTS/PM were calculated from test results performed with a model arrangement with the same pole pitch as the flywheel bearing. One bearing with a diameter of 54 cm provides a radial and axial stiffness of 2086 N/mm and 3020 N/mm, respectively, which are well suited to operate a flywheel with a flying weight of 500 kg. This flywheel, in the German Leitprojekt DYNASTORE, can provide 2 MW power for 20 ms and is ideal for power quality-type applications.

In Japan, radial-type superconducting magnetic bearings formed by an inner-cylindrical stator of YBCO bulks and an outer rotor of permanent magnets were

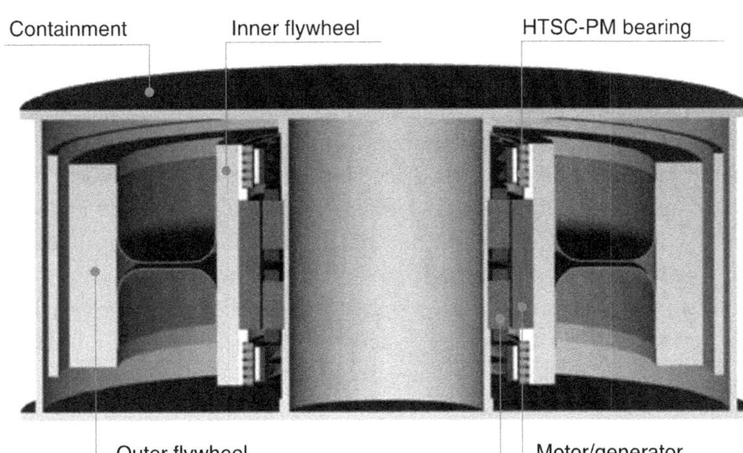

4.6 DYNASTORE flywheel energy storage system.

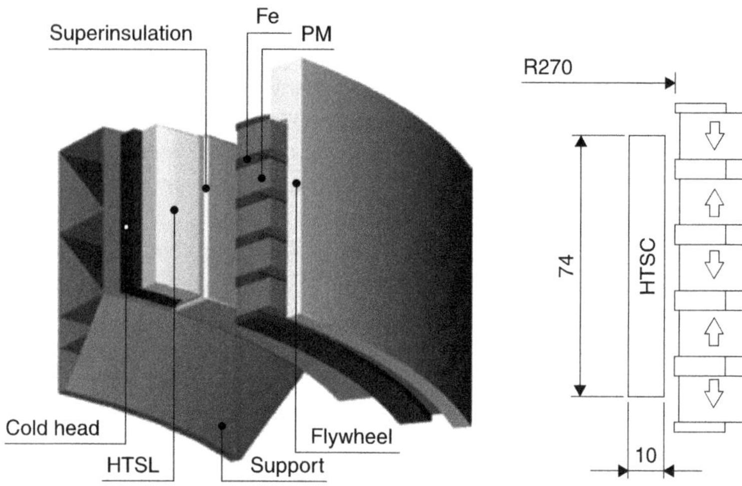

4.7 DYNASTORE flywheel: cross-section view of the bearing system with the actual dimensions of the magnetic circuit[62,63].

developed for a 10 kWh-class flywheel energy storage system in the New Energy and Industrial Technology Development Organization (NEDO) project[64–66]. In Korea, a double-evaporator thermosiphon for cooling YBCO bulks in the 100 kWh superconducting flywheel energy storage (SFES) system was designed[67,68]. In the USA, a 5 kWh/100 kW flywheel energy-storage system (FESS) using an HTS bearing suspension/damping system was developed by Boeing, as shown in Fig. 4.8[69]. Table 4.1 summarises the capabilities of HTS-levitated flywheels and provides a comparison with other forms of energy storage.

Electric utilities have a great need for efficient diurnal energy storage, such as flywheels, because the inexpensive base load capacity of the utilities is typically underutilised at night, and they must use expensive generating sources to meet their peak loads during the day. A distributed network of diurnal-storage devices could also make use of underutilised capacity in transmission lines at night and add robustness to the electric grid. These factors are expected to become more important in the forthcoming deregulation of the electric utility industry. Efficient energy storage would also be beneficial to renewable energy technologies, such as photovoltaics and wind turbines

The flywheel system is very attractive for the energy storage system because of its relatively high stored energy density. Flywheel systems are, however, now used mainly for short time usage such as load averaging or peak load energy supply because of the energy loss during rotation. One of the main losses of the flywheel system is the loss from bearings. A superconducting bearing using high temperature superconductors is very low loss.

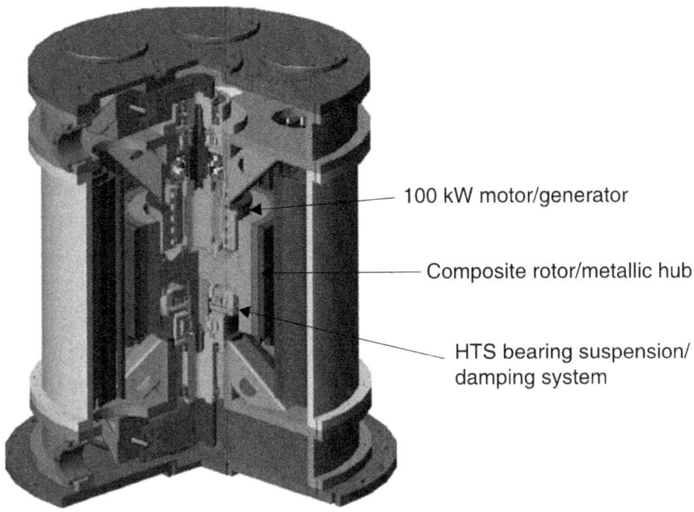

4.8 The Boeing 5 kWh, 100 kW FESS[69].

Table 4.1 Comparison of different storage systems

	Supercap	SMES	Battery	Flywheel
Availability of power E/P (s)	10–6–10–1	1–20	>1000	20–500
Energy density (kWs/kg)	5–20	< 5	100–800	150–500
Power *P* (kW)	≤10	≤7,000	≤500	≤50,000
Life time (cycles)	>1 000 000	*ca.* 1 000 000	≤1 000	~1 000 000
Efficiency (%)	>95	*ca.* 90	70–85	90–95
Stand-by-losses/self discharge (%/h)	0.1–0.2	Cooling power	< 0.01	<1–10
Stand-by-losses/self discharge (kW)				0.05–1
Cost (kEURO/kWh)	10–20	30–200	0.08	1*

* Goal.
Note: supercap refers to supercapacitors and SMES refers to superconducting magnetic energy storage. In both of these systems the energy is stored electrically as opposed to a battery where the storage is chemical and a flywheel where the energy is stored as kinetic energy

4.3.3 Motors and generators

Shortly after the discovery of HTSs, work began on HTS motors. A result of this work was experimental devices of ever-increasing size[70]. The early devices used HTS wires and were analogues of conventional synchronous and homopolar motors that had been investigated earlier with LTS wires. More recently, motor designs that use bulk HTSs have been investigated as analogues to hysteresis, reluctance, and brushless PM designs. The expected benefits of HTS motors are greater motor efficiency and smaller motor size.

The ability of bulk HTSs to trap magnetic fields[2] as high as 17 T suggests their use in superconducting analogues to brushless synchronous motors that currently use PMs. In addition, the diamagnetic and hysteretic nature of bulk HTSs have suggested several new concepts for motors. For the most part, initial experiments with bulk HTS motors have allowed the entire motor to be situated in a liquid-nitrogen bath. For one application, cryogenic pumping, this condition is adequate.

With slight modifications these motors can easily be envisioned as operating in a liquid-hydrogen infrastructure. Also, there are several industrial applications where compactness or low rotor inertia is the major parameter of interest.

For these applications, efficiency or type of cooling is not a major issue, and existing bulk HTS motor designs are already sufficient. However, in most power applications, the motor housing must be at ambient temperature, and the difficulties of cooling such motors must be addressed in the future.

The ability of a bulk HTS to trap magnetic fields allows their use as the rotor of a hysteresis motor[71]. Because the magnetisation of the HTS can greatly exceed the saturation of any conventional ferromagnetic material, such hysteresis motors are expected to have a larger power density than conventional hysteresis motors. Initial experiments seem to validate this assumption. Rotor materials used in experimental HTS hysteresis motors have included sintered BSCCO and YBCO, as well as melt-textured YBCO. Both radial and axial configurations have been investigated.

It has also been suggested that the diamagnetic property of bulk superconductors can improve the efficiency and power factor of reluctance motors[72–74]. The concept for the HTS reluctance motor is to construct the rotor with a composite of ferromagnetic and bulk HTS layers, as shown in Fig. 4.9[75,76].

The composite configuration leads to a highly anisotropic ferromagnetic structure wherein magnetic flux experiences very low reluctance in one direction and very high reluctance in a perpendicular direction. The leakage flux from such configurations should be minimal. The configuration has particular advantage for rapid start-up: the diamagnetic HTS can contribute to a positive torque even when the long radial dimension of the ferromagnetic rotor components is perpendicular to the stator field.

The use of HTS components in rotating electrical machines, such as motors and generators, will considerably reduce their weight and volume. Whereas tapes can best be used for rotor windings, bulk HTS monoliths were used for hysteresis motors, HTS-trapped field motors, but also HTS reluctance motors[75]. On the basis of an advanced reluctance principle, Oswald *et al.*[76] designed, built and tested a motor with extremely high power density and dynamics. As was suggested by Kovalev *et al.*[75] HTS motors based on melt-textured bulk YBaCuO monoliths can be operated in a synchronous regime, with HTS/ferromagnet composite rotors.

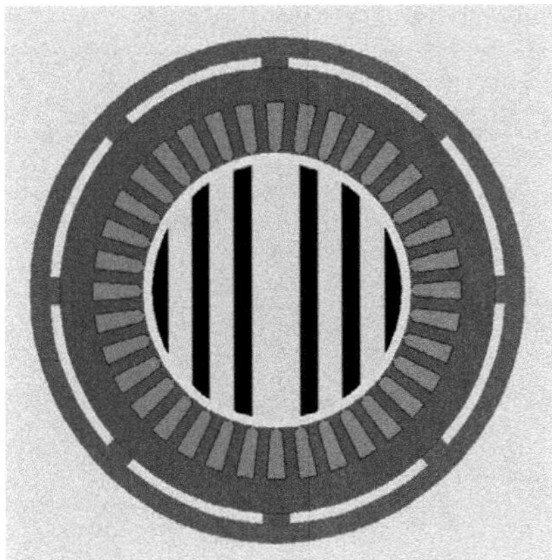

4.9 Cross-section through a ZEBRA-type superconducting reluctance motor.

From the various rotor designs, the 'Zebra' type, which consists of a stack of ferromagnetic iron and YBaCuO segments and shown in Fig. 4.10(a), proved to be particularly attractive. The two-pole rotor in Fig. 4.10(b) is immersed in liquid nitrogen and rotates on a shaft supported by roller bearings from which the oil has been removed (this is done to stop them freezing). The stator or rotating-field windings are made from Cu wire and stator laminations (teeth and yoke) from low-loss iron with a high magnetic saturation. In recent years, several of these

4.10 200 kW superconducting reluctance motor from Oswald-Elektromotoren GmbH.

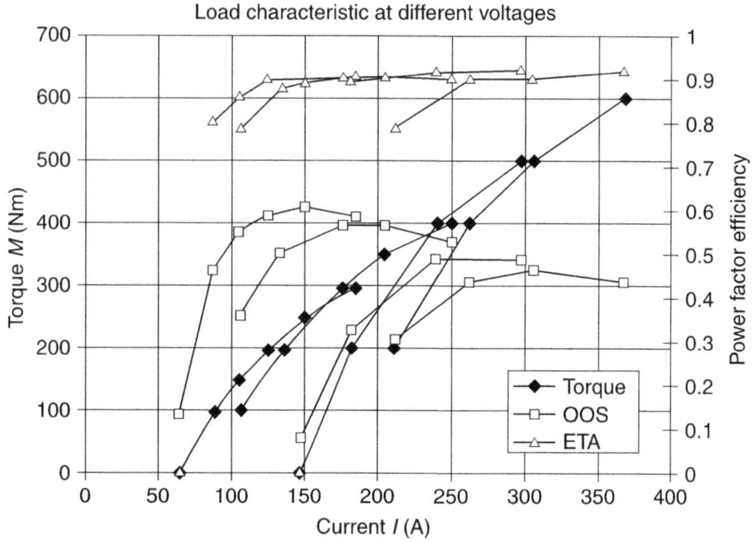

4.11 Typical load diagram for a reluctance motor of the type shown in Fig. 4.10: torque, efficiency, and power factor versus current at different voltages.

demonstrator reluctance motors were tested. The last one consisted of a stator with a 300 mm outer diameter of the lamination and a 148 mm OD rotor with an active rotor length of 320 mm. In several test cycles, the motor was tested at a fixed frequency of 50 Hz, at room temperature (RT), and at low temperatures to check the efficiency of the superconductors under different conditions. The load characteristics of the motor operated at 77 K are shown in Fig. 4.11, where the torque, the corresponding efficiency, and the power factor, cos φ, are plotted versus the current at different voltages. The efficiency exceeded 90% and the maximum power factor was determined to be between 0.7 and 0.5, whereas the maximum torque achieved a value of 600 Nm, which corresponds to 188 kW at 3000 rpm. These motors are particularly promising for industrial applications, especially as high dynamic drives. Compared with conventional high dynamic asynchronous motors, an improved performance by a factor between two and four was demonstrated; see Oswald *et al.*[76]

4.3.4 Trapped-field magnets

Bulk superconducting magnets can be used as a magnetic source similar to PMs. The benefit of the superconductor is that the field strength is much greater than that of conventional PMs, although refrigeration is necessary. An experimental magnet system has been developed and already commercialised in that the trapped-field magnets generate fields[77].

There are four known ways of magnetising superconductors:

- field cooling,
- non-field cooling,
- pulse magnetisation,
- flux pumping.

Field cooling will produce a magnetisation equal to the originally applied field. ZFC produces approximately half the applied magnetic field, as does typically pulse magnetisation. The advantage of pulse magnetisation is that the field is applied for a short time, typically using the discharge from a capacitor bank. At present, a magnetic field of the order of 3 T can be obtained at the centre of two trapped-field magnets in free space.

Flux pumping[78,79] uses a different principle. A small activation field is used to 'stroke' the surface of the superconductor, building the field up in small steps. The advantage of this technique is that the final trapped field is greater than the activation field, thus the magnetisation fixture can be made relatively small and integral to the machine, making machines such as trapped field motors and generators practical.

4.3.5 Magnetic separation

Purification of water is a very important technology. It is now common to use superconducting solenoids for magnetic separation of contaminants from polluted water. The force exerted on the magnetic materials to be removed is given by the magnetic induction field gradient. Hence, one needs a steep field gradient to achieve a large magnetic force for separation. Since the field distribution of a large superconducting solenoid is quite uniform, steel wires are usually inserted in the bore of the magnet to achieve a steep field gradient. In contrast, the field created by a bulk superconducting magnet has both a large field and a large gradient. Thus, it can be directly used for magnetic separation[77]. One commercialised magnetic separation device uses bulk HTS magnets to clean the filter of a membrane-type magnetic separator. In this separator the membrane or the filter catches contaminants of polluted water. The polluted water is treated with coagulants that can flocculate small dirt particles to make flocs larger than the mesh size of the filter, thereby allowing the filter to catch the contaminants.

In Japan, a membrane-magnetic separator using YBCO bulks was developed for water treatment with a water treatment capacity of 100 m^3 per day[80]. A high-gradient magnetic separation system (HGMS) formed by face-to-face type YBCO bulk magnets was developed to separate a slurry of mixed a-hematite (Fe_2O_3) particles. Its separation ratios of over 99% and 90% were achieved at flow rates of 0.5 l/min and 2.0 l/min, respectively[81]. A HGMS using GdBaCuO bulks was developed for purification of wastewater from a paper factory in Japan[82–86]. The developed system can treat wastewater (2000 t/day) and be continuously operated without changing magnetic filters.

4.3.6 Magnetron sputtering

Magnetron sputtering is widely used to produce thin films for various industrial applications. In the sputtering process, a target or a cathode plate is bombarded by energetic ions generated in a glow discharge plasma. The bombardment process consists of ejection of target atoms and subsequent deposition on a substrate. In this process, secondary electrons are also emitted from the target and maintain the plasma formation. PMs play a key role in the magnetron sputtering process. They are arranged such that one pole is positioned at the centre of the target material, and the other in a ring shape covers the outer edge of the target. This arrangement creates magnetic closed loops starting from the centre toward the edge. Soft magnetic materials are also used to form a ring along the periphery of the target. Secondary electrons are constrained along the loop due to the Lorentz force. If the field strength can be made higher, more dense plasma will be available near the target, which increases the deposition rate.

It is also possible to maintain plasma even in high vacuum in large fields, leading to the production of higher quality films. A further advantage is that the deposit can reach the bottom of the grooves on patterned films. In addition, ferromagnetic thin films can be made with high-field magnetron sputtering.

In conventional magnetron sputtering devices with permanent magnets, the field could not pass through ferromagnetic target materials. Hence, Nishijima and Takeda[87] propose replacing FeNdB magnets by trapped-field magnets in a magnetron sputtering device in which trapped-field magnets are installed as a magnetic source. A SmBaCuO disk, 65 mm in diameter, is used as a trapped-field magnet that can generate 6 T at 50 K.

The parallel field of 0.6 T could be generated at the target surface using this trapped-field magnet. The experimental results show that a copper deposition rate of 0.08 nm/s is achievable in an argon atmosphere at 1.06×10^{-6} Pa. It was also possible to deposit Fe particles using the high-field magnetron sputtering device. The system will be commercialised in the near future.

4.3.7 Current leads

HTS current leads (CLs) have become suitable for practical HTS applications. Both HTS tapes and HTS bulks can be applied for HTS CLs. The advantages of CLs with HTS bulks compared to CLs with HTS tapes include low thermal conductivity, a single lead handling a large current, auto quench protection, low manufacturing cost, etc. Bi-2223 bulk has been used for current leads for cryogenic devices[87]. In Japan, a pair of 1 kA current leads using Bi-2223 bulks was developed for LTS magnet systems[88]. Current leads with an available current of greater than 2.5 kA were fabricated using YBCO bulks[89]. Two types of current leads, i.e., '1 K-type' and 'B-type', were fabricated and tested using DyBaCuO bulks[90].

4.3.8 Fault current limiters

Fault current limiters (FCLs) are one of the most promising applications of HTS bulks. Bulk Bi-2223 samples are suitable to be the key element for the development of an electrical fault current limiter[87]. In France, a FCL prototype was developed at 1080 A under 1100 V with Bi-2223 bulks in 1998[91]. A 100 kVA class inductive prototype of SFCL using Bi-2212 bulks was developed within the joint collaboration between Hydro-Québec, Canada and Siemens, in Germany[92]. In Japan, a kA-class resistive SFCL device for 6 kV distribution networks was developed from YBCO bulks in 1998[93]. A three-phase 6.25 kVA (275/105 V) high temperature superconducting fault current limiting transformer (HTc-SFCLT) with functions of both superconducting transformer in normal operating condition and superconducting fault current limiter in fault condition was developed with its secondary coil consisting of MCP-BSCCO 2212/CuNi composited bulk monofilar coils in Japan[94]. A three-phase resistive current limiter based on bifilar coils of MCP-BSCCO 2212 bulk material was developed and successfully tested up to the nominal voltage and power (10 kV, 10 MVA) within the German project CURL 10[95,96].

4.3.9 Instrumentation

The low magnetic drag of an HTS bearing has suggested its use in sensitive instrumentation such as gas pressure indicators, gyroscopes and gravimeters[54]. In the latter application, subradian oscillations or small angular rotations are observed. If the motion is small enough, the HTS should remain in an elastic regime, with all of the flux lines remaining in their pinning centres, and the dissipation should be extremely small. Such bearings should provide a much higher spring constant than is possible with a torsional fibre.

HTS bearings have been proposed for use in a sensitive detector to study the polarisation of the cosmic background radiation (CMB)[97]. Information about the spatial anisotropy of the CMB has allowed cosmologists to distinguish between different evolutionary models of the universe, and information about the polarisation could lead to further refinements. The proposed device would use a rotating half-plate that modulates the CMB from a given patch of the sky passing into a polarisation analyzer. The challenge in adapting the technique to CMB polarimetry is coupling the rotation mechanism to bolometric detectors, which have high sensitivity to microphonic noise. Standard mechanical rotation mechanisms, such as gears and bearings, have stick-slip friction, which induces vibrations. The vibrations inject mechanical energy into the detectors and cause wires and other components to vibrate, thereby inducing excess electrical noise. The low rotational drag and vibration isolation characteristics of the HTS bearing make it suitable for providing a low-noise environment. Cryogenic cooling is already available in the CMB experiment as it is required to detect long wavelength radiation.

4.4 Challenges and future trends

Bulk materials can, and have been, used for current leads. Bulk materials have and are being used for fault current limiters (BSCCO – nexans cylinders etc) and they are being used as the 'active' component in an otherwise passive magnetic bearing. All of these applications are described in detail above. In addition, bulk materials have been used in reluctance machines; a range of these were developed by Oswald and others.

Current leads and fault current limiters exploit the very high current carrying capacity of superconductors. Bearings and reluctance machines exploit the diamagnetic properties of HTS machines.

None of the above, however, exploit the most unique property of bulk HTSs and that is their ability to function as very-high-field magnets. Bulk HTS magnets have found their way into water purification systems as was described earlier but apart from this there are no current applications in which HTSs are used purely as magnets.

When borne in mind that the maximum trapped field recorded in an HTS magnet was 17.24 T and that the maximum realistically available from a permanent magnet is an order of magnitude less than that, it becomes clear that there is enormous potential for devices in which HTSs are used as magnets.

To use superconductors as permanent magnets we need to overcome two problems. The first is magnetisation and the second is demagnetisation.

4.4.1 Magnetisation

Before any magnet can be used it must first be magnetised. Leaving aside the effect known as spontaneous magnetisation (which occurs in ferromagnets, is temperature dependent and occurs below the Curie temperature) this is usually achieved using an external magnetic field. In addition, the best magnetic materials such as NdFeB require a large magnetic field, far in excess of the final remnant field to be fully magnetised. It is relatively rare therefore for a permanent magnet machine to be assembled unmagnetised and for the magnetisation to be done *in situ*. This can lead to assembly problems, especially for physically large permanent magnet machines in which the lateral forces could cause flexure during assembly. However, assembly is done at room temperature, generally at least 100 degrees below the Curie point.

No-one has tried to produce a large-scale machine that relies on pre-magnetised HTSs. This is generally viewed as impractical for three reasons:

1. The need to keep the machine cold both during assembly and for the lifetime of the machine.
2. If using the full field-trapping capability of the HTS then one would be, potentially at least, dealing with magnetic fields and therefore forces of an order of magnitude greater than those experienced with a conventional permanent magnet machine.
3. If a superconductor is put into a changing magnetic field then its magnetisation can change in response.

4.4.2 Demagnetisation

There are several ways in which demagnetisation is a problem for machines based on bulk HTSs. The first and most obvious is loss of cooling. If the temperature rises above T_c then there will of course be total demagnetisation, but even a partial temperature rise may cause some demagnetisation if the persistent currents are in excess of J_c for the new higher temperature. The design of the machine therefore has to take this into account.

In addition there may be some loss of trapped field due to flux creep. This is well understood and fortunately the rate of loss decreases with time and depending on the machine will rapidly become insignificant.

Of more concern is the behaviour of the superconductor when it is subjected to a changing magnetic field. A particular manifestation of this is the so-called cross-field effect, a discussion of which follows below.

The crossed field effect is the phenomenon of magnetisation decay observed when the material is subjected to magnetic fields in a direction transverse to the original magnetisation. This could possibly result in magnets losing their magnetisation and thus becoming worthless in applications that exploit the trapped field in HTS materials. In spite of having been investigated for more than three decades, the subject remains a very intriguing topic for which many experimental facts are still far from being understood. In particular, we need to know whether large numbers of cycles of the crossed field lead eventually to total demagnetisation or saturation at a useful value.

The behaviour can be predicted by using a method first proposed by Vanderbemden *et al.*[98], which involves modelling the superconductor with a highly *non-linear E-J* constitutive law $\boldsymbol{E} \propto \boldsymbol{J}^n$, with n being a large number[99]. Such a relation can be incorporated within the numerical method introduced[100,101], in which the electric field $\boldsymbol{E}$ is always assumed to be parallel to the current density $\boldsymbol{J}$. The merit of this work is to validate of the *E-J* approach for modelling the results of the crossed-field experiments on bulk HTSs. Despite its extreme simplicity, this approach will be shown to reproduce successfully many features of the collapse of magnetic moment under transverse fields. In addition to being relevant to several applications of bulk HTS magnets, such geometry will also enable the spatial distribution of the *c*-axis magnetic flux on the top surface of the superconductor to be predicted before and after transverse fields are applied and to compare theoretical predictions to experimental data.

Bulk melt-textured $YBa_2Cu_3O_{7-x}$ (YBCO) single domains were fabricated by a top seeded melt growth (TSMG) technique[102,103]. In all crossed field experiments, the melt-textured samples were first pre-magnetised parallel to the *c*-axis by field cooling (FC). The c-axis field was then removed and a constant time interval (two minutes) allowed for magnetic relaxation. A series of transverse magnetic field cycles were then applied parallel to the ab plane of the samples. A schematic illustration of the measurement configuration is shown in Fig. 4.5. A pick-up coil wound closely around the sample was used to measure the average magnetic properties along the z-axis for the sample as described above.

The influence of a large number of transverse field sweeps on the remnant induction is examined in Fig. 4.12. It shows that successive cycles cause the magnetic induction to decrease by smaller and smaller amounts, as illustrated by the log-log plot of the data. The striking feature of Fig. 4.12 is that the induction B_z does not appear to saturate, even after a large number of field sweeps. Indeed, the B_z vs. cycle number N curves can be fitted using a power law $B_z \sim N^{-\alpha}$, with the exponent α being an increasing function of the transverse field amplitude H_{max}. Although the study shows that the continuous application of a transverse field leads to a monotonic decay of the trapped magnetic field in superconductors, it does not indicate that the superconducting bulk will completely lose the magnetisation and lose the potential to be used in rotating machines. In a rotating machine the superconductor will see a rotating field and the effect of this is much more benign, as can be seen in Fig. 4.13.

According to the results presented above, the demagnetisation effect in the superconductor can be minimised by optimising the design of the machines, especially in rotating applications. Even in the situation of a superconductor being

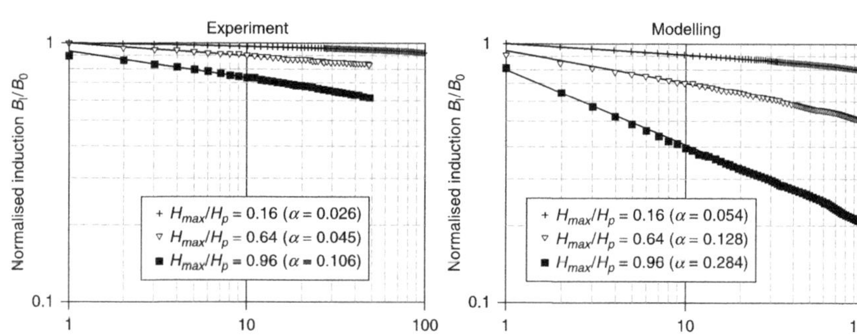

4.12 Log–log plot of the measured and modelled magnetic induction at the end of each transverse field cycle for three different transverse field sweep amplitudes H_{max}, where H_p is the field required to fully penetrate the sample and B_0 is the original trapped field.

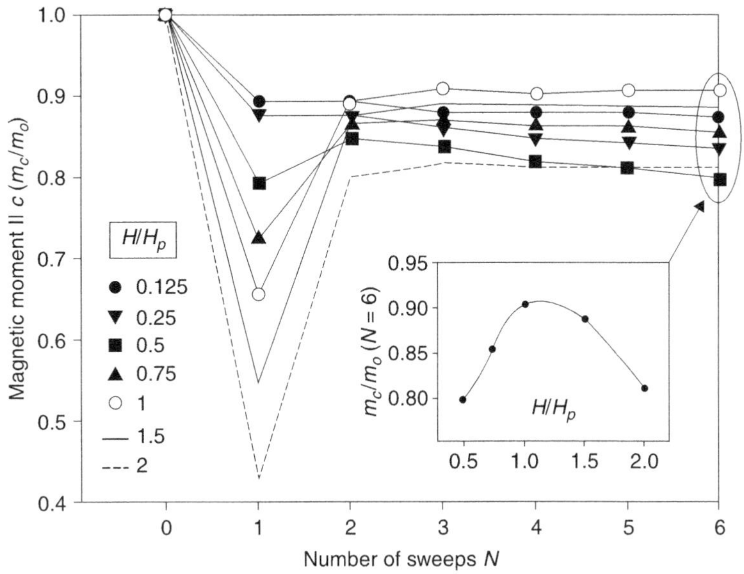

4.13 Measured remained magnetic moment at the end of each rotation cycle for several amplitudes of magnetic field ranging from 0.125 H_p to 2 H_p (where H_p is the field required to fully penetrate the sample). The magnetic moment is normalised with respect to its initial value m_0.

exposed to a pure transverse field, the demagnetisation problem would not be catastrophic either. If the power-law decay behaviour holds true, after $N \sim 5 \times 10^6$ cycles, which would correspond to the application of a 60 Hz AC field for one day, the magnetic induction should reach ~68% of its initial value B_0. A cycle time of more than 422 years would be needed to reach 50% of B_0. Therefore, bulk superconducting materials have the potential to be an alternative magnetic excitation in electrical devices to produce magnetic field in an order of 5–10 times higher than those of conventional permanent magnets. Successful application of superconducting materials in the electrical power system will open a new era in the field of power generation.

A practical machine design therefore has to consider not only conventional electromagnets and cryogenics; it has to provide a way of first developing and then maintaining the field in the machine. There are four known methods of trapping magnetic fields in superconductors:

- Field cooling. Cool the superconductor in a magnetic field and then turn the field off. This will lead to a trapped magnetic field approximately equal to the applied magnetic field.
- Zero field cooling. Cool the superconductor, apply a magnetic field and then remove it. This will lead to a maximum trapped magnetic field of half the applied field or less.

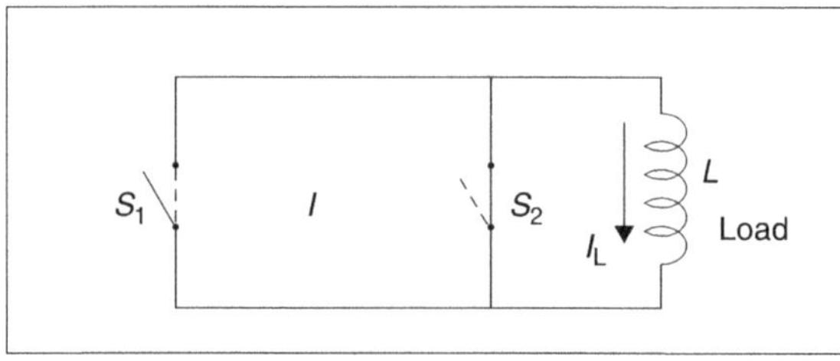

4.14 The basic circuit illustrating the flux pump principle[105].

- Pulse field cooling. Cool the superconductor and then rapidly apply a magnetic field. This is generally done using a coil and a capacitor bank that rapidly charges the coil. The first pulse will magnetise the superconductor partly and then subsequent pulses will reinforce the trapped magnetic field, increasing it a small and diminishing amount each time.
- Flux pumping. There are two types of flux pumping. The first, described in Klundert *et al.*[104], is shown in Fig. 4.14[105] and is used to charge a superconducting coil. It involves a system of superconducting switches (S1 and S2) together with a superconducting storage coil (L). Two superconducting circuits are used. Breaking one of the circuits by opening a superconducting switch (S1) allows flux to be brought inside the circuit. The first switch (S1) is then closed and a second switch (S2) opens allowing flux to enter the coil.
- The second type is for magnetising bulk superconductors. It involves passing a magnetic wave over the surface of a superconductor[78,79]. The superconductor reacts to the changing magnetic field, becoming magnetised in the process. This can then be repeated and the superconductor is gradually magnetised. This is the only process that magnetises the superconductor so that the final trapped field is greater than the original trapped field. The system is described in more detail below.

Flux pumping uses the change in permeability of a magnetic material with temperature to produce a magnetic wave. Local disturbances in the permeability of the magnetic material produce local changes in the magnetic flux density. If you take a circular puck of some ferromagnetic material and heat it on its rim then the heat will diffuse into the centre of the puck as a 'thermal wave'. Since the permeability of the puck is changing with temperature there will also be an associated 'magnetic wave'. We can predict what this waveform will look like by considering Gauss's Law:

$$\nabla B = 0 \qquad [4.1]$$

According to Gauss's Law, since the divergence of B is zero the magnetic wave will have a peak in its centre and a trough in front and behind of it. This can be easily seen in Fig. 4.15, showing a disc positioned above a superconductor.

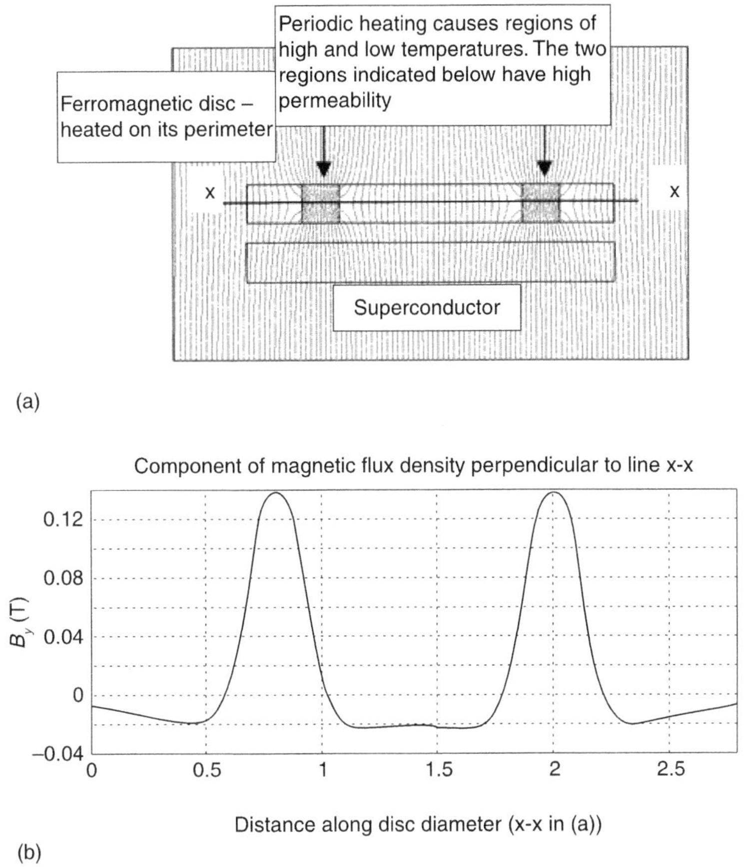

4.15 Field profile for a ferromagnetic disc heated at its perimeter and positioned above a superconductor. Note that the superconductor has been field-cooled.

Regions in the left and right hand halves of the disc have been given high permeability to signify the change in permeability due to temperature and the whole assembly has been placed in a uniform magnetic field. To the left and right of the regions of high permeability the flux lines are widely spaced and in the material itself they are bunched together. This leads to a flux profile along the top of the superconductor as is shown in Fig. 4.15(b). In the figure we have plotted the magnitude of ΔB_y, defined as the change in magnetic flux density due to the change in permeability of the magnetic circuit. Thus there may be a background field of, for example, 1 T and increasing the permeability locally raises it to 1.1 T. In this case then ΔB_y would be 0.1 T. This is done because the superconductor responds to changes in flux density.

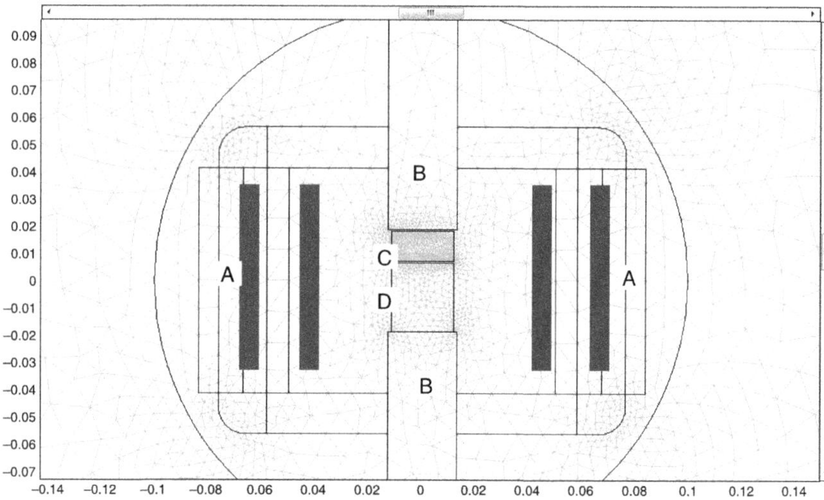

4.16 Overall arrangement. Magnetic field is provided by a pair of coils marked A via a magnetic circuit B to a superconductor C. The whole arrangement is regulated by a permeable region D.

Note this waveform has both positive and negative ΔB_y and this according to the equation system will induce currents in both the positive and the negative z direction.

A typical arrangement is shown in Fig. 4.16. This arrangement, which was created using the modelling environment Comsol, can be used to progressively magnetise a superconductor until it is fully penetrated.

4.5 Technological and commercial barriers to adoption of bulk superconductors

There are several steps along the way to commercialisation of bulk superconductors. The first is the manufacturing process itself. This has been extensively studied and many groups have reported on their ability to make high-quality samples. Initially these were made singly but as the process matured it was shown that with a suitably well-controlled furnace a batch process in which 25 samples at a time can be made. The furnace used for this is a laboratory-type box furnace and the limit of 25 samples corresponds to the size of the furnace, not the maximum number of samples that can be made. It is reasonable to assume that the process can be scaled up with the purchase of larger or multiple furnaces.

Central to the utilisation of superconductors is the overall cost of manufacture. This cost is dominated at the moment by two things; the first is the purchase of the constituent powders. Bought in small batches the cost of these is currently

£600–800 per kilogram. This price, it is felt, would be considerably reduced were the materials to be produced on an industrial scale. The constituent materials are not in themselves rare; yttrium is the 28th most common element in the earth's crust. There are large deposits of barite in various countries of the world. It is not particularly difficult to extract either metal but there are relatively few uses for both, which at least partly explains the current price. The other component in the process of manufacture is the energy required to produce good quality YBCO and (RE)BCO. The process is in two parts. The first is the growth of the grains themselves; this will typically take 5–6 days at temperatures greater than 1000 °C. This is followed by oxygenation, which takes two weeks but occurs at a much reduced temperature, typically 400 °C less. Manufacture is therefore an energy-intensive process and for this reason YBCO will never be as cheap gram-for-gram as typical rare earth magnets such as NdFeB and SmCo.

However this is not a bar to the adoption of bulk materials. A YBCO magnet can have a strength of 1–2 orders of magnitude greater than similarly sized rare earth magnets. Thus machines can be built with specifications far in excess of those made using NdFeB. A motor or a generator (such as one used for wind power) for example that uses 10 T magnets is one-tenth of the size of a machine that uses 1 T magnets. Very high fields make bulks appropriate for use in large accelerator magnets, magnetic resonance imaging (MRI) and nuclear magnetic resonance (NMR) machines, as well as facilitating machines that rely on the magnetohydrodynamic effect to produce power, brake space ships and perform magnetic cleansing of, for example, sea water. There are many other high-value machines where the use of high-power compact superconducting permanent magnets would provide a compelling economic argument.

This argument is even more compelling when it is considered that NdFeB, which, in the early days, was at a similar price to the current cost of YBCO has almost completely replaced ALNICO and ferrites, its cheaper and inferior predecessors. YBCO provides a similar step change in properties and it is a reasonable assumption that, in high-value machines at least, YBCO will replace NdFeB.

Currently the only major commercial use of superconductors is in MRI machines. It is instructive to compare the material and cooling costs of the magnet for an MRI machine constructed from bulk materials with that constructed conventionally from NbTi. The results (factoring in all of the manufacturing costs) are given in Tables 4.2 and 4.3 and Eq. 4.2.

$$\frac{Q_{out}}{W_{in}} = \frac{T_h - T_c}{T_h} \Rightarrow \frac{M_s \cdot dT \cdot C_p}{dW} = \frac{T_h - T}{T_h} \Rightarrow W = \int_{T_h}^{T_l} \frac{M_s \cdot C_p \cdot T_h}{(T_h - T) \cdot l_f} dT \qquad [4.2]$$

where Q_{out} = heat removed (i.e. useful work done), W_{in} = work expended, T_h = temperature of the heat sink, T_c = temperature of the cold sink, M_s = mass of the system being cooled, dT = fractional change in temperature, C_p = specific heat, d_W

Table 4.2 Comparison between material costs of MRI systems constructed from NbTi, MgB_2 and YBCO

	0.5 T	1.5 T	3 T
NbTi wire system	Cost: £22.5–27.5 k Temperature 4 K	Cost: £45–55 k Temperature 4 K	Cost: £90–110 k Temperature 4 K
MgBs system (1.5 × 1 × 1 m)	Cost £5 k Temp 20 K (J_c = 1.5 × 10^9 A/m^2) J = 1 × 10^9 A/m^2 Material volume 24 000 cm^2 Material weight 60 kg	Not applicable	Not applicable
YBCO system (temperature priority) (1.5 × 1 × 1 m)	Cost £80–100 k Temp 55 K (J_c = 2 × 10^9 A/m^2) J = 1.5 × 10^9 A/m^2 J_c at 77 K = 4 × 10^8 A/m^2 Material weight 120 kg	Cost £80–100 k Temp 45 K (J_c = 5 × 10^9 A/m^2) J = 3 × 10^9 A/m^2 Material volume 24 000 cm^2 Material weight 120 kg	Cost £80–100 k Temp 30 K (J_c = 8.5 × 10^9 A/m^2) J = 6 × 10^9 A/m^2 Material volume 24 000 cm^2 Material weight 120 kg
YBCO system (cost priority) (1.5 × 1 × 1 m)	Cost £20–25 k Temp 4 K (J_c = 1.2 × 10^{10} A/m^2) J = 1.2 × 10^{10} A/m^2 Material volume 6000 cm^2 Material weight 30 kg	Cost £40–50 k Temp 4 K (J_c = 1.2 × 10^{10} A/m^2) J = 1.2 × 10^{10} A/m^2 Material volume 12 000 cm^2 Material weight 60 kg	Cost £60–75 k Temp 4 K (J_c = 1.2 × 10^{10} A/m^2) J = 1.2 × 10^{10} A/m^2 Material volume 18 000 cm^2 Material weight 90 kg

Uniform region is roughly an ellipsoid of 25–35 cm diameter
The cost of YBCO bulks is estimated at £800/kg. This is a laboratory cost and would be considerably less were the materials to be manufactured on an industrial scale

= fractional work done, T_l = temperature of the cold sink, l_f = liquifier efficiency, W = work.

This equation is just the Carnot efficiency and is being used in conjunction with the liquefier efficiency to work out how much heat is required to cool the magnets. Not only is YBCO competitive on price but it is competitive on both operating temperature and, importantly, overall mass and cooling costs. The mass of the magnet will be an order of magnitude less than the NbTi system it replaces, while the cool-down costs are also greater than an order of magnitude less than the NbTi system. In addition, the YBCO system will not require the use of large amounts of costly and increasingly scarce helium. These factors lead to much lower installation and running costs. Finally, for an MRI machine it might be appropriate to use bulk MgB_2 and this is even more competitive on costs while being slightly less competitive on operating temperature and similarly competitive on overall mass.

Table 4.3 Comparison between cooling costs of an MRI system constructed from NbTi, with that constructed from YBCO

System	Cost
Conventional NbTi MRI system W_{in} = 8000–12 000 MJ	
Cold mass of magnet (kg)	2500–3500
T_h (K) (room temperature)	293
T_l (K) (cryogenic temperature)	4
C_p (J/kg K (heat capacity of cold mass)	400*
I_f Liquifier efficiency (%)	20
W_{in} (MJ)	8000–12000
YBCO bulk system (W_{in} = 338 MJ)	
Cold mass of magnet (kg)	240
Cold mass of magnet (YBCO) (kg)	120
T_h (K) (room temperature)	293
T_l (K) (cryogenic temperature)	40
C_p (J/kg K) (heat capacity of cold mass)	174*
I_f Liquifier efficiency (%)	20
W_{in} (MJ)	338

* Approximate.

4.6 Future trends

The ability to produce high-quality magnetisable bulk superconductors that can be magnetised *in situ* opens up a wealth of different potential applications. In addition, the ability to magnetise *in situ* presents the possibility that machines can be built with complicated magnetisation patterns that would have been difficult or impossible to build in any other way.

4.7 References

1 M. Murakami, M. Morita, K. Doi and M. Miyamoto, *Jpn. J. Appl. Phys.* **28**, 1189, 1989.

2 M. Tomita and M. Murakami, 'High-temperature superconductor bulk magnets that can trap magnetic fields of over 17 tesla at 29 K,' *Nature*, **421** (6922) 517–520, January 2003.

3 S. Jin, T. H. Tiefel, *et al.*, 'High critical currents in Y-Ba-Cu-O superconductors,' *Appl. Phys. Let.*, **52**(24) 2074–2076, 1988.

4 H. T. Ren, L. Xiao, Y. L. Jiao and M. H. Zheng, 'Large domain YBCO bulk superconductors,' *Cryogenics and Superconductivity*, **28**(30) 28–32, 2000.

5 M. Murakami, M. Morita and N. Koyama, 'Magnetization of a YBa2Cu3O7 crystal prepared by the quench and melt growth process,' *Jap. Journ. of Appl. Phys.*, **28**(7) L1125–L1127, 1989.

6 K. Salama, V. Selvamanickam, *et al.*, 'High current density in bulk YBa2Cu3Ox superconductor,' *Appl. Phys. Let.*, **54**(23) 2352–2354, June 1989.

7 L. Zhou, P. X. Zhang, *et al.*, 'High Jc YBCO superconductors prepared by the powder melting process,' *IEEE Trans. on Magn.*, **27**(2) 912–913, 1991.

8 M. Murakami, T. Oyama, *et al.*, 'Melt processing of bulk high T_{ic} superconductors and their application,' *IEEE Trans. on Magn.*, **27**(2) 1479–1486, March 1991.
9 Y. X. Chen, L. Xiao, H. T. Ren, Y. L. Jiao and M. H. Zheng, 'Effects of process parameters on YBCO bulks prepared by TSMTG,' *Jour. of the Chinese Rare Earth Society*, **21**(5) 526–529, Oct. 2003.
10 Y. L. Jiao, L. Xiao and M. H. Zheng, 'Large bulk YBaCuO superconductors fabricated by multiseeding melt growth methods,' *Jour. of Guangdong Non-ferrous Metals*, **15**(2–3) 196–200, 2005.
11 H. T. Ren, L. Xiao, *et al.*, 'Batch process of single domain YBCO bulk superconductors,' *Chinese Jour. of Low Temp. Phy.*, **25**(1) 11–16, Feb. 2003.
12 Y. Yamada and Y. Shiohara, 'Continuous crystal growth of YBa2Cu3O7-x by the modified top-seeded crystal pulling method,' *Phy C*, **217**(1) 182–188, Nov. 1993.
13 A. Odagawa, Y. Enomoto, *et al.*, 'Superconductive properties of a YBa2Cu3O7-x single crystal prepared by crystal pulling,' *Phy. C*, 217(1) 222–226, Nov. 1993.
14 Y. Kanamori and Y. Shiohara, 'Effect of temperature gradient in the solution on the growth rate of YBa2Cu3O7-x bulk single crystal,' *Phy. C*, **264**(3–4) 305–310, 1996.
15 J. S. Szmgd and K. Suzuki, 'Numerical analysis of transport phenomena in Y-Ba-Cu-O melt during growth of superconducting crystal Y-123 by Czochralski method,' *Phy. C*, **392–396**, 446–452, 2003.
16 X. Yao, T. Mizukoshi, *et al.*, 'Superconducting and crystal growth of YBCO grown under 1 atm oxygen pressure,' *Phy. C*, **263**, 197–199, 1996.
17 S. I. Yoo, N. Sakai, H. Takaichi, T. Higuchi and M. Murakam, 'Melt processing for obtaining NdBa2Cu3Oy superconductors with high T_c and large J_c,' *Applied Physical Let.*, **65**, no.5, 633–635, Aug. 1994.
18 M. Murakami, N. Sakai, *et al.*, 'Melt-processed light rare earth element- Ba-Cu-O,' *Sup. Scie. and Tech.*, **9**(12) 1015–1032, 1996.
19 M. Muraldihar, N. Sakai, *et al.*, 'New type of pinning structure effective at very high magnetic fields,' *Phy. Revi. Lett.*, **89**(23) 237001/1–237001/4, Dec. 2002.
20 J. X. Jin, X. K. Fu, H. K. Liu and S. X. Dou, 'Performance and applications of bulk Bi-2223 HTS bars produced using a hot-press technique', *Phy. C*, **341–348**(1–4) 2621–2622, Nov. 2000.
21 A. Polasek, L. A. Saléh, *et al.*, 'Processing of bulk Bi-2223 hightemperature superconductor,' *Mat. Res.*, **8**(4) 391–394, 2005.
22 J. G. Noudem, D. Bourgault, *et al.*, 'Possible application of bulk textured Bi-2223 for current limitation,' *Phy. C*, **349**(1) 47–52, 2001.
23 X. K. Fu, V. Rouessac, *et al.*, 'Bi-2223 bar current leads fabricated by the combination of cold isostatic pressing and hot-pressing,' *Physica C*, **320**(3) 183–188, July 1999.
24 K. H. Weck and S. Elschner, 'Development and successful testing of MCP BSCCO-2212 components for a 10 MVA resistive superconducting fault current limiter,' *Superconductor Science and Technology*, **17**(5) S122–S126, May 2004.
25 J. Bock, F. Breuer, H. Walter, *et al.*, 'Development and successful testing of MCP BSCCO-2212 components for a 10 MVA resistive superconducting fault current limiter,' *Superconductor Science and Technology*, **17**(5) S122–S126, May 2004.
26 Y. Yamada, M. Takiguchi, *et al.*, 'Transport performance of Bi-2212 current leads prepared by a diffusion process,' *IEEE Trans. Appl. Supercond.*, **10**(1) 1481–1484, Mar 2000.
27 Y. Yamada, O. Suzuki, *et al.*, 'Performance of Bi-2212 HTS current leads prepared by the diffusion process,' *IEEE Trans. Appl. Supercond.*, **12**(2) 1332–1335, Jun. 2003.

28 H. Tamura, T. Mito, Y. Yamada, K. Tachikawa and R. Heller, 'Design study of HTS current lead using reinforced Bi-2212 tubular bulk,' *IEEE Trans. On Appl. Supercond.*, **14**(2) 686–689, June 2004.
29 Y. Yamada, F. Yamashita, *et al.*, 'Structure and superconducting properties of Bi-2212 cylinders prepared by diffusion process,' *Adv. In Cryog. Engi. (Materials)*, **44** 547–554, 1998.
30 T. A. Prikhna, *et al.*, 'Formation of magnesium diboride-based materials with high critical currents and mechanical characteristics by high pressure synthesis,' *Journal of Physics*, **43** 496–499, 2006.
31 G. Giunchi, 'High density MgB_2 obtained by reactive liquid Mg infiltration,' *Int. Jour. of Mod. Phy. B*, **17** 453–460, 2003.
32 S. Ueda, J. Shimoyama, I. Iwayama, *et al.*, 'High critical current properties of MgB_2 bulks prepared by a diffusion method,' *Applied Physics Letters*, 86 (22) 222502, 2005.
33 D. Litzkendorf, T. Habisreuther, *et al.*, 'Function elements of melt textured YBCO for cryomagnetic applications,' *Phy. C*, **372–376**, 1163–1166, Aug. 2002.
34 G. Krabbes, G. Fuchs, *et al.*, '16 T trapped fields in modified YBaCuO: materials aspects,' *Phy. C*, **378–381**, 636–640, October 2002.
35 T. Wolf, A. C. Bornarel, *et al.*, 'High irreversibility fields and current densities in $NdBa_2Cu_3O7–\delta$ single crystals and melt-textured samples,' *Phy. Rev. B*, **56**(10) 6308–6319, Sep. 1997.
36 M. Matsui, N. Sakai and M. Murakami, 'Effect of cooling rate on the trapped-fields of Nd–Ba–Cu–O bulk superconductors,' *Phy. C*, **378–381** 732–737, Oct. 2002.
37 S. Nariki, N. Sakai and M. Murakami, 'Processing of high-performance Gd–Ba–Cu–O bulk superconductor with Ag addition,' *Supercond. Scie. and Tech.*, **15**(5) 648–652, May 2002.
38 M. Muralidhar, M. Jirsa, N. Sakai and M. Murakami, 'Progress in meltprocessed (Nd–Sm–Gd)Ba2Cu3Oy superconductors,' *Superconductor Science and Technology*, **16**(1) R1–R16, Jan. 2003.
39 G. Desgardin, I. Monot and B. Raveau, 'Texturing of high-T_c superconductors,' *Superconductor Science and Technology*, **12**(7) R115–R133, Jan. 1999.
40 H. Teshima, M. Sawamura, M. Morito and M. Tsuchimoto, 'Levitation forces of a single-grained Y-Ba-Cu-O bulk superconductor of 48 mm in diameter,' *Cryogenics*, **37**(9) 505–509, Sept. 1997.
41 W. Hennig, D. Parks, R. Weinstein and R. P. Sawh, 'Enhanced levitation forces with field cooled YBa_2Cu_3O7-x,' *Applied Physics Letters*, **72**(23) 3059–3061, Jun. 1998.
42 H. Fujimoto and H. Kamijo, 'Superconducting bulk magnets for magnetic levitation systems,' *Phy. C*, **335**(1) 83–86, 2000.
43 A. Yahara and H. Matsuba, 'Magnetic shielding effect on oxide superconducting plates,' *IEEE Tran. on Magn.*, **25**(2) 498–500, Mar. 1989.
44 M. J. Hurben, O. G. Symko, *et al.*, 'Characteristics of YBCO magnetic shields,' *IEEE Tran. on Magn*, **27**(2) 1874–1876. Mar. 1991.
45 M. Murakami, 'Novel application of high-T(C) bulk superconductors,' *Appl. Supercond.*, **1** 1157–1173, 1993.
46 F. Werfel, U. Floegel-Delor, T. Riedel, R. Rothfeld, D. Wippich, *et al.*, 'Progress toward 500 kg HTS bearings,' *IEEE Trans. Appl. Superconduct.*, **13** 2173–2178, June 2003.
47 J. S. Wang, S. Y. Wang, Z. Y. Ren, H. Jiang, M. Zhu *et al.*, 'Experiment results of high temperature superconducting Maglev vehicle,' *Physica C*, **386** 431–437, 2003.

48 H. Walter, S. Arsac, J. Bock, S. O. Siems, W. R. Canders *et al.*, 'Liquid hydrogen tank with cylindrical superconducting bearing for automotive application,' *IEEE Trans. Appl. Superconduct.*, **13** 2150–2153, June 2003.
49 H. Minami and J. Yuyama, 'Construction and performance test of a magnetically levitated transport system in vacuum using high-*T*. superconductors,' *Jpn. J. Appl. Phys.*, **34** 346–349, 1995.
50 H. Kamijo, T. Higuchi, H. Fujimoto, H. Ichikawa and T. Ishigohka, 'Flux-trapping characteristics of oxide superconducting bulks in array,' *IEEE Trans. Appl. Superconduct.*, **9** 976–979, June 1999.
51 J. S. Wang, S. Y. Wang, *et al.*, 'A scheme of maglev vehicle using high T_c bulk superconductors,' *IEEE Tran. On Appl. Supercond.*, **9**(2) 904–907, June 1999.
52 S. Y. Wang, J. S. Wang, *et al.*, 'The man-loading high-temperature superconducting maglev test vehicle,' *IEEE Trans. On Appl. Supercond.*, **13**(2) 2134–2137, 2003.
53 Z. G. Deng, J. S. Wang, *et al.*, 'Performance advances of HTS maglev vehicle system in three essential aspects,' *IEEE Trans. On Appl. Supercond.*, **19**(3) 2137–2141, June 2009.
54 E. H. Brandt, 'Rigid levitation and suspension of high-temperature superconductors by magnets,' *Amer. J. Phys.*, **58** 43–49, 1990.
55 J. Hull, 'Superconducting bearings,' *Supercond. Sci. Technol.*, **13** R1–R14, 2000.
56 F. C. Moon, *Superconducting Levitation*. New York: Wiley, 1994.
57 H. Walter, J. Bock, Ch. Frohne, K. Schippel, H. May *et al.*, 'First heavy load bearing for industrial application with shaft loads up to 10 kN,' *IOP J. Phys.: Conf. Ser.*, **43** 995–998 (2006).
58 J. Hull, 'Flywheels on a roll,' *IEEE Spectr.*, **34** 20–25, July 1997.
59 S. Nagaya, N. Kashima, M. Minami, H. Kawashima and S. Unisuga, 'Study on high temperature superconducting magnetic bearing for 10 kWh flywheel energy storage system,' *IEEE Trans. Appl. Superconduct.*, **11** 1649–1652, Mar. 2001.
60 K. Matsunaga, M. Tomita, N. Yamachi, K. Iida, J. Yoshioka *et al.*, 'Fabrication and evaluation of superconducting bearing module for 10 kWh flywheel,' *Physica C*, **378–381** 883–887, 2002.
61 A.C. Day, J. R. Hull, M. Strasik, P. E. Johnson, K. E. Mccrary *et al.*, 'Temperature and frequency effects in a high-performance superconducting bearing,' *IEEE Trans. Appl. Superconduct.*, **13** 2179–2184, June 2003.
62 J. R. Hull, T. M. Mulcahy and J. F. Labataille, 'Velocity dependence of rotational loss in Evershed-type superconducting bearings,' *Appl. Phys. Lett.*, **70** 655–657, 1997.
63 W.-R. Canders, H. May and J. Hoffmann, 'Contactless Magnetic Bearings for Flywheel Energy Storage Systems,' (2006). Talk, private communication; also German BMWi project DYNASTORE. Now coordinated by the Institute of Electrical Machines, Traction and Drives, Technical University, Braunsehweig, Germany.
64 S. O. Siems, W.-R. Canders, H. Walter and J. Bock, 'Superconducting magnetic bearings for a 2 MW/10 kWH class energy storage flywheel system,' *Supercond. Sci. Technol.*, **17** S229–S233 (2004).
65 T. Ichihara, K. Matsunaga, *et al.*, 'Application of superconducting magnetic bearings to a 10 kWh-class flywheel energy storage system,' *IEEE Trans. On Appl. Supercond.*, **15**(2) 2245–2248, 2005.
66 I. Masaie, K. Demachi, T. Ichihara and M. Uesaka, 'Numerical evaluation of rotational speed degradation in the superconducting magnetic bearing for various superconducting bulk shapes,' *IEEE Trans. On Appl. Supercond.*, **15**(2) 2257–2260, June 2005.

67 R. Shiraishi, K. Demachi, M. Uesaka and R. Takahata, 'Numerical and experimental analysis of the rotation speed degradation of superconducting magnetic bearings,' *IEEE Trans. On Appl. Supercond.*, **13**(2) 2279–2282, June 2003.

68 H. S. Choi and H. M. Park, 'The optimum design for magnetic flux distribution of a superconducting flywheel energy storage system,' *IEEE Trans. On Appl. Supercond.*, **19**(3) 2116–2119, June 2009.

69 S. Jung, J. Lee, *et al.*, 'Double-evaporator thermosiphon for cooling 100 kwh class superconductor flywheel energy storage system bearings,' *IEEE Tran. On Appl. Supercond.*, **19**(3) 2103–2106, 2009.

70 M. Strasik, P. E. Johnson, *et al.*, 'Design, fabrication, and test of a 5-kWh/100-kW flywheel energy storage utilizing a high-temperature superconducting bearing,' *IEEE Trans. On Appl. Supercond.*, **17**(2) 2133–2137, June 2007.

71 D. Driscoll, V. Dombrovski and B. Zhang, 'Development status of superconducting motors,' *IEEE Power Eng. Rev.*, **20** 12–15, May 2000.

72 L. K. Kovalev, K. V. Ilushin, S. M. A. Koneev, K. L. Kovalev, V. T. Penkin *et al.*, 'Hysteresis and reluctance electric machines with bulk HTS rotor elements,' *IEEE Trans. Appl. Superconduct.*, **9** 1261–1264, June 1999.

73 A. Oswald, M. Krone, M. Soll, T. Strasser, J. Oswald *et al.*, 'Superconducting reluctance motors with YBCO bulk material,' *IEEE Trans. Appl. Superconduct.*, **9** 1201–1204, June 1999.

74 A. Oswald, M. Krone, T. Strasser, K. J. Best, M. Soll *et al.*, 'Design of HTS reluctance motors up to several hundred kW,' *Physica C*, **372–376** 1513–1516, 2002.

75 L. K. Kovalev, K. V. Ilushin, V. T. Penkin, K. L. Kovalev, A. E. Larinoff *et al.*, 'High output power reluctance electric motors with bulk high-temperature superconductor elements,' *Supercond. Sci. Technol.*, **15** 817–822, 2002.

76 A. Oswald, K.-J. Best, M. Setzer, M. Söll, W. Gawalek *et al.*, 'Reluctance motors with bulk HTS material,' *Supercond. Sci. Technol.*, **18** S24–S29 (2005).

77 T. Oka, Y. Itoh, Y. Yanagi, M. Yoshikawa and H. I. Mizutani, 'Construction of a 2–5 T class superconducting magnetic field generator with use of an Sm123 bulk superconductor and its application to high-magnetic field demanding devices,' *Physica C*, **335** 101–106, 2000.

78 T. Coombs, Z. Hong and X. Zhu, 'A thermally actuated superconducting flux pump', *Physica C*, **468** (2008) 153–159.

79 Yan Y., Hsu C. H., Hong Z., Xian W., Yuan W. *et al.*, 'A novel design of thermally actuated magnetization flux pump for high temperature superconducting bulks', *IEEE Trans. On Appl. Supercon.*, **21**, No. 3, 2011, 1568–1571.

80 J. H. P. Watson and I. Younas, 'Superconducting discs as permanent magnets for magnetic separation,' *Mater. Sci. Eng.*, **B53** 220–224, 1998.

81 N. Nishijima, N. Saho, *et al.*, 'Magnetization method for long high-T_{ic} bulk superconductors used for magnetic separation,' *IEEE Tran. On Appl. Supe.*, **13**(2) 1580–1583, Jun 2003.

82 K. Yokoyama, T. Oka, *et al.*, 'Solid-liquid magnetic separation using bulk superconducting magnets,' *IEEE Tran. On Appl. Supercond.*, **13**(2) 1592–1595, Jun. 2003.

83 Y. Kakihara, T Fukunishi, *et al.*, 'Superconducting high gradient magnetic separation for purification of wastewater from paper factory,' *IEEE Trans. On Appl. Supercond.*, **14**(2) 1565–1567, 2004.

84 S. Nishijima and S. Takeda, 'Superconducting high gradient magnetic separation for purification of wastewater from paper factory,' *IEEE Tran. on Applied Superconductivity*, **16**(2) 1142–1145, 2006.
85 S. I. Takeda and S. Nishijima, 'Development of magnetic separation of water-soluble materials using superconducting magnet,' *IEEE Tran. On Appl. Supercond.*, **17**(2) 2178–2180, Jun. 2007.
86 S. Nishijima and S. I. Takeda, 'Research and development of superconducting high gradient magnetic separation for purification of wastewater from paper factory,' *IEEE Trans. On Appl. Supercond.*, **17**(2) 2311–2314, June 2007.
87 U. Mizutani1, H. Hazama, T. Matsuda, Y. Yanagi, Y. Itoh *et al.*, 'Magnetron sputtering activated by a 60 mm diameter superconducting bulk Magnet,' *Supercond. Sci. Technol.* **18** (2005) S30–S33.
88 J. X. Jin, X. K. Fu, H. K. Liu and S. X. Dou, 'Performance and application of bulk Bi-2223 HTS bars produced using a hot-press technique,' *Physica C*, vol, **341–348** 2621–2622, Nov. 2000.
89 K. Ueda, T. Bohno, *et al.*, 'Design and testing of a pair of current leads using bismuth compound superconductor,' *IEEE Tran. on Applied Superconductivity*, **3**(1) 400–403, March 1993.
90 H. Teshima, M. Morita and H. Hirano, 'Critical current of Y-Ba-Cu-O bulk current leads,' *IEEE Tran. On Appl. Supercond.*, **14**(2) 1778–1781, Jun. 2004.
91 H. Fujishiro, M. Ikebe, *et al.*, 'Low-thermal-conductive DyBaCuO bulk superconductor for current lead application,' *IEEE Tran. On Appl. Supercond.*, **16**(2) 1007–1010, Jun. 2006.
92 J. Noudem, J. M. Barbut, *et al.*, 'Current limitation at 1080 A under 1100 V with bulk Bi-2223,' *IEEE Tran. On Appl. Supercond.*, **9**(2) 664–667, Jun. 1999.
93 J. Cave, D. Willtn, R. Nadi and Y. Brissette, 'Development of inductive fault current limiters up to 100 kVA class using bulk HTS materials,' *IEEE Tran. On Appl. Supercond.*, **9**(2) 1335–1338, Jun. 1999.
94 D. Ito, C. Yang, *et al.*, 'kA class resistive fault current limiting device development using QMG HTS bulk superconductor,' *IEEE Tran. On Appl. Supercond.*, **9**(2) 1312–1315, Jun. 1999.
95 C. Kurupakorn, N. Hayakawa, *et al.*, 'Development of high temperature superconducting fault current limiting transformer (HTc-SFCLT) with Bi-2212 bulk coil,' *IEEE Tran. On Appl. Supercond.*, **14**(2) 900–903, June 2004.
96 S. Elschner, F. Breuer, *et al.*, 'Manufacturing and testing of MCP 2212 bifilar coils for a 10 MVA fault current limiter,' *IEEE Tran. On Appl. Supercond.*, **13**(2) 1980–1983, June 2003.
97 R. Kreutz, J. Bock, *et al.*, 'System technology and test of CURL 10, a 10 kV, 10 MVA resistive high-T_c superconducting fault current limiter,' *IEEE Tran. On Appl. Supercond.*, **15**(2) 1961–1964, 2005.
98 S. Hanany, T. Matsumura, B. Johnson, T. Jones, J. R. Hull *et al.*, 'A cosmic microwave background radiation polarimeter using superconducting bearings,' *IEEE Trans. Appl. Superconduct.*, **13**, 2128–2133, June 2003.
99 P. Vanderbemden, Z. Hong, T.A. Coombs, S. Denis, M. Ausloos *et al.*, *Phys. Rev. B* **75** (2007), 174515
100 N. Amemiya, K. Miyamoto, N. Banno and O. Tsukamoto, *IEEE Trans. Appl. Supercond.* **7** (1997), 2110.
101 Z. Hong, A.M. Campbell and T.A. Coombs, *Supercond. Sci. Technol.* **19** (2006), 1246.

102 Z. Hong, A.M. Campbell and T.A. Coombs, *IEEE Trans. Appl. Supercond.* **17** (2007), 3761.
103 R. Cloots, T. Koutzarova, J.P. Mathieu and M. Ausloos, *Supercond. Sci. Technol.* **18** (2005), R9.
104 N. Hari Babu, M. Kambara, Y.H. Shi, D.A. Cardwell, C.D. Tarrant *et al.*, *Supercond. Sci. Technol.* **15** (2002), 104.
105 L. J. M. Van De Klundert and H. H. J. Ten Kate, 'Fully superconducting rectifiers and flux pumps', *Cryogenics*, 21, 195–206, 1981.

5
High temperature superconductor (HTS) magnets

H. JONES, University of Oxford, UK

Abstract: At the time of writing, high temperature superconducting magnets have not fulfilled their early promise, mainly because of the difficulties in getting these reactive and brittle ceramics into wire form and, consequently, their expense. However, for some niche applications, HTS magnets have been developed. In this chapter, the author outlines his experience of building four such systems after introductory discussions about superconducting magnets in general and design considerations. The recent commercial availability of so-called second-generation (2G) coated conductors opens up a more promising scenario, provided the cost can come down. This scenario is discussed and some conclusions are drawn.

Key words: high temperature superconductor (HTS) magnets, first generation P-i-T conductors (1G), second generation coated conductors (2G), use of iron in HTS magnets, superconducting wires and tapes.

5.1 Introduction

Around 1987, the year following Bednorz and Muller's Nobel Prize-winning discovery of high temperature superconductors (HTS), many of us who worked in the area of superconducting magnets, whether academics or industrialists, were concerned that their jobs were going to be de-skilled as the prospect of easy-to-use and build magnets operating in low cost and easy-to-handle liquid nitrogen would be the norm in a couple of years or so. We shouldn't have worried. A quarter of a century on, superconducting magnets, low or high temperature, are still a formidable challenge both intellectually and technically.

This chapter is concerned with electromagnets made wholly, or partly, with high temperature superconductors in wire form. In other words, windings that generate magnetic fields *per se* as opposed to motors or transformers, etc. In general, the magnetic field will need to be accessed to be used for research, measurement or some process.

At the time of writing, the position regarding high temperature superconducting magnets is dominated by the cost of the conductor. The cost is such that only high-value-added, niche applications attract interest. Even the matter of cost isn't straightforward. On the face of it, using the now accepted 'unit' of 'kiloamp metre', NbTi is some two orders of magnitude cheaper than second-generation (2G) YBCO coated conductor (CC). However, this unit doesn't reflect the fact that NbTi has to be used at liquid helium temperatures whereas YBCO CC can be used at much higher temperatures. Liquid helium is expensive and difficult to handle/contain but, at the upper limit, YBCO CC can be used in cheap and easy liquid nitrogen, so surely that's worth the extra cost? The arguments come back that (a) at LN_2 temperatures, YBCO isn't particularly good from a critical current standpoint and, (b) these days, 're-condensing' technology, whereby helium is re-liquefied *in situ* using a cryocooler, is so advanced that the need for these lower temperatures is no obstacle and the cheapness of NbTi more than makes up for the lower temperature cryogenics. So, in spite of everything, why should we bother with HTS magnets?

Figure 5.1 shows some cutting edge condensed matter physics research by Paul Goddard, a young Oxford Researcher. It is on organic magnetism. It is beyond the scope of this chapter to go into that here but note the field levels required.

LTS is only good for the low 20s of tesla. In Fig 5.1, the fields were obtained using pulsed techniques. Although pulsed magnets are useful tools to access very high fields, it is generally desirable to have steady fields for many experiments if possible. The record for LTS magnets, which are the only realistic means of achieving widespread access to steady high fields in research laboratories, stands at 23.5 T (Bruker Corporation[1]). At a handful of international magnet laboratories it is possible to access higher fields by using tens of megawatts of electrical power in water-cooled magnets or a combination of these and superconducting magnets: *hybrid* magnets. The record stands at 45 T at the National High Magnetic Field Laboratory, Tallahassee, Florida.

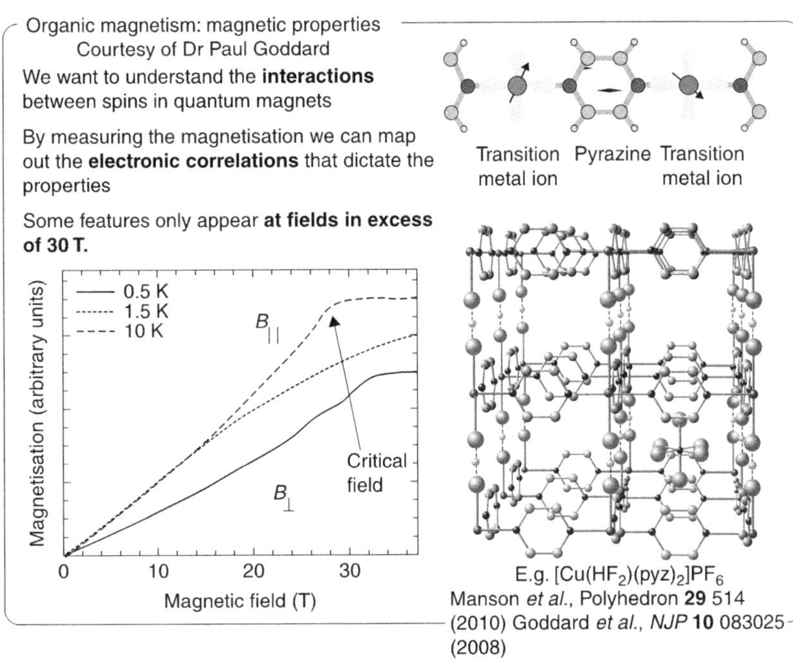

5.1 High steady fields for research. (Source: P. Goddard.)

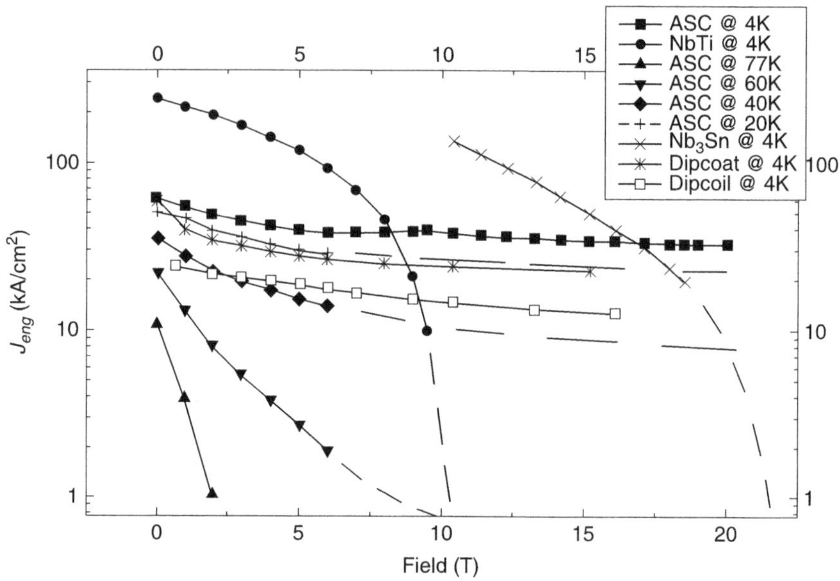

5.2 The engineering current density of various superconductors.

Figure 5.2 shows the engineering current density of both high and low T_c superconducting wires as a function of magnetic field. From this, it can be seen that at the higher temperatures, HTSs struggle to compete with NbTi and certainly Nb_3Sn. As we start to approach the liquid helium regime though, high temperature superconductivity exhibits the remarkable property whereby J_e becomes almost independent of field and, on paper at least, the prospect of very-high-field superconducting magnets arises.

Before getting into any detail on HTS magnets, it is worth looking at some background on magnets in general, or at least electromagnets – not necessarily superconducting.

5.2 Electromagnets

Magnets impact on the lives of mankind more than most of us ever realise. More specifically, electromagnets have revolutionised the world since the mid-nineteenth century although, more recently, advances in permanent magnets have been remarkable and the energy products of modern materials have risen exponentially. However, here we are concerned with electromagnets and specifically with superconducting electromagnets. For a treatise on magnets in general, the reader is referred to Herlach and Jones[2].

Electromagnets are a manifestation of Ampère's law and its developments such as the Biot Savart law[3]. Conventional electromagnets feature everywhere although not that often in everyday life to give a magnetic field *per se*. More likely they will feature as motor windings in, for instance, washing machines. Only in very specialist applications is the generation of a field for its own sake of interest but that is what we will concentrate on here. In particular we will dwell on high-field magnets for research as this is one area where superconductors have had a great impact.

Following electromagnets fall broadly into three classes[4]:

1 Dissipative windings surrounding a ferromagnetic core,
2 Dissipative windings with air cores,
3 Superconducting windings usually with air cores but possibly with ferromagnetic cores.

The first case isn't of great interest for high fields as iron is useful only to about 2 T. Later on however, we will see iron being used with HTS windings for more modest fields at higher temperatures. It is worth giving some consideration to the second category, however.

For a copper winding, the field generated by a simple solenoid in its bore is given by the well-known relation

$$B_0 = G(\alpha,\beta)\ [(W\lambda)/(\rho a_1)]^{1/2} \qquad [5.1]$$

B_0 = field in teslas at the centroid of the bore; $G(\alpha,\beta)$ = a geometry factor; W is the electrical power required in watts; λ = the packing factor of the windings; ρ = the resistivity of copper; a_1 = the radius of the bore (Montgomery's convention).

The provenance of this equation can be found, for example in Montgomery[5] and we won't dwell on it here except to say that, for a winding of constant current density, a packing factor of 0.7 (low but water cooling channels have to be allowed for), a bore of radius 2.5 cm and using the most efficient value of $G(\alpha,\beta) = 0.179$, we find that for a B_0 of 10 T, we require 1.9 MW of power. This assumes a copper resistivity of $1.7 \times 10^{-6}\ \Omega$ cm at 20 °C. In fact it is more likely that the copper will be somewhat hotter, say, 70 °C in which case ρ will be $2.05 \times 10^{-6}\ \Omega$ cm and the power requirement then will be 2.3 MW. Just to emphasise the point, 10 T, by no means a high field in absolute terms, requires the power that would satisfy the domestic requirements of some hundred houses – a typical village.

5.3 Superconducting magnets

5.3.1 Magnet design

Basics

Inserting $\rho = 0$ into Eq. 5.1 implies that this would give infinite field, but this is clearly nonsense and so some other approach must be found. B depends on J, or I (Klauder[3]). So, we must explore the relation with field and current that makes sense in the case of a solenoidal magnet wound with a conductor that has zero resistance.

Inductance

It should be established at this point that superconducting properties have the result that there is no ohmic resistance but all the reactive parameters, e.g. inductance, are unaffected. Consequently, a wound solenoid, which has inductance L, needs a voltage in order to energise it because

$$V_{ps} = -L dI/dt \quad [5.2]$$

Where V_{ps} is the minimum power supply voltage necessary.

So, taking the parameters of a typical compact, laboratory superconducting magnet, obtainable commercially, capable of generating 18 T in a 40 mm bore at 100 A, we find that $L = 40$ H. This means that to energise the magnet fully in 10 minutes, the voltage required would be 6.66 volts. In practice, it is normal to adjust the sweep rate, and therefore the charging voltage downwards as the field increases. So, at the lower fields a voltage of 10 V or more would be necessary. It is worth making the point that a voltage is required because we now go on to develop relations that assume zero resistance and therefore that field is purely a function of current and not power. Implicitly, the uninitiated might assume that a power supply needs no voltage.

Field–current relations

Referring to Fig. 5.3(a) we can write

$$\mathbf{dB} = \mu_0 I/4\pi r^3 (\mathbf{dl} \times \mathbf{r}) \quad [5.3]$$

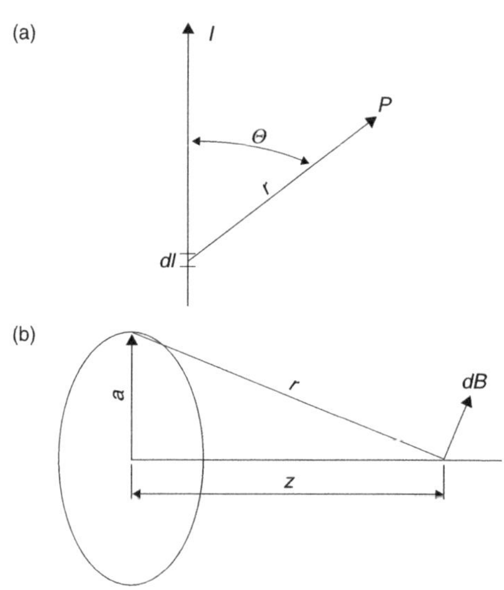

5.3 (a) The derivation of Eq. 5.3. It shows a conductor carrying a current *I* and the effect of an element *dl* on *P*. (b) See Eq. 5.4.

It is now possible to derive an expression for the on-axis field due to a current loop at a position z along its axis – see Fig. 5.3(b), it is

$$dB_z = (\mu_0 Ia/4\pi r^3)\, dl \qquad [5.4]$$

where dB_z is the axial component of field (radial components disappear owing to circular symmetry), I = the current flowing in the loop element dl, a is the radius of the loop, r = the distance from the current element to the point z and μ_0 is the permeability of free space.

$$B_z = (\mu_0 Ia/4\pi r^3) \int_0^{2\pi a} dl = Ia^2/2r^3 \qquad [5.5]$$

As $r = (a^2 + z^2)^{1/2}$, B_z may be written in the convenient form

$$B_z = (\mu_0 Ia^2)/2(a^2 + z^2)^{3/2} \qquad [5.6]$$

In Fig. 5.4, a thick solenoid of practical proportions is shown. Equation 5.6 can be integrated over a cylindrical sheet of current loops $2b$ long, and a number of such sheets between the inner radius a_1 and the outer radius a_2. So integrating we get

$$B_z(0,0) = \mu_0 I/2 \int_{a_1}^{a_2} \int_0^b (a^2/(a^2 + z^2)^{3/2}) dZda \qquad [5.7]$$

The double integral yields a logarithmic function

$$B_z(0,0) = Ja_1\{\mu_0 \beta \ln[[\alpha + (\alpha^2 + \beta^2)^{1/2}]/[1 + (1 + \beta^2)^{1/2}]]\} \qquad [5.8]$$

Where $\alpha = a_2/a_1$ and $\beta = b/a_1$ and J is the overall current density in the winding. The factor in curly brackets is known as the 'F factor', strictly in this case $F_0(\alpha, \beta)$. Hence the centroid, on-axis field of a finite solenoid can be written as

$$B_z(0,0) = Ja_1 F_0(\alpha, \beta) \qquad [5.9]$$

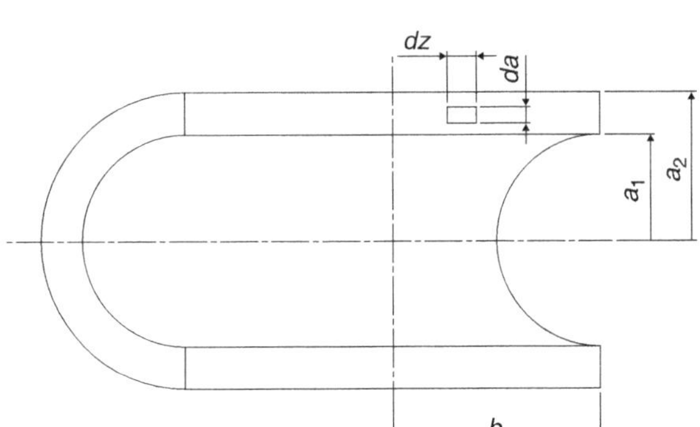

5.4 Definition of parameters for a thick solenoid of finite length.

This is the defining relation of superconducting magnet design. It is relatively trivial to compute and plot the F factor as a function of α and β. One such plot is shown in Fig. 5.5.

Optimising volume and B(I) efficiency

A consideration in designing a magnet is the quantity, and therefore the cost, of superconductor required. It can be shown, see for example Montgomery[5], that the minimum value of the winding volume is $V_{min} = 2\pi\alpha^3(\alpha^2 - 1)\beta$. The line generated is plotted on Fig. 5.5. So, at first sight, minimising the volume of superconductor is the thing to do. Not necessarily. For a superconducting magnet, the principal limiting factor is $I_c(B)$. That is to say a superconducting magnet only remains a superconducting magnet provided no part of its windings is subject to a field that exceeds the critical field for the (super) current carried in the windings. This means that we have to look for that area of the electromagnet that experiences the highest magnetic field. The highest field, B_{max} is at the inner layer on the mid plane in a simple solenoid. The ratio B_{max}/B_0 cannot be expressed in analytical terms but is numerically computed after, for instance, Boom and Livingstone[6], so some possibilities for more efficient design emerge. The problem with simply

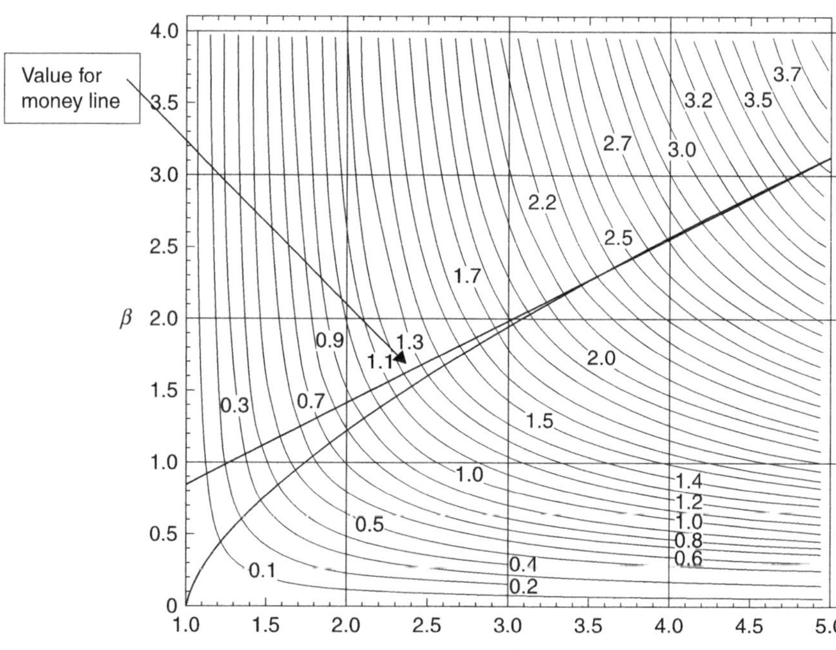

5.5 The F factor curve family. The F values shown are multiplied by 10^{-6}.

opting for minimum volume is that, roughly speaking, one ends up with a relatively short, fat solenoid. This has two consequences. Firstly the ratio B_{max}/B_0 could be high enough that, in spite of minimising outlay on conductor, one still has a lower useful field than might otherwise be possible for the same current. Secondly, minimum volume usually gives relatively poor homogeneity of field. This latter consideration may or may not matter.

It is possible to compute a 'value for money' line on Fig. 5.5. This is computed by minimising the product of volume and B_{max}/B_0. The latter ratio is plotted as a function of the geometric parameters in Fig. 5.6. Generating the function from scratch is time-consuming and will not be addressed here save to say that Wilson[7] treats it quite comprehensively as does Montgomery[5] and, of course, the reader can consult the original, Boom and Livingstone[6].

5.3.2 Design examples

Simple solenoid

Now we have access to practical materials and some elements of design, we can attempt to design a real magnet. Although this chapter is about HTS magnets, here we will use data and examples for LTS as they are well established. The principles are exactly the same for HTSs although, these days, finite element (FE) modelling would be more likely to be used. The first example will be a simple solenoid

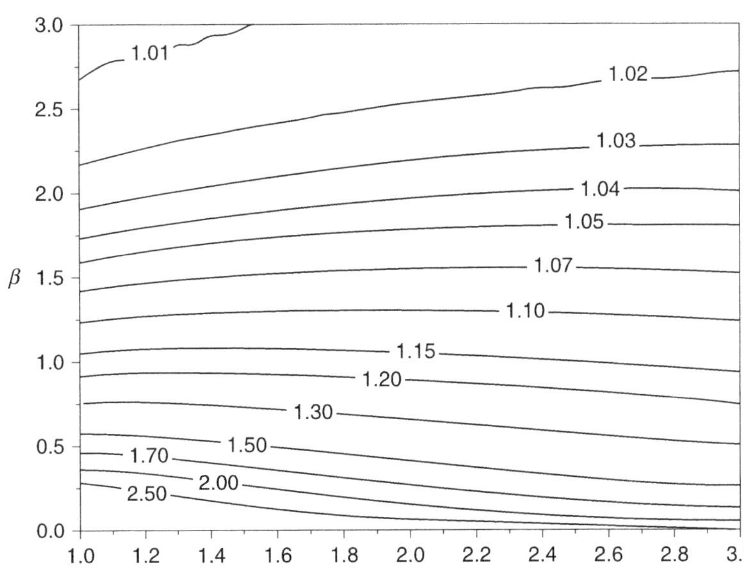

5.6 Ratio of B_{max} to B_0.

wound from NbTi filamentary material. In Fig. 5.7 there is shown a family of representative $I_c(B)$ characteristics for commercial NbTi filamentary wires of different diameters. From these characteristics the limitations of any magnet become apparent.

Magnet specification

Figure 5.7 shows immediately that, depending on which wire is chosen, various options emerge. For example one could choose to have a large current for a given field or a small current for the same field. For the former, one chooses a large diameter wire, for the latter, a small diameter. If we just think in rough terms of ampére-turns, then both choices would give the same result. However, the many turns/few amps version implies a high inductance magnet ($L \propto n^2$, where n is the number of turns) and *vice versa* would yield a low inductance and high current. Both these have their technical pros and cons; a high inductance implies that a high voltage is needed for rapid sweeping whereas a high current demands heavy current leads and the associated poor helium economy. So, we are faced with a design choice and that means more information is needed, i.e., *a specification.* The following is suggested.

Field:	7 T
Bore:	35 mm
Current (max):	100 A
Homogeneity:	Not critical

Regarding the current, a rather smaller current could easily be used but this is a good compromise for illustrative purposes. Writing again the defining equation

$$B_z(0,0) = Ja_1F_0(\alpha, \beta) \qquad [5.10]$$

on the assumption that this is a tutorial exercise, some practical quantification needs to be injected . . .

The current density J in Eq. 5.10 can be re-jigged as $I\nu$, where ν is the turns density, i.e. the number of winding turns of superconductor per unit area. For round wires, such as we will restrict ourselves to here, typical values of ν are given in Table 5.1. The values are based on those recorded over many years at the Clarendon Laboratory, Oxford, and reflect insulation, interleaving and other constructional considerations.

Most experimentalists prefer low currents for compactness of cryostat, cryogen economy and easily handleable external circuitry. As mentioned earlier, this usually implies high inductance so, to keep the power supply more manageable, modest voltages mean relatively long sweep times. It is here that a necessary digression is appropriate.

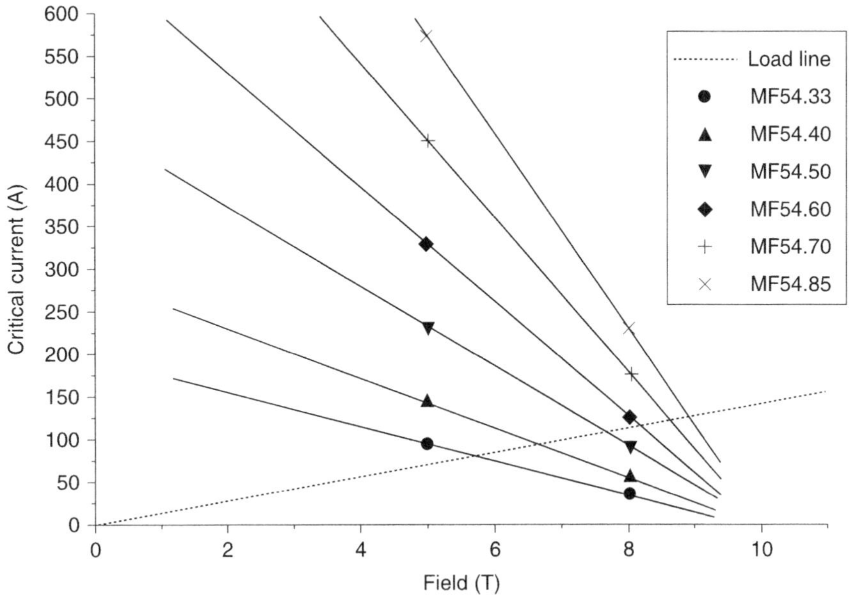

5.7 Short sample characteristics of different sizes of NbTi filamentary superconductor with 54 filaments. The load line of the first design example is shown also. The size of the wire is given by the 54/xx figure. For example 54/40 is 0.4 mm.

Persistence versus driven

Talk to any non-specialist and the one thing that they will know about superconducting magnets is that they can be 'persistent'. A superconducting magnet can be energised, shorted and, provided the LHe is topped up, will act, more or less, like a permanent magnet. We will not go into the specifics of how this is done here but, suffice to say, in most research applications in solid state physics, persistent operation is unnecessary as most operations require frequent

Table 5.1 Turns density for different wire diameters

Nominal wire diameter (mm)	ν (turns cm^{-2})*
0.25	1200
0.33	700–800
0.40	485
0.50	295
0.75	182

*Strictly this should be turns m^{-2} but this choice has more practical relevance and the conversion for calculations is trivial.

changes of field or continuous sweeping for the production of spectra. In these circumstances the ability for persistence, with the requirement for a superconducting switch, can actually be a nuisance.

Final design

In what follows, the simple on-axis centroid fields will be calculated using expressions developed already. The ratio of B_m/B_0, for simple solenoids, can be estimated from Fig. 5.6. However, for the design examples set out below, the detailed on-axis and off-axis calculations will be done using computer software. It is worth repeating that it is reasonably straightforward to write homemade software for on-axis field profiles for any number of coils on the same axis using formulae and expressions given herein. It is rather less trivial to take account of the off-axis fields, although it is possible. As also mentioned earlier, some very good commercial software can be obtained[8]. Computer programs will not be listed here, as provenance and authorship can be complicated.

Referring to the specification, it is possible from Fig. 5.7 to get a rough estimate of which wire to use. We can plot a *load line*, the gradient of which is 100/7 AT^{-1}. Allowing a suitable margin, it appears that the 0.5 mm diameter conductor would be appropriate. The characteristics shown in Fig. 5.7 are known as *short sample curves*. This is because the $I_c(B)$ characteristics are determined by measuring the critical current as a function of field of a short length of the conductor. For a discussion of critical current measurements, see Jones and Jenkins[9]. In Fig. 5.7 the load line for the specified magnet is drawn. Note the intersection with the 54/50 short sample curve.

From Eq. 5.9 with $I\nu$ inserted for J and setting $I = 90$ A,

$$F(\alpha,\beta) = 7/(90 \times 1.75 \times 10^{-2} \times 2.95 \times 10^{6}) = 1.506 \times 10^{-6} \qquad [5.11]$$

Figure 5.5 shows that we could choose α from ~2.5 to >5. and β from 1 to >4. However, as discussed earlier, we could use the minimum volume line or the value for money (VfM) line. Without any particular homogeneity constraints, it would make sense to opt for a geometry based on these. The values of α and β are around 2.7 and 1.7 respectively. In fact in this part of the $\alpha(\beta)$ characteristic, detailed calculation shows that it makes very little difference whether V_{min} or V_{VfM} is chosen. Only when α and β are in the 1.5 and <1 region, respectively, are the differences important. However, although marginal, there are benefits in using the VfM line so that is what we will use here.

Detailed design

Using the VfM line, we find that for $F(\alpha,\beta) = 1.51 \times 10^{-6}$, $\alpha = 2.75$ and $\beta = 1.8$. Thus the final parameters of the magnet are shown in Table 5.2.

Table 5.2 Parameters of a 'value for money' magnet

a_1 (cm)	a_2 (cm)	b (cm)	υ (T cm^{-2})	B_m/B_0	Turns	Length (km)
1.75	4.81	3.15	295	1.051	5687	1.172

Low $F(\alpha,\beta)$ coils

Before leaving simple solenoids, it is instructive to examine, albeit briefly, a design example where the V_{min} versus V_{VfM} issue is significant. Obviously low $F(\alpha,\beta)$ values yield the biggest differences between the V_{min} and V_{VfM} lines. So, consider the following specification.

Field:	4 T
Bore:	3.5 cm
Current:	100 A
Conductor:	0.33 mm
Turns density:	750 t cm^{-2}

From this $F(\alpha,\beta) = 0.3$. From Fig. 5.5 we can see that the V_{min} values of α and β are 1.47 and 0.75 respectively. For V_{VfM} the corresponding values are $\alpha = 1.4$ and $\beta = 1.02$. The consequences of both these approaches are shown in Table 5.3.

To sum up, using VfM one needs 12% more conductor but the ratio B_m/B_0 is reduced by 10% and thus the maximum *useful* field is 4.1 T compared to 3.7 T for a V_{min} coil.

The above simple examples serve only to illustrate the general principles of solenoid magnet design. Things can be a lot more complicated: split magnets, nested coils, asymmetric coils, compensated coils, bucking coils, and so on and so forth. However, magnet design is becoming easier with the advent of finite element packages from a number of companies although their use can be quite a skill in its own right. At the Clarendon Laboratory, we use packages from Vector Fields (Cobham technical Services Ltd)[10] and COMSOL[11]. Information on the use and availability of these packages can be found from the relevant websites.

Table 5.3 Comparison of parameters; minimum volume vs value for money for a low $F(\alpha,\beta)$ coil

	a_1	$F(\alpha,\beta)$	α	β	Volume (cm^3)	B_m/B_0
V_{min}	1.75	0.3	1.47	0.75	58.681	1.287
V_{VfM}	5	0.3	1.40	1.02	65.95	1.165

5.4 Superconducting wires

5.4.1 Introduction

A selection of practical superconductors is shown in Fig. 5.8. At the forefront is a selection of LTSs, either NbTi in single strand filamentary form or in Rutherford cable, developed for accelerator applications. Further up is a length of 'bronze route' Nb_3Sn. This has a glass weave insulation as Nb_3Sn filamentary wires are wound into coils using a composite of bronze and niobium filaments, with Cu island stabilisers. After winding, the coil is heated to several hundred degrees for tens of hours to promote solid diffusion of Sn from the bronze into the niobium to form Nb_3Sn, which is very brittle and cannot be wound. This is the wind and react or 'W&R' process. Next to it is a loop of Nb_3Sn tape. Here a thin layer (~10 μm) of Nb_3Sn is on a substrate with a copper stabilising tape/solder sandwich. This can be wound, usually into flat pancakes, such that the thin layer is at the neutral bending axis so react and wind can be used. This latter gives rise to a sense of *déjà vu* when using similar techniques for HTS tapes to practitioners over a certain age!

Further up, on the right, is a length of MgB_2 wire. Although MgB_2 is technically an HTS, it is unlikely to find any uses other than niche applications so it won't be discussed further here. At the top of Fig. 5.8, various HTS conductors are shown. It is appropriate to discuss these in more detail here.

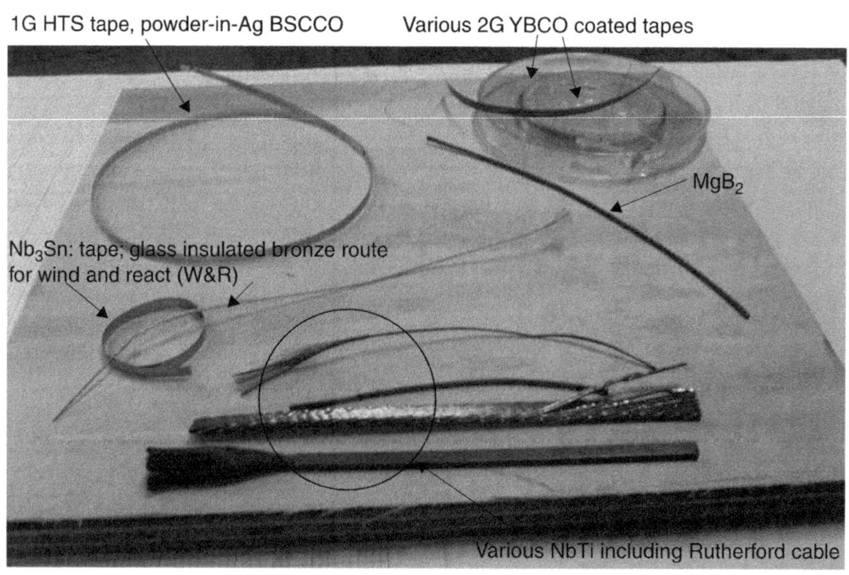

5.8 Some practical superconductors.

5.4.2 HTS conductors

A whimsical observation but based on hard experience is Jones's Postulate: *The higher the temperature at which a superconductor works, the more impossible it is to actually use.* It applies throughout the temperature range and has held for over 40 years in the author's experience. NbTi, once developed, has proved a very tractable, and inexpensive, superconductor up to ~10 T and ~10 K. It is virtually strain-proof and can easily be handled using react and wind. Nb_3Sn, on the other hand, with a T_c of only 8 K higher than NbTi, is much more difficult, as described above. It is also at least an order of magnitude more expensive. However, the highest fields available in LTS magnets ($B \leq 23$ T) rely on its use.

High temperature superconductors, that is, those that can operate in liquid nitrogen, are many times more difficult to use than Nb_3Sn even. After the discovery of the main materials of interest, i.e, YBCO and BSCCO (2223 and 2212 compounds) there began a big effort world-wide to overcome the very disappointing characteristics of these materials from a magnet maker's point of view.

Dip coat and WRAT

To illustrate the problems in getting these difficult materials to perform in magnets it is worth discussing briefly a technique, wind react and tighten (WRAT) that was developed and is now purely of historical interest[12,13,14] but it highlights the sort of unique obstacles to conductor and magnet building thrown up by HTSs. For all practical purposes there are only three HTS: YBa_2Cu_3O (YBCO 123), $Bi_2Sr_2CaCu_2O$ (BSCCO 2212) and $Bi_2Sr_2Ca_2Cu_3O$ (BSCCO 2223). The granularity of these materials is the limiting factor to the critical currents that could be achieved, especially at higher temperatures. Some form of 'texturing' is needed. This can be achieved by mechanical deformation by rolling into tapes or, as we shall see later, epitaxial deposition. In the case of BSCCCO 2212, it lends itself to a partial melting technique that greatly improves the connectivity of the material and, at lower temperatures (<~30 K), exhibits the best I_cs in high magnetic fields. Thus, it is a good candidate for magnets, particularly as inserts for LTS coils working at liquid helium temperatures. The T_c of 2212 is somewhat lower than for Bi-2223, as low as 80 K so Bi-2223 remains the material of choice for LN_2 operations.

Bi-2212 also lends itself to being dip coated, in a slurry of powder and organic binders, onto a thin Ag tape. This immediately flags up another drawback of these materials. They are highly reactive and will only exist in chemical harmony with platinum, gold and silver. Consequently, all support/matrix material has to be one of these noble metals. Obviously, Ag is the cheapest but it also is permeable to oxygen, which allows getting the O_2 stoichiometry correct during reaction. So, the prospect suggests itself of having a 'green' tape, which must then be reacted at temperatures of around 800 °C in an oxygen atmosphere to give the 2212 crystal form. This is, on first sight, a candidate for wind and react, whereby pancakes could be wound from the green tape, but this format, although giving a much higher superconductor to substrate ratio than, say, powder-in-tube, has a big disadvantage, namely the superconductor is exposed, unlike in a powder-in-tube, and this highly reactive film cannot be allowed to touch itself or anything else. Consequently the conductor is co-wound with narrow Ag corrugations at either edge and so the turns are kept slightly apart. Figure 5.9 shows this 'ventilated' pancake with a winding density of ~3 turns per mm. The tape itself is some 150 μm thick (the substrate is 50 μm thick) so the spacing from the corrugations is ~250 μm. Ventilated is quite a good word as the corrugations don't just keep the turns apart but allow the egress of vapours during the 'burn off' (of the binders) stage and free access of O_2 during the reaction stage. Another problem is that the reaction window can be a little as ±2 °C in 800 °C for optimised reaction and in a typical furnace keeping this tolerance over a large volume can be very difficult. After reaction, the pancake has to be handled very carefully while it is potted in paraffin wax. After potting, the two edges of the pancake are machined off. This gets rid of the corrugations. The wax is then melted out and what is left is a loosely wound pancake somewhat similar to a clock spring. This has polyester film painstakingly inserted between the turns, which are carefully tightened at the same time. The result is a useful insert-sized pancake, which has a high turns density and therefore good engineering current density. In very small (<15 m of tape) pancakes, 0.3 T was added to a background of 15 T.

So, we have just looked at a very complicated process to make a magnet using 2212 BSCCO. It does serve to show, however, that HTS materials are difficult to use for magnets. Just to complete the story, there was some good progress

5.9 The WRAT process.

on producing long lengths of Bi-2212 dip-coated tape using continuous reaction furnaces[15,16]. Also, Bi-2212 is not entirely defunct as it is still of interest for round-wire HTS[17] which is desirable for quite a few applications. A good, general account of Bi-2212 conductor development is given by Duperray and Herrmanns[18].

5.4.3 First generation conductors

For the most part, until the early years of the twenty-first century anyway, the biggest effort went into developing Bi-2223 powder-in-tube conductor, a typical cross-section of which is shown in Fig. 5.10. Essentially, this is made by packing the relevant powder, usually Bi-2223, into a silver tube, which is then drawn down. Several of these are then rebundled and the bundle is drawn again. The process can be repeated at will. The final multifilamentary package is then rolled flat, into a tape to impart texture to the BSCCO in an attempt to overcome the *weak link* inter grain limitation on critical current. The tape has then to be heat-treated in an oxygen atmosphere. These *first generation* or *1G* tapes, or wires, have been commercially available for some time but are very expensive. One of the drawbacks of 1G tape is the relatively high proportion of Ag. Not only does this make for a high price tag but the engineering current density is rather diluted.

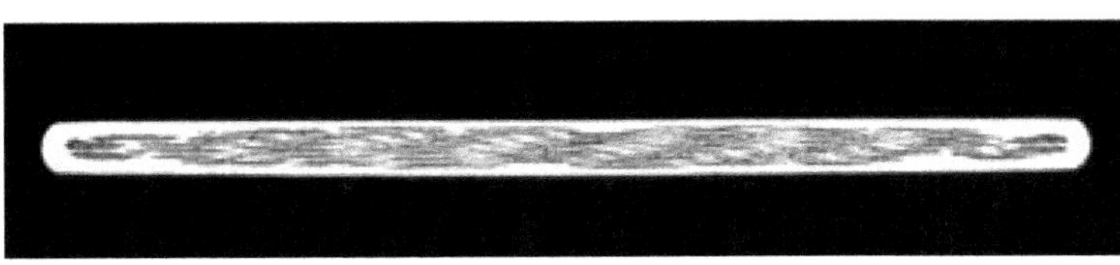

5.10 Cross section through a 1G P-i-AgT conductor.

5.4.4 Second generation conductors

There is a certain irony in that the first truly HTS, in the sense that it worked in liquid nitrogen, was YBCO discovered by Wu *et al.* in January 1987[19], with a T_c of 90 K, but it soon got kicked into the long grass when the Bi compounds were discovered by Maeda *et al.* on Christmas Day 1987[20]. The Bi compounds had a T_c in excess of 100 K and, at that time, temperature was everything so the effort swung towards developing BSCCO. As already mentioned though, these granular ceramics would not carry current much better than cold copper because of the grain boundary problem. By epitaxially depositing material using, for example, laser ablation, on a textured substrate, the grain boundary problem went away. YBCO lent itself very well to this process and extremely good intrinsic current densities have been achieved. So YBCO is in the ascendancy again. The problem is that the superconductor layer is only ~1 μm thick in an overall tape thickness of ~100 μm so the engineering current density could be much better. Also this 2G conductor, which has a very complex architecture (visit the SuperPower Inc. website[21] for some good diagrams) is even more expensive than 1G. However, the balance is that these *coated conductors* are superior to 1G powder-in-tube and commercial production is almost, but not entirely, concentrating on these second generation, (2G), materials. For a good account of the materials science behind YBCO 2G, see McManus-Driscoll[22].

Several manufacturers world-wide are working on these second generation coated conductors but possibly the best known is SuperPower Inc.[21]

5.5 Examples of HTS magnets

There follow accounts of a series of four HTS magnet systems that were developed in the Clarendon Laboratory, Oxford in the period approximately 1992–2007. They are all made from 1G conductor and are all of pancake pattern although one is a racetrack geometry. With the exception of one, all have iron cores and are essentially low-field high (er) temperature magnets.

5.5.1 Maglev demonstrator

The first illustrative example was built some time ago now by Jenkins *et al.*[23–30] but it serves as an example of what simplicity is possible using HTSs. At the same time, using early, commercially available, 1G, powder-in-Ag-tube conductors, there were unique challenges. The project was in collaboration with Loughborough University and it was to see if HTSs could, in principle, be used for attraction levitation transport. The focus of this was the maglev train at Birmingham Airport, which was used to transport passengers between the airport and Birmingham international airport. At Loughborough, a demonstrator had been built to show how feedback technology could maintain a fixed air gap for the operation of a levitated train. This used conventional copper/iron technology as did the Birmingham train itself. The idea of the project was to see if we could retro-fit an HTS magnet to the demonstrator and see if the feedback signal at ~10 Hz would work with a superconductor. The basic arrangement is shown in Fig. 5.11.

An HTS version of the electromagnet was needed to replace it in the demonstrator. It was hoped that an air-cored coil would do the job but, for simplicity, we wanted to work in liquid nitrogen. The performance of the 1G superconductor at 77 K, especially then, is inadequate to do without iron. In the event, the much higher current density in the nine double pancakes that provided the ampère turns for the magnetomotive force to the laminated iron circuit meant that, overall, there was a significant gain in efficiency. In Table 5.4 the gains are extrapolated to a magnet that might replace the conventional magnets on the Birmingham Maglev, and a comparison is made.

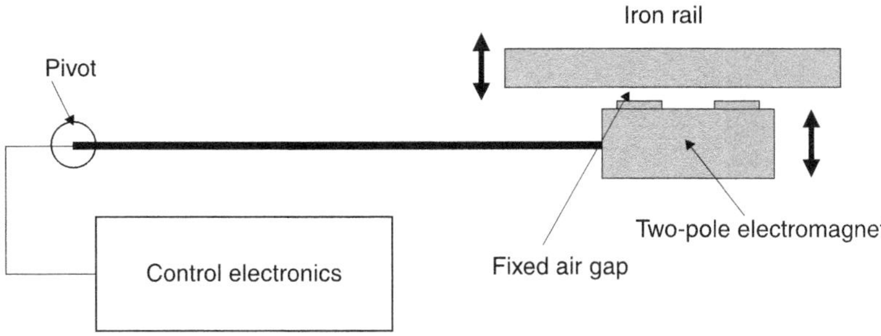

5.11 The Loughborough Demonstrator. The iron rail could be moved up and down. The control could be configured so that the electromagnet on a pivoted arm could follow the rail, with a fixed air gap or remain static.

Table 5.4 Comparison of normal and HTS coils with reference to the Birmingham maglev

	Normal	HTS
Current density (A mm^{-2})	2–3	10*
Packing factor	0.5	0.5
Slot size (mm)	230 × 115	100 × 50
Mass of core (kg)	59	37
Total magnet mass (kg)	112	63†
Lift/weight ratio	9:1	16:1

*Conservative figure but makes some allowance for cryostat.
†Estimate assuming coil is entirely Ag.

The system is shown in Fig. 5.12. A concession that had to be made was that although the magnet would have been more efficient with the windings on the poles, the immersion cryogenics would have been too complex. Instead, the windings were placed on the back iron so that a simple cryostat, made from glass-fibre reinforced plastic (GRP) cylinders produced in-house, could be gravity-fed with liquid nitrogen through flexible tubing. The demonstrator underwent extensive dynamic testing and was successfully made to levitate and maintain a fixed air gap using suitably configured feedback control. Table 5.5 shows the main parameters of the magnet system and further details can be found in a number of publications[23–30].

Interestingly, Richens[31] took advantage of the extensive testing and experimental data obtained on this magnet to develop and check his FE modelling used in designing the magnet, which is the subject of the next section.

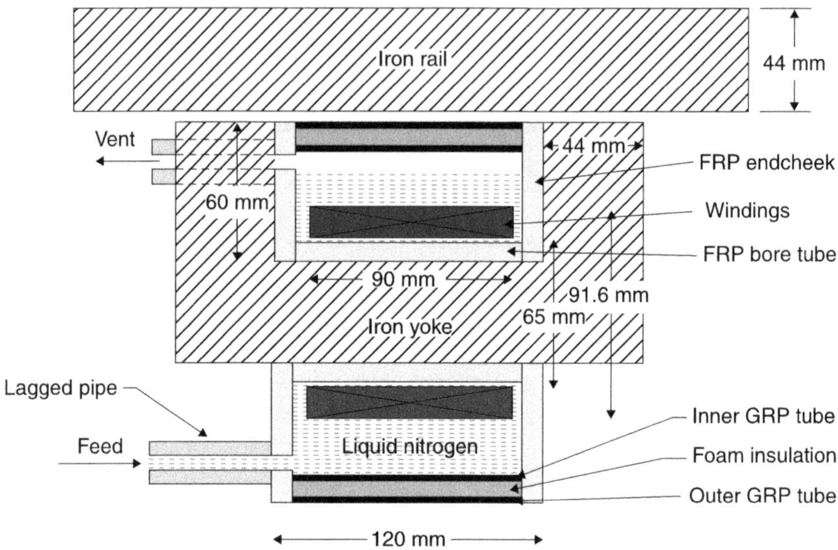

5.12 The HTS magnet/cryostat for the Maglev demonstrator.

Table 5.5 Coil data and performance for Maglev demonstrator

Coil data		Performance ($I = 5$ A)	
Number of double pancake modules	9	Ampere turns (nI):	4770
Total no. of turns	954	B_{gap} from 3D model (T)	0.54
Inner winding diameter (mm)	65	B_{gap} from simple calculation ($B = \mu_0 nI/g$) (T)	0.60
Outer winding diameter (mm)	95.6		
Length (mm)	90	Lift force ($B^2A/2\mu_0$) (N)	449
Overall turns density (cm^{-2})	80		
Operating temperature (K)	77.3		
Max. operating I (I_c) (A)	5		

5.5.2 A magnet for cryocooler operation

Despite all the inherent drawbacks of high temperature superconductivity, the fact remains that it will work, albeit not perfectly, at higher temperatures than had been dreamed of prior to 1986. One of the possibilities that opened up was the operation of magnets using cryocoolers, e.g., Gifford McMahon 'fridges'. In 1996, Richens undertook to design, build and characterise a magnet that would be capable of operating, conduction cooled, on a cryocooler. Also, he carried out a thorough study of the effects of anisotropy of superconducting tape and devised software that embraced the principle of the critical current of the whole magnet taking into account the limiting effect of the radial field component at the ends of a coil. Techniques to reduce this undesirable feature by using iron were successful[31,32]. Again, this is a pancake magnet with seven pancake pairs but its construction incorporated aspects that facilitated conduction cooling. The structure is shown in Fig. 5.13.

The magnet was successfully operated both in liquid cryogens and on a GM cryocooler where it was operated in the range 12–15 K. The stability of the magnet was thoroughly investigated and the results are shown in Fig. 5.14.

It is instructive to show the equivalent picture with a LTS magnet. In Fig. 5.15 the voltage across a small LTS magnet is shown as the current is swept up at 0.125 A s^{-1}. The voltage is purely inductive and is ~12.5 mV, which makes the magnet inductance 0.1 H. At 162 A (~1300 seconds) the magnet quenches, with no prior indication. The quench event is over in considerably less than 1 s. From Fig. 5.15 we can estimate that quench propagation in HTS is some two or three

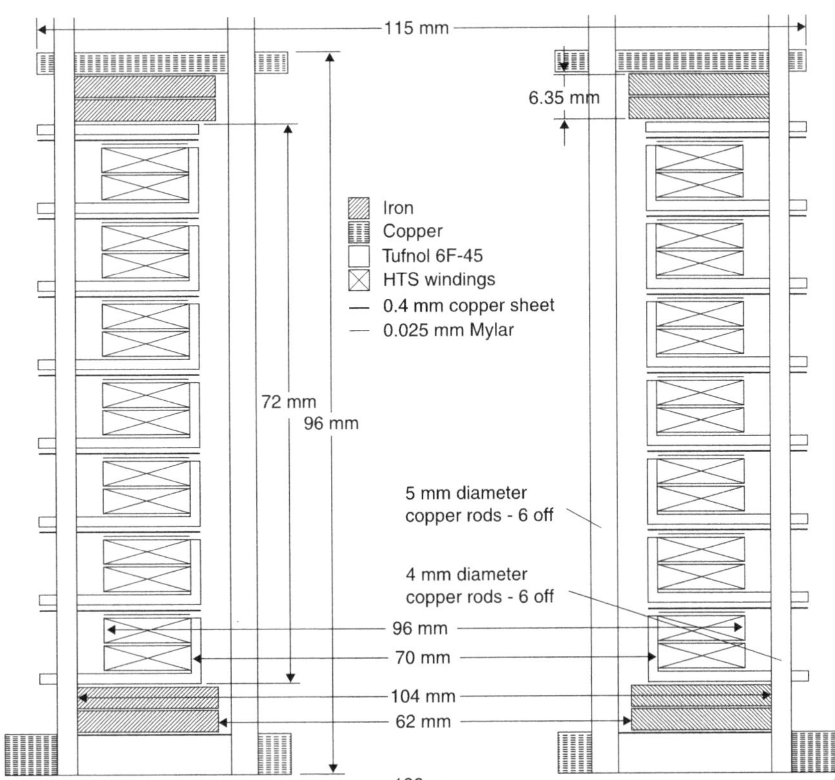

5.13 Cryocooled magnet. Note the use of copper tie bars and cheek liners to facilitate conduction cooling. Note also the iron at the end cheeks. This feature enabled a 20% increase in the critical current of the magnet. (Source: Richens[31].)

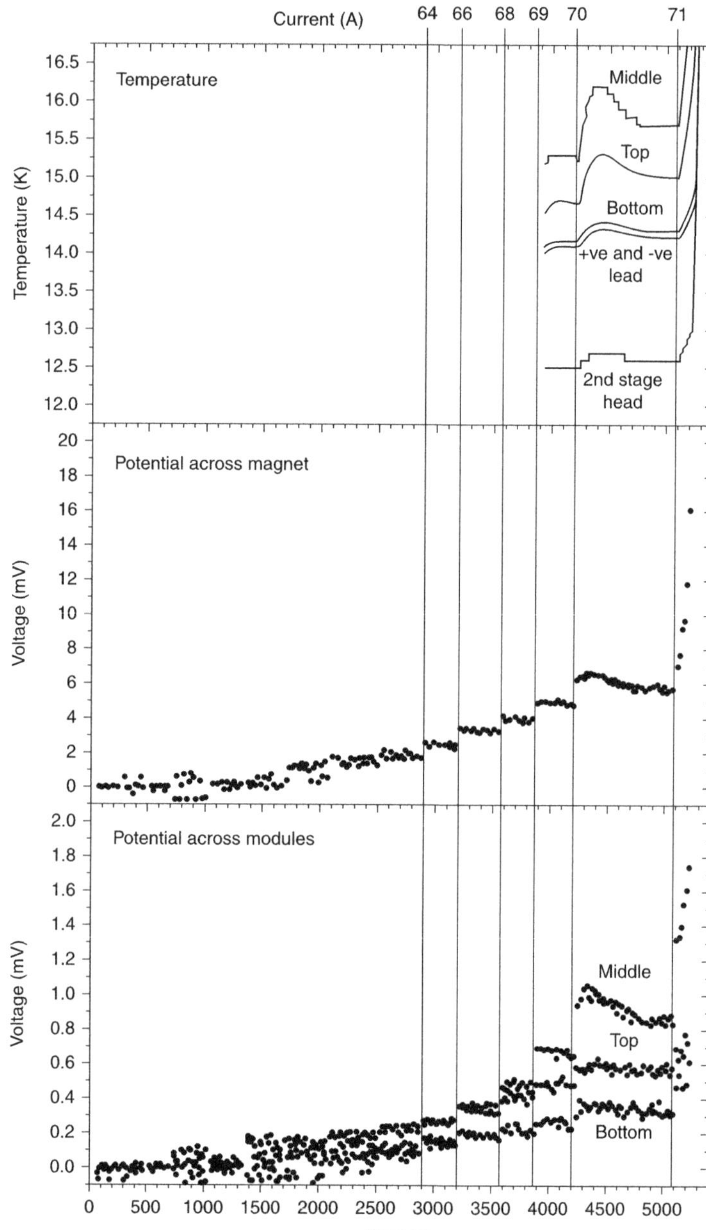

5.14 Monitoring the temperature and voltages of various parts of the coils as the current is stepped. The annotation is self-explanatory but note the gradual slope to 'quench'. (Source: Richens[31].)

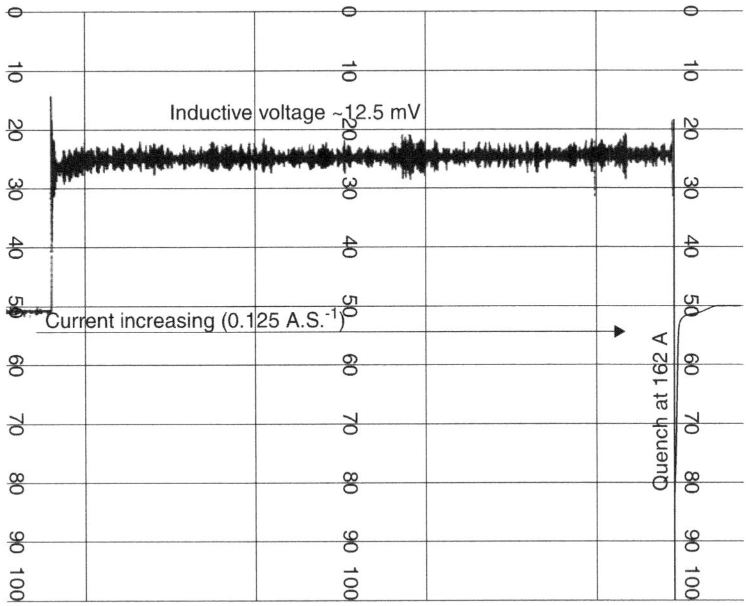

5.15 Showing the quench characteristic of a LTS magnet.

orders of magnitude slower in HTS. Further detail of this 'global' quench behaviour of HTS is given by Richens *et al.*[31,32].

5.5.3 A magnet for magnetohydrodynamic (MHD) marine propulsion

MHD

The magnetohydrodynamic (MHD) effect has been known for some time. In fact, as long ago as 1832 Faraday tried to measure a voltage across the river Thames induced by the water flow in the earth's magnetic field. The author started his professional life working on MHD power generation at AERE Harwell nearly 50 years ago. On being told, in about 1964, that the process would only ever be economic if a superconducting magnet were used, he said, 'what's a superconducting magnet?'! He has since found out . . .

Basically, if a conducting fluid is passed through an orthogonal magnetic field, mutually orthogonal to both fluid flow and *B*, an EMF is generated and power production is possible. Analogous to the dynamo, the reverse process gives a motor. Although this is a somewhat sweeping, loose analogy, by passing a current through (conductive) salt water (the sea) orthogonal to a magnetic field generated on a boat, the motor effect gives a force, which propels the boat forward, or astern, depending on the relative polarities of *B* and the current. In fact, MHD is scientifically rather more complex and good accounts are given by Witte[32] and Hales[33]. There is also a good introductory work by Davidson[35].

An LTS MHD boat, *Yamato – 1*, was built and successfully tested in Kobe harbour in 1992[36] but, thus far, the only HTS MHD boat has been produced by Hales at the Clarendon Laboratory as part of a Teaching Company Scheme in collaboration with Space Cryomagnetics Ltd (now Scientific Magnetics Ltd)[34,37]. Somewhat more modest than *Yamato – 1*, this is only a model but, in some ways, small-scale MHD throws up its own unique challenges that wouldn't be there in a full-scale version.

The pattern chosen was the 'open field' design successfully used for a non-superconducting model MHD submersible by Way and Devlin in the late 1960s[38].

In order to access the largest mass of water, the key point in maximising thrust, the optimum magnet geometry proved to be 'racetrack'. This was a departure from our experience, which hitherto had been usually circular.

Solid N_2 refrigeration

There needed to be an on-board method of refrigerating the magnet. Machinery such as cryocoolers was out of the question on the grounds of size and weight, so there needed to be a 'thermal battery' into which cooling could be 'stored' for the duration of a voyage. It was decided to refrigerate the magnet by conduction from a reservoir of solid nitrogen[37]. There are several advantages to this method. If the solid is cooled sufficiently, as well as the latent heat associated with melting at 63 K, there is a further solid–solid phase change at 35.6 K, the latent heat of which can also be exploited as the solid N_2 warms from, say, 30 K to its melting point.

The racetrack winding

This magnet used 1G powder-in-Ag-tube, 4 mm-wide conductor obtained from American Superconductor Inc. Again, it was decided to incorporate some iron, for two reasons. Firstly, the overall field is enhanced. The addition of 3 mm-thick plates of iron with a 6 mm overlap increased the critical current of the magnet by some 30% compared to using no iron. This is mainly because of the reduction in radial component 'hotspot' which is a limiting factor in anisotropic HTS tape conductors.

The initial design of the racetrack is shown in Fig. 5.16. This is a plan view and there are six pancakes, making the width of the winding pack 27.9 mm. As the magnet is to be conduction cooled, the 'former' on which the racetracks are to be wound is copper with suitable radial slits to prevent eddy currents. The inter-pancake connections are as shown in the magnet system described in Section 5.5.4. Also as in Section 5.5.4 the interleaving between turns is silk ribbon as these pancakes must be well vacuum-impregnated with epoxy resin.

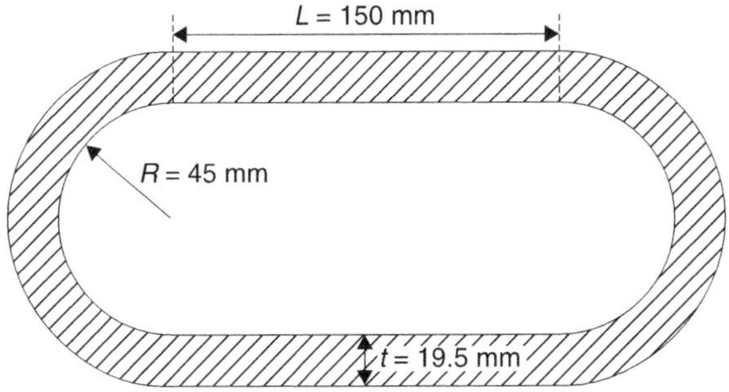

5.16 The initial design of the racetracks.

The reason Fig. 5.16 is entitled 'initial design' was that on trial windings there was a problem with the extent of lift on the straight sections. The lift was as much as 4 mm. This was a worry as, unlike a LTS such as NbTi, we couldn't use techniques such as wedging as this conductor is so fragile. It is important that no separation occurs as much of the cooling is done by radial conduction, and gaps could seriously impair this. So, it was decided to reshape the former such that it had 'curved straights' as shown in Fig. 5.17 below. Figure 5.18 shows one of the racetracks being wound.

Although the reshaping of the former made a considerable improvement on the lifting problem, there was still a degree of separation at the point where the conductor left the semicircular ends and went on to the curved straight sections. This problem was solved by attaching a bracket to the corners of the coil and pressing each pancake back onto the former using six grub screws as shown in Fig. 5.19.

The clamping arrangement is left in place until after the racetrack coil pack was impregnated, whereupon the turns were fixed in place. Impregnation comprises evacuating the coils and resin separately, introducing the resin under vacuum and then cycling to atmosphere and pressurising to 10 bar as described by Jones and Hickman[39].

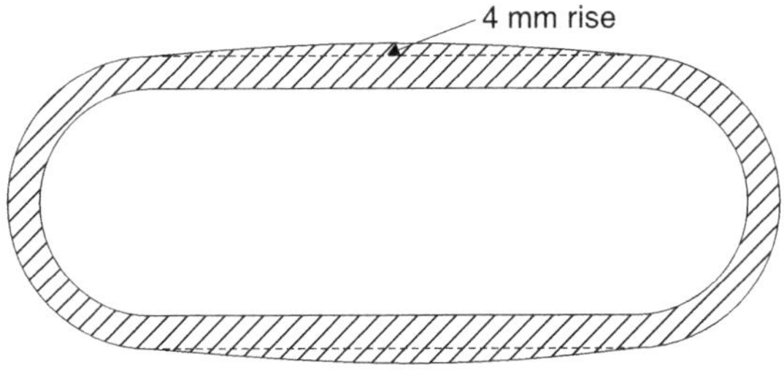

5.17 New shape of former to counteract separation.

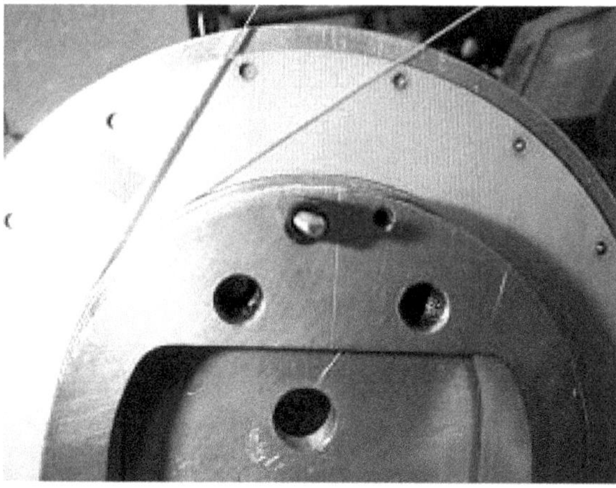

5.18 Racetrack winding. Note the slightly curved copper former and the co-winding of the HTS and the silk ribbon.

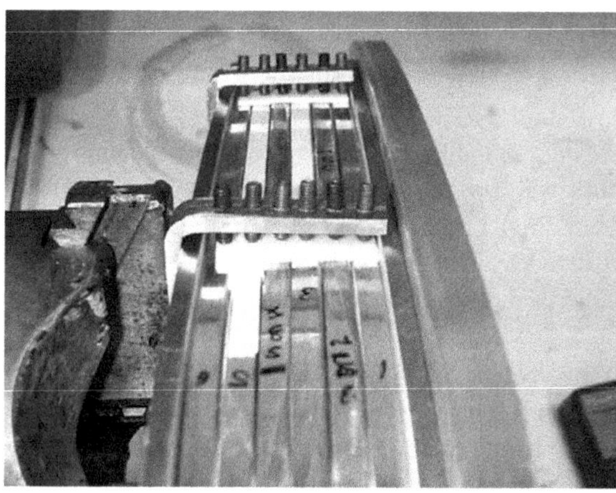

5.19 Showing the clamping arrangement to prevent separation of windings until impregnation is complete.

Assembly of system

Figure 5.20 shows the whole system. The nitrogen reservoir has in it a dense series of discs that act as a heat exchanger. The fine divisions of the heat exchanger reflect the fact that the thermal diffusivity of solid nitrogen is very poor – approximately that of wood! At the edge of the discs is the continuous coil of tubing through which liquid helium is forced through to solidify the nitrogen and reduce the temperature to ~30 K. This is done at the start of the 'voyage'. This

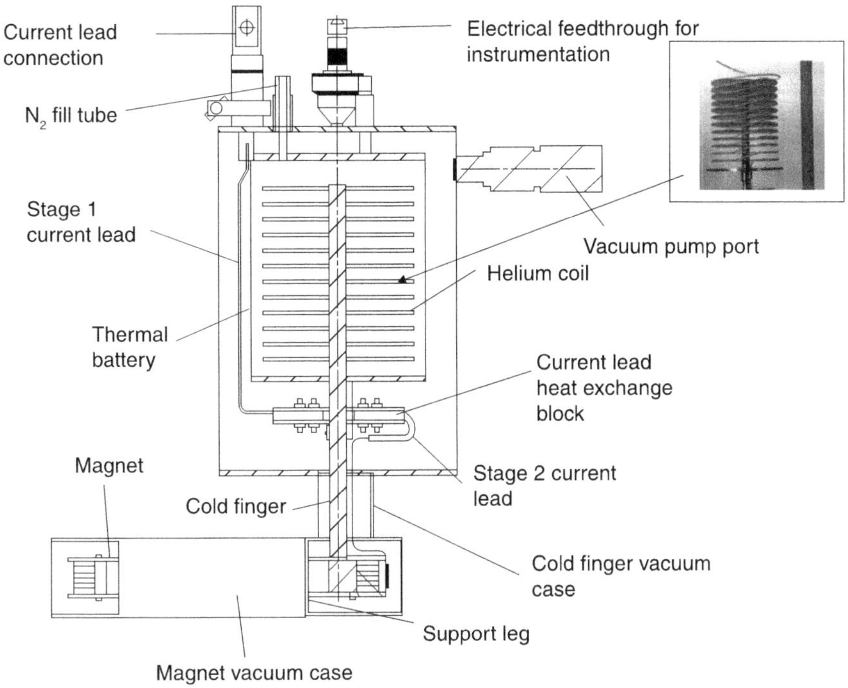

5.20 A side section of the system.

thermal battery proved to be very successful and, for example, over the period of 1 hour, a typical 'voyage', the temperature of the magnet rose only 7 K from 38 K. For simplicity, the magnet current was limited to 76 A, which is appropriate to operating at 63 K, the melting point of N_2. Obviously, at lower temperatures, a higher current would have been possible. The complete system is shown in Fig. 5.21.

5.5.4 A magnet for industrial sputter coating

The last magnet system we will discuss in more detail. It was the most recent for one thing, apart from small 2G prototypes being developed at the time of writing. This, too was the work of Patrick Hales[40,41], it was again in collaboration with industry, and it was to replace the permanent magnet magnetrons used in industrial sputter coating systems with electromagnetic magnetrons. Sputtering is an atomistic process that uses a glow discharge/plasma to generate a flux of ions incident on the target surface. This knocks off atoms of the target material (the cathode) and these are accelerated to the substrate to be coated, in effect the anode, by a high voltage. There are many applications of this coating method but a familiar example is the coating of machine cutting tools, e.g., drill bits, with

5.21 The HTS MHD propulsion unit. The magnet was powered through the current leads visible at the top of the unit by on-board batteries. Radio control was used to manipulate current levels and direction (forward and astern). The electrodes for sea water current (not shown) surrounded the straight sections of the racetrack cryostat and were made from aluminium. The boat achieved an average speed of 3 cm s^{-1}. (Source: Hales[34].)

substances like TiN, (TiAl)N, (TiZr)N and (CrZr)N. This increases the microhardness of the surface and leads to impressive increase in lifetimes.

Sputter coating

It is inappropriate to go into any depth on sputter coating, which is a complex subject in itself and there are many excellent works to consult, for example Kelly and Arnell[42]. Suffice to say, the addition of magnetic fields using magnetrons to confine and increase collision probabilities is an essential feature. Permanent magnets are the traditional sources of fields and, using NdBFe magnets, many compact and neat arrangements are possible. However, electromagnetic arrangements do bestow some advantages. The fields are controllable and can be increased, decreased and profiles can be adjusted. In principle, higher fields ought to be achievable. As the industry took a conservative view of using superconductors and cryogenics, the first thing Hales did was to investigate whether conventional electrical technology using copper-iron magnets, with water cooling if necessary, might be suitable. It transpired that it was very difficult to achieve the field levels in an envelope as small as that which NdBFe systems gave. High temperature superconductivity, on the other hand, in combination with iron and cooled from a cryocooler, seemed to offer promise and it was this option that was pursued. The industry accepted this step change in technology and, after all, they already used cryogenic pumping to achieve high vacuums so it wasn't a totally new scenario.

The system arrangement

The arrangement chosen was the 'facing targets' sputtering system as shown in Fig. 5.22 reproduced from Hales[40]. The figure shows that a pair of magnets forming N and S poles are required. So, an arrangement such as shown in Fig. 5.23 was needed. More detail of one of the magnets is shown in Fig. 5.24.

A good summary of the design of the system can be found in Hales *et al.*[37] and for full detail see Hales[40]. Essentially, the optimum number of pancake windings

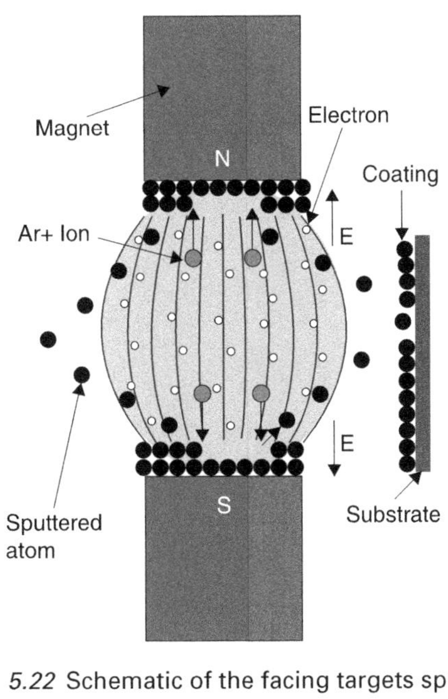

5.22 Schematic of the facing targets sputtering scheme (plan view).

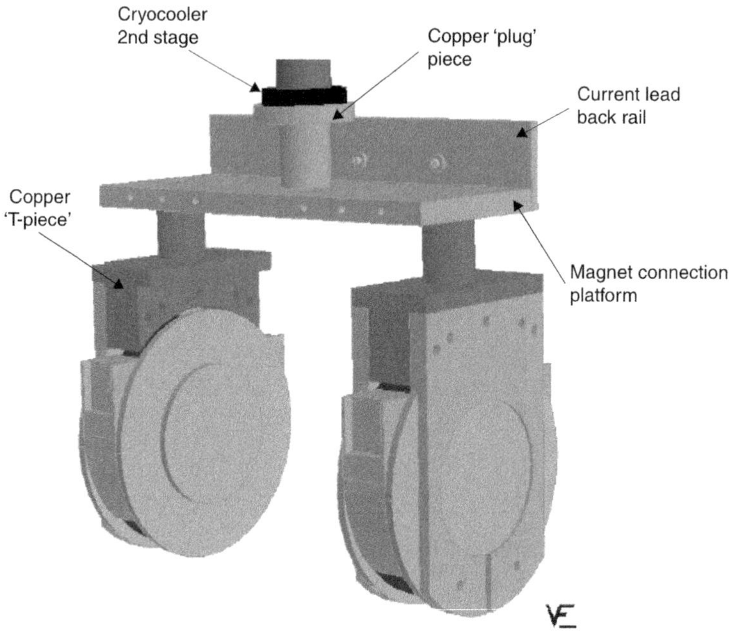

5.23 The arrangement for a facing-targets HTS cryocooler refrigerated magnet system.

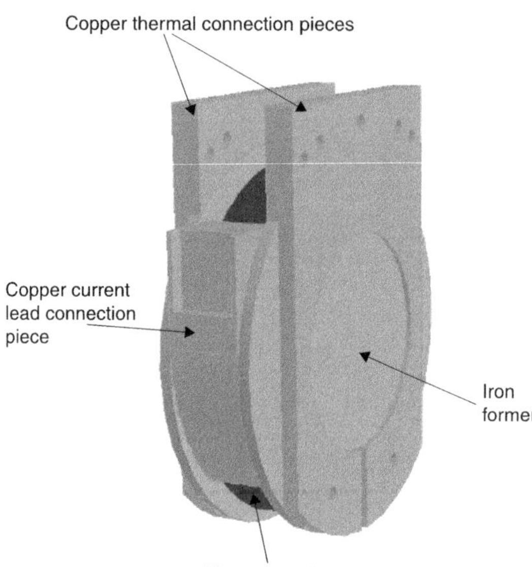

5.24 Detail of one of the 'poles'. Note that the windings are cooled by conduction finally through the iron.

proved to be six on each pole, i.e. three pancake pairs. The winding information and operating characteristics are shown in Table 5.6.

As in the MHD racetrack, the inner inter-pancake connection was by a helical tail, shown in Fig. 5.25. The pancakes were wound and interturn insulation was a 4 mm-wide woven silk ribbon obtained from the doll-making industry. Figure 5.26 shows aspects of the pancake winding.

Radial field component

As has been mentioned above, the way to minimise the B_r 'hotspots', which are at the outer corners on the inner bore of the end pancakes, is to add iron end cheeks

Table 5.6 Key parameters of HTS magnets

Parameter		Value
Conductor type		BSCCO–2223/Ag reinforced tape
Coil construction		
	Number of pancake coils	6
	Inner diameter (mm)	77
	Outer diameter (mm)	138
	Number of turns per pancake	78
	Length of conductor per pancake (m)	160
Operating conditions		
	Operating temperature (K)	20
	Maximum predicted operating current (A)	224
	Maximum centroid field (T)	3.18
	Inductance at maximum current (mH)	18.2

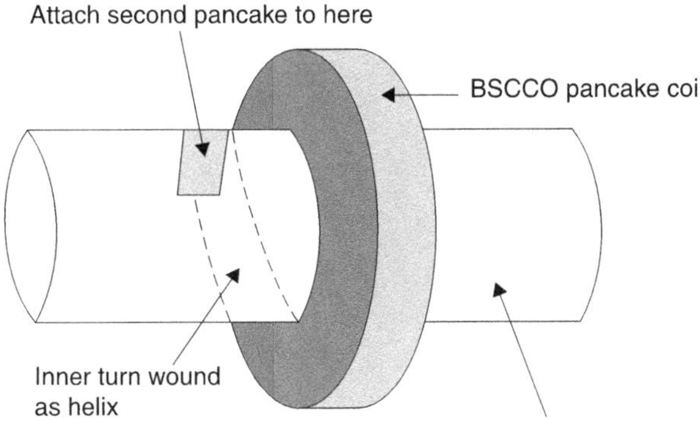

5.25 Inner interconnection arrangement.

(a) (b) (c)

5.26 (a) Co-winding of conductor and silk ribbon, (b) two soldered lap joints between pancake coils (c) view of the cross-section of one of the pancakes.

to the iron bore, and preferably for the end cheeks to protrude radially beyond the windings. The flux lines preferentially permeate the iron rather than the winding pack. This arrangement is shown in Fig. 5.27. The effect of this construction on the radial field component can be seen in Fig. 5.28.

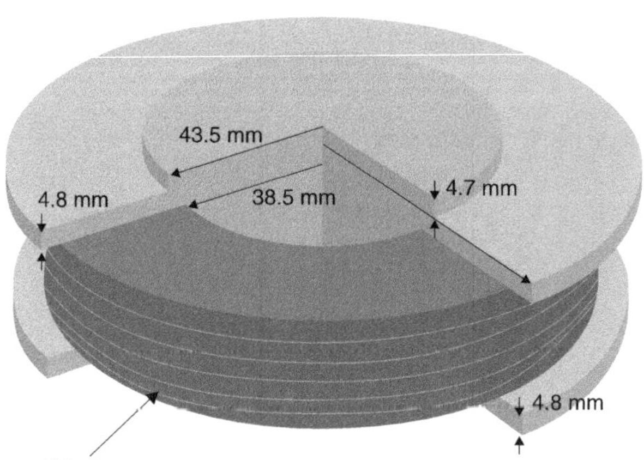

5.27 Showing the iron, which amplified the field and reduced the radial field component at the ends of the winding and thus increased the critical current.

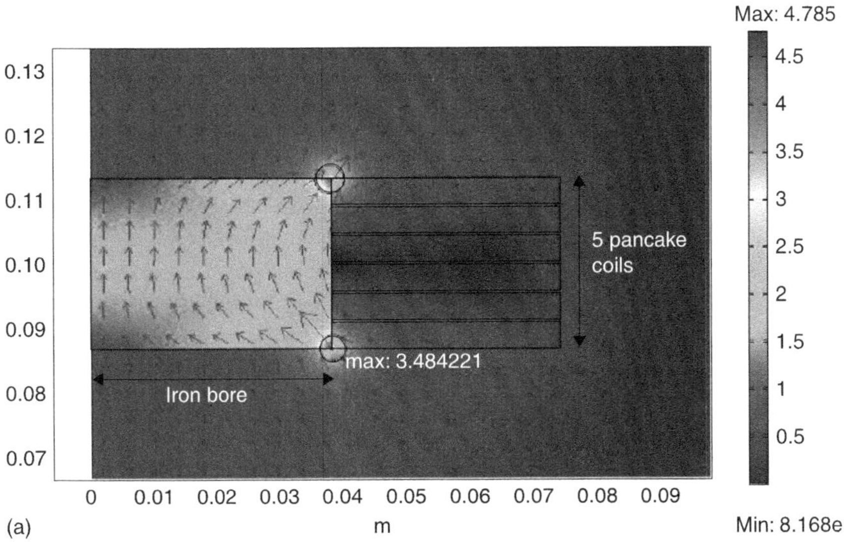

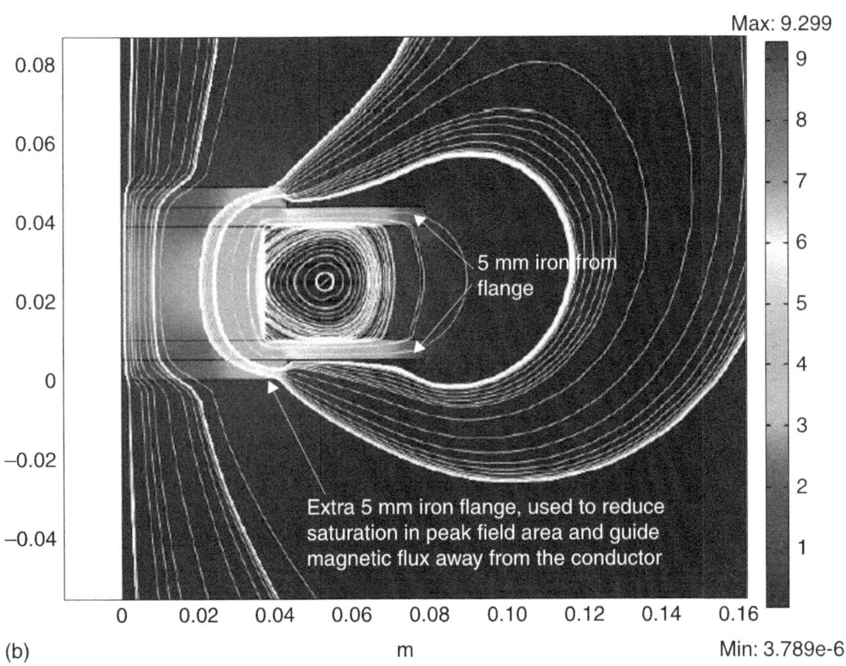

5.28 (a) Magnetic 'hot spots' in a six-pancake assembly, (b) the effect of adding iron end cheeks that extend radially.

Final arrangement

Figures 5.29, 5.30 and 5.31 show the final arrangement in the factory at Teer Coatings Ltd (West Stone House, Berry Hill Industrial Estate, Droitwich, Worcestershire WR9 9AS UK). In Fig. 5.29, several features are worthy of note. The water-cooled 'facing targets' can be seen. The Gifford-McMahon cryocooler has two stages. The first stage is connected to a radiation shield for operation at ~80 K and the second stage is connected to the magnet for operation down to 20 K. The incoming current leads are heat sunk at the first stage. From the first stage to the magnet, HTS current leads are used. These can carry up to 200 A and have virtually zero heat leak[43].

Figure 5.30 shows a wider view of the coating chamber with the door open. The radiation shield and the superinsulation are in place as is the water-cooled target shield. Figure 5.31 shows an external view of the coating cabinet.

Performance

Figure 5.32 shows the field at the targets as a function of radius from the centre of the target for several different magnet currents. Also shown are the performance

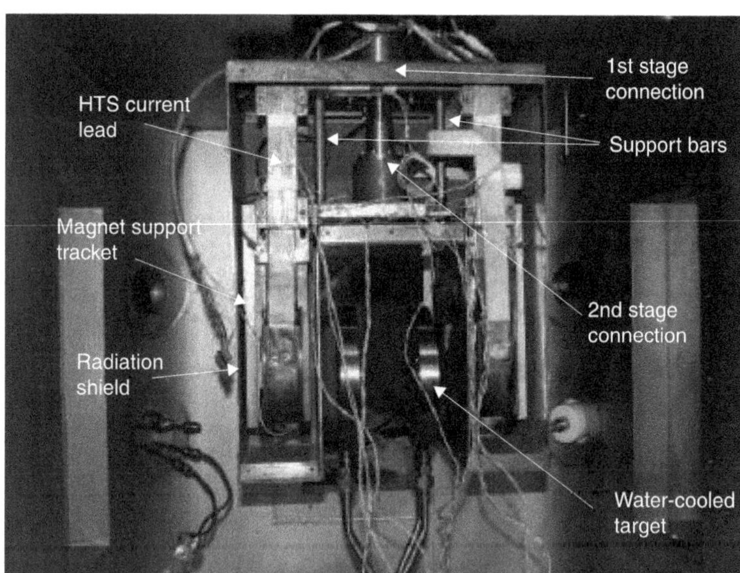

5.29 Magnet system inside the coating chamber. The inner surfaces of the poles of the magnet are 160 mm apart. The inner surfaces of the water-cooled targets are 42 mm apart. A feature of many HTS cryocooled systems is the use of HTS current leads, which can also be seen.

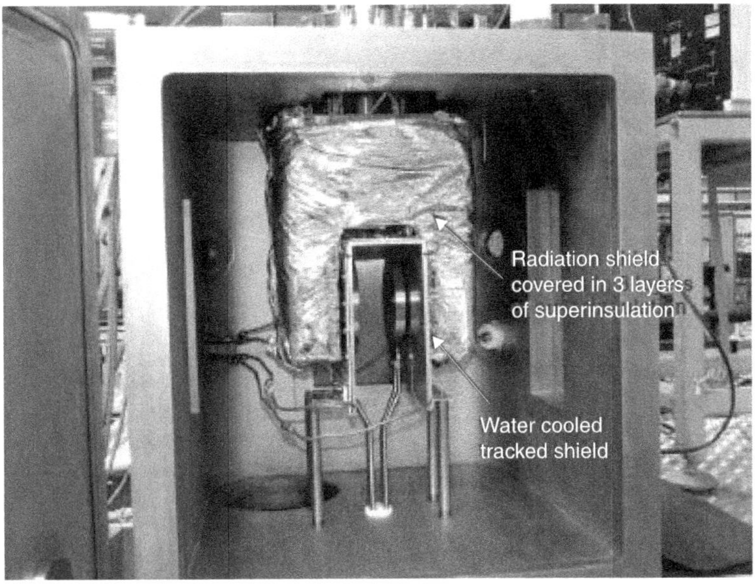

5.30 System with radiation shield and superinsulation in place. The water-cooled target shield can be seen.

5.31 Showing an external view of the coating cabinet at the premises of Teer Coatings Ltd. Behind is the compressor, which feeds the G-M cold head that is mounted on the magnet system top plate. To the front, the heavy-duty current leads that energise the magnet are going to the power supply. In the picture is Tony Hickman (see Section 5.8).

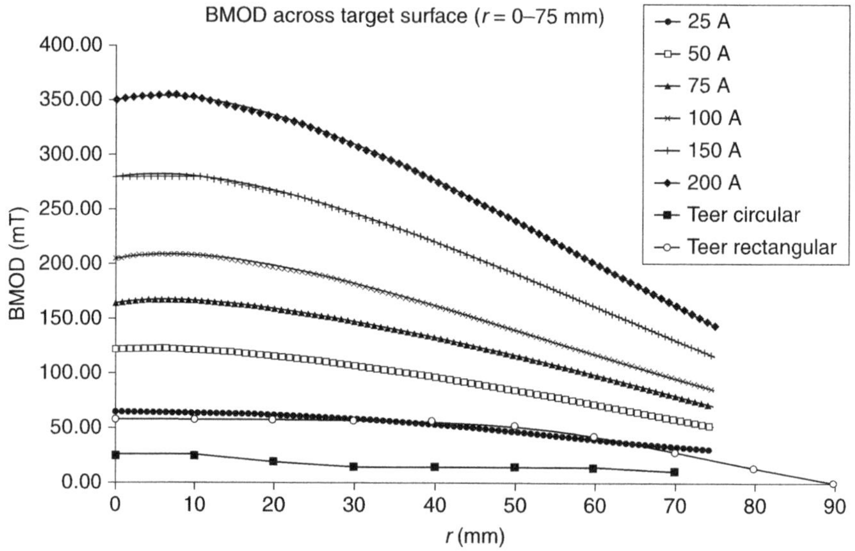

5.32 Field level at the target as a function of radius of the target at different magnet current levels.

of two permanent magnet magnetrons, 'Teer circular' and 'Teer rectangular'. It can be seen that the HTS/Fe magnet system can achieve fields some five times greater than these. Indeed, a magnet current of only 25 A matches the best of the permanent magnet systems.

Magnet stability

HTS magnets have a much shallower EJ characteristic than LTS magnets. It is difficult to discern a clear boundary between superconducting and normal. In LTS magnets, a quench, once initiated, will spread very quickly through the whole winding. As discussed in Section 5.5.2 above, the much greater operating margins and very much higher specific heat capacities mean that a quench spreads much more slowly in an HTS magnet, which can have the disadvantage that a localised hot spot can lead to a burn out. On the other hand, it should be much easier to detect the onset of a quench. A quench situation was modelled for this coil. Details can be found in Hales[40] but the relevant figure is reproduced here as Fig. 5.33.

Finally, Fig. 5.34 shows the coating plasma. This is rather remarkable as this very hot phenomenon is only a few cm from a superconducting magnet working at cryogenic temperatures!

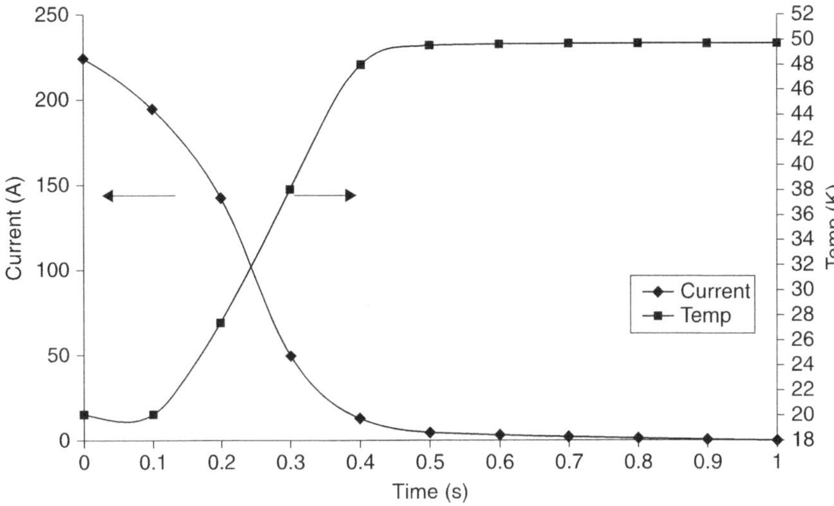

5.33 Characteristics of the coil during a quench simulation. The temperature stays within safe bounds, not exceeding 50 K.

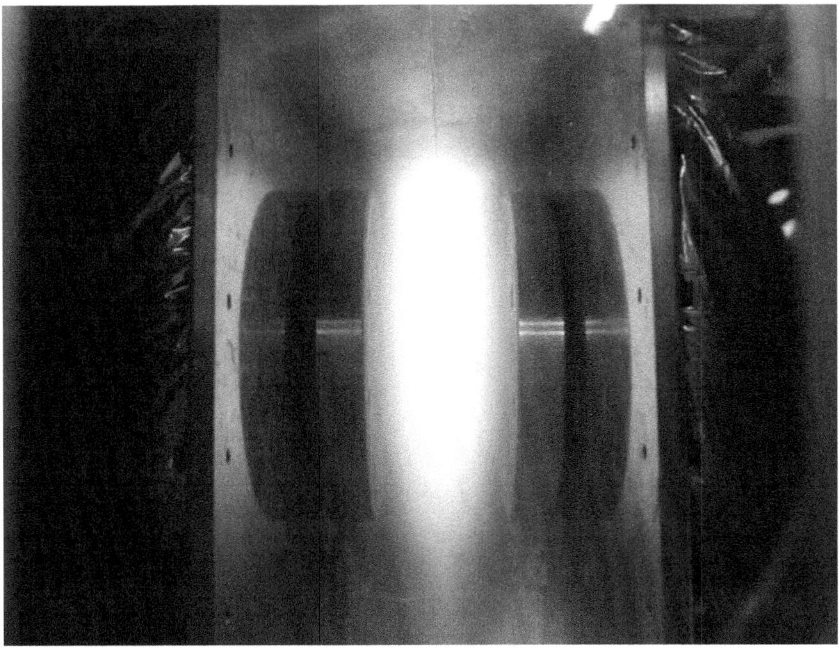

5.34 The coating plasma between the poles of the magnet.

5.6 Future trends

Much has been invested in second generation coated conductor and, in spite of the cost as it stands now, efforts to produce high-field magnets are in train. At present, the world record for a partly superconducting magnet insert is 27 T. A small coil made from SuperPower 2G tape successfully added 7 T to a background of 20 T provided by one of the National High Magnetic Field Laboratory's (NHMFL) water-cooled magnets. The coil is shown in Fig. 5.35.

Probably the flagship example is the 32 T user magnet project at NHMFL at Tallahassee, Florida, USA[44] (Fig. 5.36). Here, a magnet is being designed and built for users at the laboratory so it will be a serious application, not just a prototype. It will easily be the highest field superconducting magnet in the world. It comprises a commercially sourced LTS 'outsert' which itself is in two sections. The LTS section has two NbTi coils and three NB_3Sn coils. The LTS bore is 250 mm. The inner section will be developed in-house and will be a stack of YBCO-coated conductor pancakes. The working bore will be 32 mm.

Whilst the cost of 2G conductor remains at the present level, magnet applications will be only high value-added, niche projects such as the Tallahassee magnet. However, if large quantities of 2G conductor can be sold, e.g. for power transmission, the costs should fall. There are signs that this may happen. Several projects are in train world-wide whereby utilities are looking seriously at incorporating HTSs in their distribution networks.

5.35 A 2G HTS insert coil that added ~7 T to a background of 20 T at NHMFL Tallahassee. The coil is wound from conductor supplied by SuperPower Inc. (Source: SuperPower[21].)

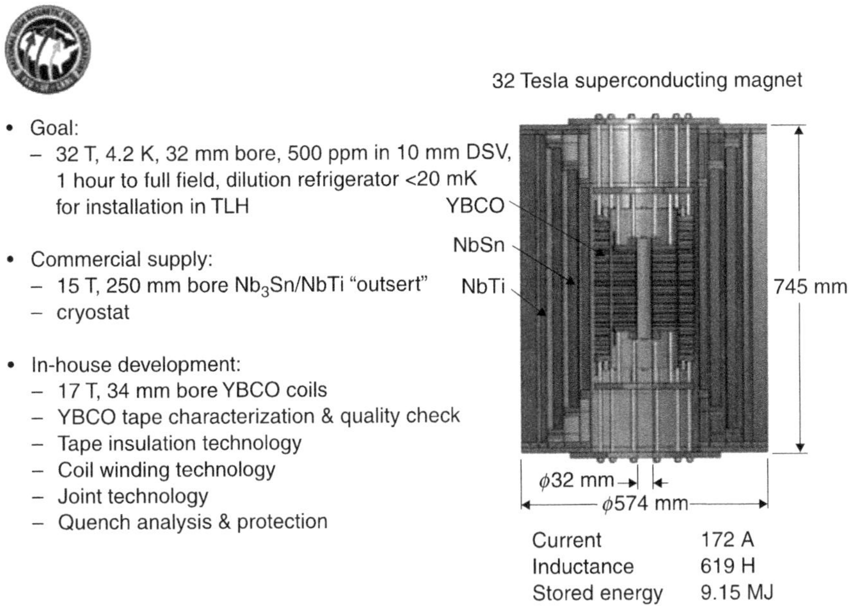

5.36 Summary of the 32 T magnet project. (Source: D. Markiewciz[40].)

5.7 Conclusions

Part of the excitement caused by the discovery of high temperature superconductivity was the prospect of very-high-field superconducting magnets. The feature of HTS that stimulated popular perception was its high critical temperature T_c: ~100 K. Coupled to that was the high upper critical field B_{c2} ~100 T. By working at lower temperatures, this field range should have been accessible. However, the first generation of conductors of the powder-in-Ag-tube type, proved to be disappointingly intractable from the point of view of getting a high-enough critical current (I_c). Consequently, HTS high-field magnet technology, so far, has not lived up to its early promise. However, the fact is that these materials could work at temperatures above the liquid helium regime, and they were applicable in the higher temperature, lower field areas, especially when used with iron.

In this chapter, using illustrations of four such systems, the author has described his own experience of HTS magnets and their difficulties/limitations. When the prospects of the recently commercially available 2G HTS conductors are reviewed, the future looks distinctly optimistic.

5.8 Acknowledgements

I have drawn freely on the work of some of my past graduate students in writing this chapter. They are: Dr Richard Jenkins, Dr Peter Richens and Dr Patrick Hales.

I am particularly indebted to Patrick as he was entirely responsible for both the MHD marine magnet and the sputter coating magnet and without the generous access to his electronic versions of his theses, for diagrams, tables and some text, the chapter would be being handed in even later! I would also like to pay tribute to the excellent technical input over the years from Tony Hickman and Robert Storey. Rachel Fell also helped considerably with figures and tables. I am grateful to Dr Denis Markiewicz for the information on NHMFL's 32 T magnet. Finally, much of my work over the last two decades was influenced by my student, the late Dr David Ryan, who died at a tragically early age in a road accident in 2004, in the USA, where he had begun a brilliant career. I dedicate this chapter to him.

5.9 References

1. Bruker Corporation, http://ir.bruker.com/phoenix.zhtml?c=121496&p=irol-newsArticle&ID=1294089&highlight accessed June 2011.
2. F Herlach and H Jones, *Magnets: Encyclopedia of Applied Physics*, Ed. GL Trigg, VCH publishers 9 (1994) 245–259.
3. Klauder LT Jnr, *Encyclopedia of Physics* eds RG Lerner and GL Trigg (VCH) p 36.
4. Friedlander FJ, *Encyclopedia of Physics* p 288.
5. Montgomery DB, 1969, *Solenoid Magnet Design*, (New York: Wiley)(Reprinted 1980 by RE Kreiger Inc).
6. Boom RW and Livingstone RS, 1962, 'Superconducting solenoids', *Proc IRE* March 275.
7. Wilson MN, 1983, *Superconducting Magnets* (Oxford: Oxford University Press).
8. For example, Cobham, Vector Fields Ltd, 24 Bankside Kidlington, Oxford OX5 1JE, UK.
9. H Jones and RG Jenkins. 'Transport critical currents' In *High-Temperature Superconducting Materials Science and Engineering New Concepts in Technology*, Ed. D Shi, Pergamon 1995, 259–304.
10. Cobham PLC, http://cobham.vectorfields.com/ accessed June 2011.
11. Comsol INC., http://www.comsol.com accessed June 2011.
12. M Newson, DT Ryan, MN Wilson and H Jones, 'High T_c insert coils for high field superconducting magnets – the Oxford' *IEEE Trans. on Applied Superconductivity*, 10(1) (March 2000) d.o.i. 10.1109/77.828273.
13. H Jones, WJ Siertsema, PE Richens, M Newson, PM Saleh *et al.* 'An update on the Oxford high magnetic field facilities in the year 2000' *Physica B*, 294–295 (2001) 598–604.
14. MS Newson, DT Ryan, MN Wilson and H Jones 'Progress in the design and operation of high-T_c Coils using dip-coat BSCC–2212/Ag tape' *IEEE Trans. on Applied Superconductivity*, 12(1) (March 2002) 725–728.
15. JW Burgoyne, C Eastell, CG Morgan, D East, RG Jenkins, *et al.* 'A novel continuous process for the production of long lengths of BSCCO–2212/Ag dip coated tape' Inst. Phys. Conf. Ser. No. 148, *Applied Superconductivity*, 1, (1995) 335.
16. CG Morgan, BM Henry, CJ Eastell, MJ Goringe, CM Grovenor *et al.* 'Continuous melt processing of Bi–2212/Ag dip coated tapes' Proc. ASC 1996, *IEEE Trans. Appl. Supercon.*, 7(2), 1711–1714 (1997) d.o.i. 10.1109/77.620909.
17. M Okada *et al.* 1998 'A new symmetrical arrangement of tape shaped multifilaments for Bi(2212) round shaped wire.' *IEEE Trans. Appl. Supercoad.* 9, 1904.

18. G Duperray and PF Herrmann 'Processing of high T_c conductors: the compound Bi(2212)' in '*A Handbook of Superconducting Materials' Vol. I: Superconductivity, Materials and Processes*', (eds. David A Cardwell and David S Ginley) IOP Publishing Ltd. (2003) 449.
19. MK Wu, JR Ashburn, CJ Torng, PH Hor, RL Meng *et al.* (1987) 'Superconductivity at 93 K in a new mixed phase Y-Ba-Cu-O compound system at ambient pressure' *Phys Rev Lett* vol 58(9), pp 908–910.
20. H Maeda, Y Tanaka, M Fukutomi and T Asano (1988) 'A new high T_c oxide superconductor without a rare earth element' *Jap J Appl Phys* 27(2) L209–210.
21. Superpower Inc., http://www.superpower-inc.com/ accessed May 2011.
22. JL Macmanus-Driscoll 'Processing of high T_c conductors: the compound YBCO' in *A Handbook of Superconducting Materials Vol. I: Superconductivity, Materials and Processes*, (eds. David A Cardwell and David S Ginley) IOP Publishing Ltd. (2003) 565.
23. AA El-Abbar, RM Goodall, CJ Macleod, AM Campbell, RG Jenkins *et al.* 'Dynamic characteristics of iron-cored magnets using high T_c superconductors' Inst. Phys. Conf. Ser. No. 148, *Applied Superconductivity*, 1, (1995) 703–706.
24. AM Campbell, RM Goodall, H Jones, RG Jenkins, CJ Macleod *et al.* 'Control aspects of superconducting magnets for magnetic suspensions and bearings' *IEEE Trans. Appl. Supercond.*, 5, (1995) 650–653 d.o.i. 10.1109/77.402632.
25. RM Goodall, CJ Macleod, AA El-Abbar, H Jones, RG Jenkins *et al.* 'The potential for EMS Maglev using high T_c superconductors' *Proc Maglev '95*, Bremen, VDE-VERLAG, (1995) 209–216.
26. RG Jenkins, AM Campbell, RM Goodall, and H Jones 'Designing HTS coils for magnetic circuits' Proc. MT 14, Tampere, Finland, June 1995, *IEEE Trans. Mag.*, 32, (1996) 2683–2686 d.o.i. 10.1109/20.511427.
27. H Jones, RG Jenkins, RM Goodall, C Macleod, AA Abbar *et al.* 'High temperature superconducting magnets with active control for attraction levitation transport applications' *NASA Conference Publication 3336*, Part 2, (July 1996), 535–550.
28. RM Goodall, C Macleod, A El-Abbar, H Jones and A Campbell 'The use of iron-cored HTS magnets for EMS Maglev' *Proc 9th Int Symp on Superconductivity (ISS '96)* Springer-Verlag (1997) 1393–1398.
29. H Jones, PE Richens, DT Ryan and RM Goodall 'Progress in high-T_c magnet development in Oxford' *Physica B*, 246–247, 333–336, (1998).
30. PE Richens, H Jones, RM Goodall, AM Campbell, D Cardwell *et al.* 'Progress in iron cored high-T_c magnet development for electromagnetic actuator applications' Proc. MT–15, Beijing, China, October 1997, *Proc. of Fifteenth International Conference on Magnet Technology*, (Eds. L Liangzhen, S Guoliao, and Y Luguang), Part 1, 785–788, Science Press, Beijing, China, (1998) d.o.i. 10.1109/TASC.2002.1018504.
31. PE Richens 'High temperature superconductors in electromagnetic applications', DPhil thesis Oxford 2000.
32. H Jones and PE Richens 'A detailed study of the design, construction and cryo-operation of an HTS magnet' *IEEE Trans. on Applied Superconductivity*, vol. 12(1) (March 2002) 1741–1746 d.o.i. 10.1109/TASC.2002.1018744.
33. H Witte 'Scaled down magnetohydrodynamics for a marine propulsion demonstrator', MSc thesis Oxford 2002.
34. PW Hales 'Magnetohydrodynamics using superconducting magnets', MSc thesis Oxford 2004.
35. PA Davidson *An Introduction to Magnetohydrodynamics*, Cambridge University Press 2001.

36. Y Sasakawa. *Yamato – 1, World's First Superconducting Magnetohydrodynamic Propulsion Ship*, Ship and Ocean Foundation 1996.
37. P Hales, H Jones, S Milward and S Harrison, 'Investigation into the use of solid nitrogen to create a thermal battery for cooling a high-temperature superconducting magnet', *Cryogenics* 45 (2005) 109–115.
38. JD Doss *Engineer's Guide to High-Temperature Superconductivity*, John Wiley and Sons 1989 p 208.
39. H Jones and AL Hickman 'Insulation and impregnation procedures used in magnet technology at the Clarendon Laboratory, Oxford' *IEEE Trans. on Applied Superconductivity*, 10(1) (March 2000) 1345–1346 d.o.i. 10.1109/77.828485.
40. PW Hales 'Resistive and superconducting magnets for magnetron sputtering', DPhil thesis Oxford 2007.
41. P Hales, H Jones, A Hickman, K Cooke and D Teer, 'A wide split, cryocooler refrigerated high temperature superconducting magnet system, for industrial sputter coating applications' *IEEE Trans. Appl. Supercond.* 18(2) (June 2008) 812–815 d.o.i.10.1109/TASC.2008.921246.
42. PJ Kelly and RD Arnell, 'Magnetron sputtering: a review of recent developments and applications'. *Vacuum* 1999 56 pp 159–172.
43. D Ryan, 'Current leads' in '*A Handbook of Superconducting Materials' Vol. I: Superconductivity, Materials and Processes* (eds. David A Cardwell and David S Ginley), IOP Publishing Ltd. (2003) 1601–1608.
44. Denis Markiewicz, NHMFL, private communication, Jan 2011.

6
Cryogenics for high temperature superconductor (HTS) systems

P. K. GHOSHAL, Oxford Instruments NanoScience,
Abingdon, UK

Abstract: This is an overview of cryogenics in relation to the advances in superconducting engineering and technology. This chapter will discuss the fundamentals of a cryogenics system and its requirements for HTS, wet and dry system cooling methods, cryogenics issues with cooling sources, heat sources and measurement, cryostat construction, practical thermometry in dry/cryogen-free systems, development and applications using cryogenics and HTS, and associated challenges.

Key words: cryogen-free, HTS, heat transfer, cryocoolers.

6.1 Introduction

Everything should be made as simple as possible, but not simpler.
Albert Einstein

Cryogenics is an essential but often undesirable part of a superconducting system because it calls for an increase in cost and complexity, and reduces reliability and safety. Cryogenics is not just an ancillary, but a driver and centre of advanced technology in the development of ideas and practical achievements. This is a field that is wide and open, technically demanding and challenging, that requires a variety of engineering and scientific disciplines.

Cryogenics is a specialised branch of applied physics and engineering involving technology associated with temperatures below 120 K. This is mainly focused towards the study of low temperature and studying material behaviour at low temperature. Developments in technology have been mostly attributed to liquefying the 'non-condensable' gases of the atmosphere. The technological challenges for engineers and applied physicists have been in regard to the properties of pure chemical substances and their mixtures, compressor machinery, fluid flow and heat exchange and thermal insulation techniques.

Cryogenics for high temperature superconductors (HTS) does not always mean operation at the temperature of liquid nitrogen or near its critical temperature. For any application, the power density and engineering current density is important since it is clearly most efficient to operate a superconductor near its highest permissible temperature. The following question comes to mind, 'what is the best operating temperature for the system?' The real scenario emerges with extremely important factors that need to be addressed and evaluated, particularly for HTSs and the magnet system. This opens a whole topic of discussion in relation to volume and economics, e.g. an HTS magnet operating at <10 K will require substantially less superconductor than a magnet operating at ~80 K, but the cost and saving with cryogenics and superconductor needs close evaluation.

Cryogenics is a combination of scientific and engineering disciplines, applied to realise low temperatures that make process analysis, equipment design and machine construction more challenging, thus demanding knowledge from the cryogenic engineer, but also providing him with particular insights in many aspects of these diverse disciplines. The study of cryogenics will make physicists and engineers fully conversant with phase transitions, supercritical fluids, two-phase flows, non-linear heat transfer regimes, flow instabilities and fields in advanced science and technology. Technically, cryogenic system construction requires advanced knowledge of structural materials, assembly and joining techniques, leak detection and safety along with quality assurance, knowledge of risk analysis (hazard operation – HAZOP and hazard analysis – HAZAN).

This chapter aims at presenting a general view of cryogenics in relation to advanced engineering and technology. Additional reading on the related subjects is given in Timmerhaus and Flynn[1] and White and Meeson[2]. This chapter broadly discusses the fundamentals of a cryogenics system and its requirements for HTSs, wet and dry system cooling methods, cryogenics issues with cooling sources, heat sources and measurement, cryostat construction, cooler performance, thermometry, structural materials, developments and applications using cryogenics and HTSs and associated challenges.

6.1.1 Cryogenic liquid

Basic understanding of cryogenic liquids is necessary for understanding any applications using cryogenics as described in this chapter. The commonly used cryogenic fluids are given in Table 6.1. This chapter does not say anything about commercial cryogenic refrigerators (cryocoolers) and liquefaction techniques. For greater depth and detail please refer to Section 6.11 Sources of further information and advice.

There are primarily two methods of cooling a sample for cryogenic measurement.

- The first method uses direct immersion of the sample in a cryogenic liquid. This method of cooling is ideal where sample heating at high current densities is a concern. This method allows for easy and efficient cooling independent of shape (in wire, tape or films, crystals or block) because this direct cooling has a very high heat transfer coefficient between the cooling medium and heat removal surface.
- The second method is through an intermediate medium (primarily conduction cooling) where the sample is attached to the cold head of the refrigerator as

Table 6.1 Properties of common cryogenic fluids

Cryogenic fluid property	^4He	H_2 (normal)	Ne	N_2	Ar	O_2	CH_4
Molecular weight	4.00	2.02	20.18	28.01	39.95	32.00	16.04
Critical temperature (K)	5.20	33.18	44.49	126.20	150.7	154.60	190.6
Critical pressure (atm)	2.25	12.98	26.44	33.51	47.99	49.77	45.39
Boiling point (K)	4.23	20.27	27.10	77.35	87.30	90.20	111.7
Melting point (K)	4.2*	13.95	24.56	63.15	83.81	54.36	90.72
Density of liquid at b.p. (kg/m^3)	124.7	70.8	1207	806.1	1395	1141	422.4
Density of gas at 273 K and 1 atm (kg/m^3)	0.1785	0.0899	0.8998	1.25	1.784	1.429	0.7175
Thermal conductivity of liquid at b.p. (mW/(m.K))	18.66	103.4	155	145.8	125.6	151.6	183.9
Latent heat of vaporization (h_{fg}) at b.p. (kJ/kg)	20.75	445.4	85.75	199.2	161.1	213.1	510.8

* At 140 atm pressure.

Source: Timmerhaus and Flynn.[1]

shown in Fig. 6.1. The advantage of this method is that this minimises handling of cryogenic fluids. The disadvantage of this method is that this provides limited cooling power, limited transport current (against joule heating) and thermal link to provide adequate cooling.

The basic immersion-type cooling and measurement, the sample and the insert probe is submerged in cryogenic liquid contained in a vacuum-jacketed Dewar flask.

The most commonly used liquid cryogens for superconducting applications are liquid helium and liquid nitrogen.

- Liquid helium (LHe) is used to cool and operate superconducting magnets fabricated from low temperature superconductors,
- Liquid nitrogen is used:
  - to cool the intermediate shield to intercept heat in liquid helium cryostats,
  - for pre-cooling liquid helium magnets to 80–100K before cooling to 4.2 K is economical from the point of view of LHe consumption compared to the magnet being cooled directly from room temperature to 4.2 K with LHe.

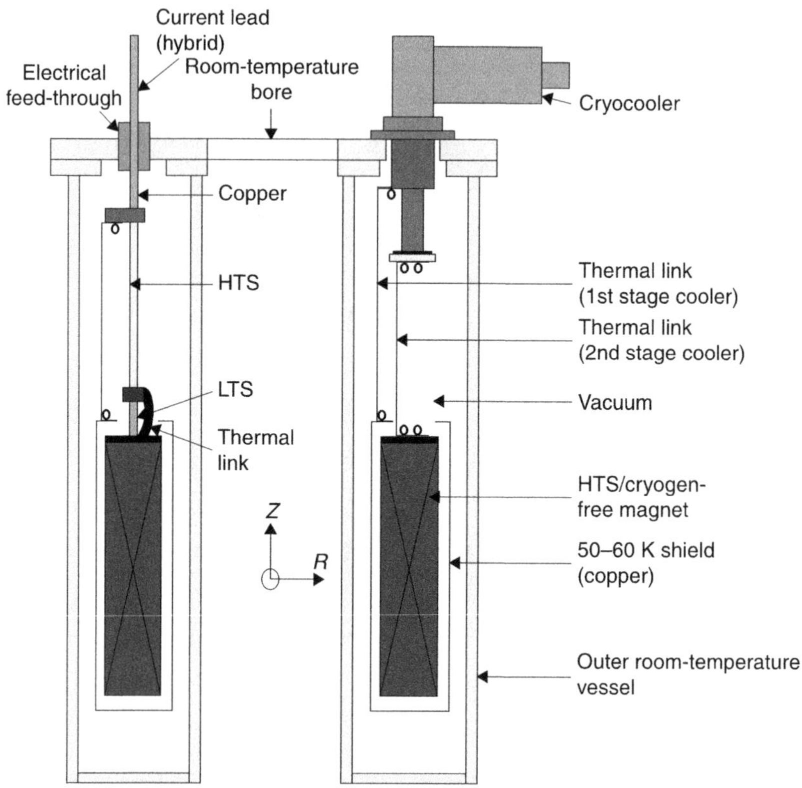

6.1 Schematic of conduction cooling magnet system in cryostat.

There are only a few cryogens that have an atmospheric saturation temperature (boiling point) below 100 K. Liquid helium still dominates the majority of the high-field superconducting magnets utilising low temperature superconductors (LTS), LTS/high temperature superconductor (HTS) hybrids and HTS materials[3]. From the prominent cryogens (helium, hydrogen, neon, nitrogen and oxygen), helium is the only cryogen suitable for LTS magnets. For HTS magnets, it is always best to operate them dry. If operated wet, hydrogen, neon and nitrogen all cover the operating temperature range with limitations to specific critical parameters, but safety needs close attention.

For HTS applications, there are three primary solid cryogens that could be used: solid nitrogen (SN_2), solid neon (SNe) and solid argon (SAr). Detailed analysis of the thermodynamic and mechanical properties of the above solid cryogens suggests that SN_2 is the preferred candidate for cooling HTS. SN_2 remains solid up to 64 K with good electrical insulation properties and is inexpensive, which makes it preferable. Hence, for HTS magnets operating at a temperature in 20–60 K using BSCCO or YBCO and MgB_2 operating in the range 10–30 K, SN_2 gives an enhanced thermal capacity and acts as a thermal reservoir. In general, solid cryogens have good heat capacity and are a substitute for impregnants. Solid cryogens are preferred over epoxy in certain applications for the following reasons: thermal conductivity, mechanical strength and electrical insulation.

6.2 Cooling methods for superconducting magnets

The various cooling methods used over the years for superconducting devices are:

- Bath-cooled cryostat. LHe in direct cooling (mode of heat transfer is both conduction and convection),
- Forced cooled. LHe is forced to flow under pressure inside the coil winding as an in cable in conduit conductor (CICC) (mode of heat transfer is both conduction and convection),
- Cryocooled. The system is cooled using cryocoolers (mode of heat transfer is conduction).

A block diagram of a typical cryogenic refrigerator is shown in Fig. 6.2. There is a heat load at the cold temperature (critical temperature T_c) at which the HTS device wants to operate. This heat load can be from the environment, typically radiation, solid/gas conduction and fluid convection from adjacent regions or structures at higher temperatures (for example, a liquid nitrogen heat shield at 77 K or the warm end of a current lead at 300 K). The heat load can also be internally generated in the component from sources such as ac losses and eddy currents in conductors. The heat from the load is transferred to the cold region of the refrigerator by solid conduction or fluid convection. Therefore, there has to be a small temperature difference (typically a few degrees Kelvin or less) between the cold region of the refrigerator and the slightly warmer HTS load. The working fluid (typically but not exclusively helium gas) in the cold section of the refrigerator absorbs this heat and flows next to a heat exchanger where this fluid is further heated by exchange with a counter-flow gas stream. From there it enters a compressor where external work is done on the fluid, compressing it to high pressures and further increasing the fluid temperature. After this the fluid flows to an ambient heat exchanger where the heat of compression is rejected to the ambient near room temperature (at T_A). Next the fluid goes to the heat exchanger discussed above where it is further cooled by rejecting heat to the counter-flow gas stream or regenerative matrix. From here the cold fluid can go to the load or be further cooled by expansion of the fluid to remove energy. If the working fluid temperature is below the inversion temperature (~45 K for helium), a Joule–Thompson expansion can further reduce the temperature. The efficiency of an ideal process (called the coefficient of performance, COP) for such a refrigerator can be shown as

$$\eta_{Carnot} = \frac{T_C}{T_A - T_C} \qquad [6.1]$$

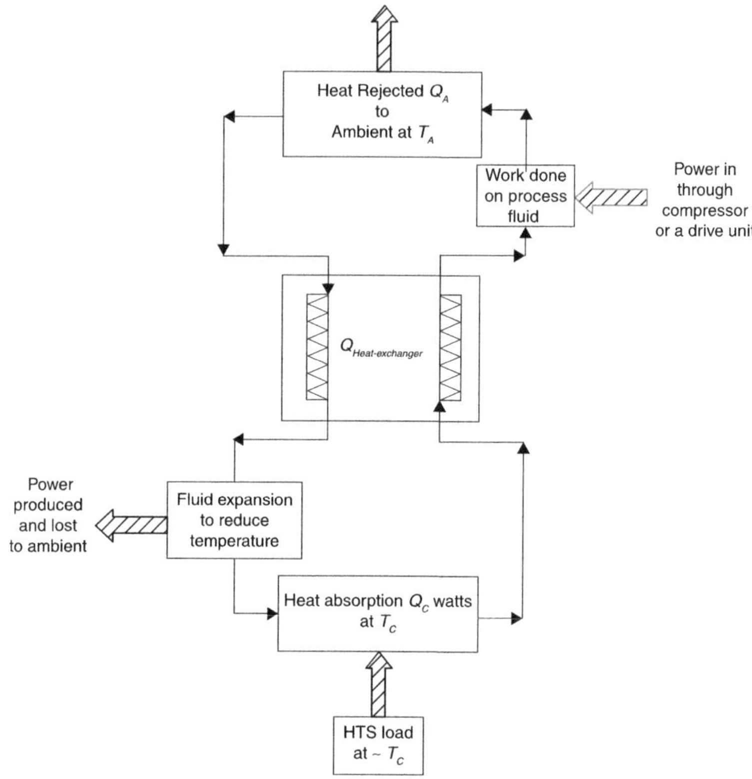

6.2 A typical cryogenic refrigerator.

The basic law of thermodynamics, ideal Carnot equation, gives

$$W_{Carnot} = Q_C\left(\frac{T_A}{T_C} - 1\right) \qquad [6.2]$$

where η_{Carnot} = Carnot efficiency, T_A = higher operating temperature (ambient), T_C = lower operating temperature, W_{Carnot} = Carnot work done, Q_C = heat required to be removed at T_C.

Several observations can be made. To begin with, part of the attraction of HTS electrical applications is the fact that T_C is in the range 25–80 K compared to 4–10 K typical of LTSs; this results in higher Carnot efficiencies and means that the cryogenic refrigerators for HTSs can be smaller (about 10 times compared to LTS). Also, heat should be rejected at as low a temperature as practical; usually this is room or ambient temperature. Real refrigerators typically run at a fraction (less than 10% and up to 30%) of Carnot efficiency due to losses in compressors, heat exchanger effectiveness, etc. The inverse of the Carnot efficiency is called the Carnot specific power and is the number of watts required at ambient to provide 1 W of refrigeration at the lower operating temperature T_C.

For a low temperature of 80 K and an ambient temperature of 300 K, the η_{Carnot} is 0.364 and the specific power is 2.75 W/W is calculated with Eqs 6.1 and 6.2. Typically, small cryocoolers have a COP of about 10% of Carnot, so the specific power would be 27.5 W/W

6.2.1 Cryostat and measurement

Designing a cryostat that will vary with application depending on the type of measurement envisaged may be broadly divided into five stages:

1. Basic requirement: commercial or a research measurement, dry or wet (LHe or liquid nitrogen (LN_2)), cooling power needed for operation, temperature range, vibration isolation, associated pumping systems, dimensions and weight.
2. Thermometry and temperature control: appropriate thermometer for the system (sensitivity, accuracy, effect of magnetic field, etc.), location and mounting of thermometer, temperature control and accuracy.
3. Access within cryostat: sample or the measurement system mount, measuring environment (wet, dry or gas-cooled), access to the sample environment without warming up the whole system (for research magnet), service access requirement (e.g. wiring, windows, mechanical drivers, etc.)
4. Magnet system: magnet specification (strength, homogeneity, persistence requirement), operating current, operating temperature, magnetic materials (cryostat material, magnet support), effect of eddy current due to magnet quench, effect of the cooling components/links due to eddy current heating near the magnet.
5. Thermal design: conduction heat loads (support and spacers), convection heat load (difficult to calculate) and radiation heat load (shield requirement basis for both wet and dry systems).

6.2.2 High temperature superconductors (HTS): an overview

HTS power equipment has the potential to become a key twenty-first century technology for improving the capacity, efficiency and reliability of the electricity delivery system. Possibilities include higher capacity HTS power lines, synchronous machines, transformers, radio frequency cavities, superconducting magnetic energy storage (SMES), nuclear magnetic resonance (NMR) and magnetic resonance imaging (MRI) magnets, future fusion and accelerator magnets and other possibilities that will reduce size and allow additional capacity to be placed.

6.2.3 Current and voltage contacts: Ag sheathed HTS materials and thin films

Designing any superconducting system comprising transport measurement has always been difficult, especially making the electrical contact directly to the superconductor. If contacts are resistive, this is realised as joule heating and becomes difficult to control and measure, and often cascades very fast. Many materials can be extended to low temperature superconductivity, where contacts are reasonably straightforward but this is not the same for high temperature superconductivity because of typical architecture and structure.

Before looking at HTS contacts[4,5,6], it is necessary to describe the resistance of a contact that is independent of contact area. The contact resistivity $\rho_{contact}$ (measured in Ω/cm^2) can be represented in Eq. 6.3.

$$\rho_{contact} \rightarrow R_{contact}A_{contact} \qquad [6.3]$$

where $R_{contact}$ is the contact resistance and $A_{contact}$ is the equivalent contact area. In order to achieve high-quality contacts on oxide-film superconductors[7], contact pads are normally made from a material with a low oxygen affinity, such as gold or silver (sprayed or deposited). There is a distinct difference between bulk and thin film cases, because in films optimum oxygen annealing temperatures are less and annealing time is shorter. That is why for most HTS, the films of gold or silver are deposited *in situ* and helps to improve the quality of contact interfaces but this increases the risk of overheating the HTS film when soldering or making fixed contact.

For HTSs, lead-tin solders are difficult to use because the HTS does not normally wet, and using flux is strictly prohibited in bare oxide superconductors. Therefore, indium-based solders are used, which have a lower melting point ~120°C (with Sn) and ~145°C (with Ag) and found to be stronger than pure indium.

The wetting surface and the contact sizing for HTS in LN_2 are different from LHe. In the case of HTS films, the contact pad that is available on the YBCO conductor is usually thin, therefore the spreading resistance within the contact pad area is limited and poses a problem. The basic design criterion of all joints is to limit the joule heating, and generally for HTS films this is limited to a temperature rise <0.5 K (this normally reduces the critical current in YBCO at 77 K by ~5%). As a guideline[8], the contact area is scaled as seen in Eq. 6.4.

$$\dot{q} \approx 5 \times 10^2 . \Delta T^{2.5} \times A_{contact} (\mathrm{W.m^{-2}.K^{-2.5}}) \qquad [6.4]$$

where $\dot{q}$ is the power generation at the point of contact. The requirement on the contact area is greater when we want to limit the temperature difference ΔT between the sample and the LN_2 to <0.5 K. The contact area resistance increases by power of 2.5 as the heat transfer rate scales by $\Delta T^{2.5}$.

This takes a new dimension when HTS is in a dry environment in vacuum, where the contact cooling is primarily by conduction (cold finger) from a refrigerator depending on the available cooling power available. The heat that needs to carried away dry is calculated based on the heat sink that will be available through the cooling path or the link available and is estimated using Fourier's Law (Eq. 6.5).

$$\dot{q}_{conduction} = k(T).A.(dT/dx) \quad [6.5]$$

where $k(T)$ is the temperature dependent thermal conductivity, dT/dx is the temperature gradient along the sample, and A is the area of cross-section.

If the sample and the sample holder with a low thermal conductivity are reasonably long, then the temperature on the other end (away from the cooler of the sink) gives a high temperature gradient. Therefore in conduction cooling, intermediate thermal anchoring is encouraged within the sample or the cold finger for cooling with a high thermal conduction to the sink. The selection of suitable sample holders is critical to the measurement that is envisaged. For example, ac loss measurement could be challenging when this is targeted for a rotating field environment[9]. In order to address very challenging measurements, engineering and manufacture of composite sample holders is critical towards the thermal properties with a balance of ac heat and temperature distribution with non-metals. Therefore, based on the measurements that need to be carried out under various cooling environments and materials (thin film, bulk, round or rectangular), a detailed consideration of the contact resistance and cooling paths needs to be engineered and qualified prior to assembly and test.

6.2.4 Voltage–current (VI) characteristics

The voltage–current curve is critical for any superconductor to find the critical current (I_c). I_c depends on the magnetic field, the temperature and the strain. Obtaining good accurate I_c data on short and high current samples is always a problem. Generally, VI characteristics obtained at different magnetic fields, voltages across a superconductor rise abruptly with a very narrow current region. It is very difficult to predict and evaluate I_c but it is normally defined as where the voltage takes off quickly. In practice, I_c is determined by common criteria: resistivity and electric field. The details of each criterion are beyond the scope of this section. Readers are encouraged to read the books referred in Section 6.11: Sources of further information and advice for in-depth information.

Most magnet application engineers normally prefer to have the resistivity criteria suitable because this helps to select and compare different conductors over a limited magnetic field range where I_c does not vary much. But this is limited to a point where current varying range is small. Therefore, for a wide range of current and application, the resistivity criteria of obtaining I_c is not suitable. The voltage generated depends on the distance between the voltage taps within the superconductor. Therefore the critical level is normally expressed in terms of the electric field (E_c) along the conductor for engineering design application. I_c values obtained from a VI curve using electric field criteria is labelled as $E_c = 0.1\ \mu V/cm$. This criteria gives a relatively constant I_c (as resistance becomes linear) in a high magnetic field and is normally not preferred for HTSs or in high-field application.

The commonly used I_c evaluating criteria are similar to electric field criteria and covers a wide range of current[10]. Here the critical current is obtained by drawing a tangent to the VI curve at a fixed electric field level $E_c = 0.1\ \mu V/cm$ to the intersection of the lines with the current axis. For most of the practical superconductors, the empirical equation for VI is found over a wider range of voltages and is represented as Eq. 6.6.

$$V = c.I^n \tag{6.6}$$

where c is the proportional constant and n is an index of the degree of nonlinearity (also referred to as a quality index number used for evaluating a superconductor). For multifilamentary superconducting wire, n is normally high (10–50) and as soon as $n \rightarrow 1$, this shows that the superconductor tends to be resistive.

6.2.5 Strain and I_c criteria

It is observed both in low-T_c[11,12] and high-T_c superconductors[13] that I_c degrades with strain (both hoop and bending strain). Therefore for practical superconductors, it is very important to determine (a) the minimum bending radius that a superconductor may be put through while coiling and bending cables and (b) the maximum strain that it can withstand due to electromagnetic stress and thermal stress (generally hoop stress when cooled and energised). Strain is an important parameter, measured at zero magnetic field, and to a certain extent may be used within the design at zero field (free from Lorentz force) by knowing the irreversible bending strain limit between room temperature and low temperature under controlled thermal cycle.

6.2.6 Cryogenics and safety

Handling cryogenics and associated accessories may be safe having learnt the hazards and ways to mitigate them without learning the hard way. This topic in itself is vast, so the hazards that are normally seen while handling cryogens are briefly listed below (noting that we can protect ourselves against all of these hazards by simply following correct procedures):

- Extreme cold and risk of cold burns (frostbite),
- Asphyxiation (oxygen level is reduced),
- Fire and explosion hazard (through oxygen enrichment),

- High magnetic fields (large attractive forces and affecting medical implants),
- Electrical hazards,
- Vacuum hazards,
- High-pressure hazards,
- Radioactive hazards (for example, cobalt-60 in thermometers),
- Lifting and moving heavy equipment,
- Transferring and moving cryogens in cryostats,
- Liquid air and liquid oxygen (this is not flammable but materials coming in contact combust spontaneously).

6.3 Cryogenics issues: cooling sources, heat sources and measurement

Any superconducting system with superconducting magnets or associated with superconductivity demands proper understanding of the cooling technology. In order to address the issues in cryogenics or in general cooling the following needs careful attention: cooling sources, heat sources and measurement technology.

Cooling sources for any magnet system could be liquid cryogens or cryocoolers to cool the magnet system and subsequently maintain the operating temperature. Cryogens have enormous cooling power and are mostly used for direct cooling, but with cryocoolers, and their limited cooling power, the system is cooled by conduction. This could give a temperature gradient between the source and the medium. The cooling source should be sufficient to balance the heat load from various heat sources within and into the system. The primary heat sources within a magnet system and cryostat is through conduction (support for magnet and cryostat, wiring, external leads, etc.), convection (contamination within the vacuum space), radiation (between cryostat walls and shield if present) and, for peak heat load, additional heat as the system or magnet experiences while ramping up the magnet (associated with ac losses from the magnet and joule heating in the leads).

6.3.1 Heat transfer at low temperature

As mentioned above, heat conduction mechanisms are of major concern when designing a cryostat for test and commercial application. Designing a dry system for good heat transfer from the magnet or test environment across solid–solid links requires good thermal anchoring to the sample, wiring and temperature sensors, interfaces for cryocoolers and variable temperature inserts. Sometime the sample environment cooling is enhanced with solid–gas interfaces.

6.3.2 Conduction heat transfer: solid, liquid and gases

Heat conduction is a phenomenon that occurs through the interaction of neighbouring atoms and molecules, transferring their energy/heat (partially) to their neighbours. This is the most significant means of heat transfer within a solid and between solid objects in thermal contact. Fluids are less conductive, with the exception of helium. Therefore the thermal contact conductance between solid bodies in contact is vital for cryogenic design and applications that need to cool the system with cryocoolers. Steady-state conduction can only happen when the temperature difference driving the conduction is constant when temperatures in the conducting object do not change and are said to be in quasi-equilibrium. Therefore, in steady-state conduction the amount of heat entering a system is equal to the amount of heat removed. Transient conduction is the process where the temperature within the conducting object changes with time. Analysis is complex and carried out mostly with empirical equations, and limited measured data is available. Conduction loss is normally calculated using Fourier's law.

Heat conduction through gases is normally used for cooling samples (as exchange gas) and/or for shielding (vapour cooling). Calculations for gas conduction are not as accurate as those for solids. Exchange gas used for conduction is normally used to enhance cooling and achieve homogenous temperature distribution in a low-thermal-conducting sample. The heat transfer rate between the solid and gas having a temperature gradient (ΔT) is expressed by Newton's law of cooling (Eq. 6.7), where the rate of heat transfer is proportional to the difference in temperatures between the body with its surroundings and heat transfer co-efficient:

$$\dot{q}_{Convection} = h_{Convection}.A_{Surface}.\Delta T \qquad [6.7]$$

where $h_{Convection}$, is the convective heat transfer coefficient and $A_{Surface}$ is the surface area of the solid. The heat conduction/transfer rate under forced flow is much greater than for conducting gas free-flow (also referred to as convection) that is proportional to $h_{Convection}$[14]. Discussion on an analytical solution is beyond the scope of this section. Please refer to suggested reading for further information and details[1,2,8].

6.3.3 Radiation heat transfer

This is a significant mode of heat transfer experienced at cryogenic temperatures because of large thermal difference between the various parts of the cryostat. The radiation heat flow is between two surfaces having a temperature difference. The radiation heat flow is represented by the Stefan–Boltzmann equation (Eq. 6.8).

$$\dot{q}_{Radiation} = \sigma.\varepsilon.A.(T_{hot}^4 - T_{cold}^4) \qquad [6.8]$$

where ε is the emissivity of the surface at operating temperature, A is the surface area of the incident heat flux, and σ is the Stefan–Boltzmann constant (5.67×10^{-8} W/m^2.K^4).

Heat flow is proportional to emissivity and the temperature difference between the two surfaces (ε is critical and is about 1.0 for an oxidised surface and lower for polished metallic surfaces). In order to reduce the radiation heat flow ε a temperature gradient is introduced using an additional 'line of sight' shield (sometimes referred to as a radiation shield). Multiple radiation shields break the temperature gradient into smaller steps thereby significantly reducing radiation heat transfer into the low temperature surfaces. In principle, with N floating (rigid to semi-rigid) radiation shields, the radiation heat transfer could be reduced by[15] ~ $(N+1)^{-1}$.

Convective heat transfer by residual gas

Within cryogenics, generally convective heat transfer means the transfer of heat from one place to another by the movement of residual gases within the vacuum space (mostly due to out-gassing from the surfaces of the cryostat and the components within the vacuum space). Convection is actually a combined effect of conduction and residual gas flow.

Free or natural convection occurs in the absence of vacuum where the free gas molecules' movement is caused by the density variations due to the temperature in the gas molecules (that do not condense to a cold surface). This is minimised by continuously evacuating the space thereby removing the number of molecules that take part in convection in free space. In order to further reduce the convective heat transfer and overall effective emissivity, multilayer insulation (MLI) such as aluminised Mylar™ is placed between the hot and cold surfaces within the vacuum space. For further details on convection mechanism and kinetic theory of gases at low temperature see text books in cryogenic engineering[1,15].

Vacuum pumping

In order to achieve better thermal stability and reduce heat load into a cryogenic environment (such as a magnet or sample), the volume of space isolating the cold environment from room temperature is evacuated to a level where there is a minimum of residual or free conducting molecules (depending on specific requirements), without contaminating the working environment. In order to achieve a high level of vacuum, vacuum pumps (depending on the throughput, maximum vacuum and flow required) are selected that are fit for purpose. At the time of writing, turbo molecular pumps with integrated mechanical pumps (both oil and oil-free) are widely used, replacing rotary and diffusion pumps with cold traps, achieving a vacuum $<10^{-6}$ mbar. For a deeper understanding of vacuum and its associated risks, see the books on vacuum technology[16] since it is beyond the scope of this section.

6.3.4 Joule heating due to transport current

There are more heat sources that are important for a cryostat design for both wet and dry systems that significantly affect the thermal design and performance.

Resistive heating, also referred to as joule heating, is experienced primarily from (a) the sample contacts, which can be minimised by good sample cooling and making low-resistance sample contacts[2,8] and (b) current leads – this is normally optimised based on the operating cycle, cross-section area, cooling power available and protection for the lead itself.

Therefore, systems that require long hold time should use de-mountable current leads in order to avoid any conduction loss. For dry systems and limited wet systems, HTS leads significantly reduce the heat leak into the system and are preferred over an optimised copper wire lead where demountable leads are not used[17,18,19].

6.4 Cryostat construction

A cryostat is generally a double-walled vessel (mostly evacuated) that is cold within and stationary, similar to a Dewar flask, to maintain cold cryogenic temperatures. Low temperatures are maintained within a cryostat by using mechanical refrigeration method or cryogenic fluids, depending on specific application. These vessels are also partially qualified as pressure vessels during risk and hazard analysis. Typical sizes vary from cryocans used in laboratories for medical research, a few litres in volume (for preserving biological cells), to fusion reactor vessels in order to keep the magnet cold at its operating temperature, and which may be many thousands of litres.

Bath cryostats are similar in construction to vacuum flasks filled with cryogens. A cold plate is placed in thermal contact with the cryogen, which is replenished as it boils away, depending on the volume and construction of the cryostat towards reducing heat leak. In case of LHe, helium vapour that boils away from the bath is used to cool thermal shields around the outside of the LHe bath. The boil-off rate may be reduced further by using several concentric layers of shielding, with gradually increasing temperatures.

Closed-cycle cryostats consist of a chamber through which cold helium vapour is pumped, and an external mechanical refrigerator that extracts the warm helium exhaust, which is cooled and recycled. Closed-cycle cryostats demand large amounts of electrical power, do not need refilling with helium and, in theory, can run continuously for an indefinite period. Within these cryostats, magnets or the system is cooled by attaching a link that is in thermal contact with the helium vapour chamber by conduction cooling through to a metallic cold plate (cooler stage) inside a vacuum chamber.

Continuous-flow cryostats are attractive for laboratory experiments where cryostats are cooled by LHe from a storage Dewar flask. The cryostat is continuously replenished by a steady flow of LHe from the storage Dewar flask, with LHe boiling within the cryostat. Continuous-flow cryostats do not require electrical power, but large quantities of expensive LHe are consumed during operation. Therefore, facilities having continuous-flow cryostats are encouraged to capture and recover the helium as it escapes from the cryostat; this can be expensive. The volume of the storage Dewar flask defines the time over which cooling can be maintained before replacement with another storage Dewar flask for continuous operation.

6.4.1 Material selection

The choice of materials for cryostat construction is a challenge in its own right. With known material properties, we are limited to what may be used between cryostat construction and heat transfer medium at cryogenic temperatures. The following properties are critical to selection for thermal, mechanical and electrical design (structural, support, cooling, etc.) and are not presented in any particular order.

Copper, in reality, does not provide great cooling power compared to a ceramic at cryogenic temperatures but pure copper has excellent thermal conductivity at low temperatures which makes this material suitable for sample holders in order to carry heat to the sink[20]. Most materials (both metals and non-metals) become brittle at cryogenic temperatures with the exception of face-centred-cube (FCC) crystal structures (copper, brass, aluminium) and a few other non-metals that do not embrittle, such as Kapton™ and Teflon™. FCC materials generally show an increase in their yield strength at cryogenic temperatures, which is even seen in solders.

Thermal properties

- Thermal conductivity is an important parameter that plays a critical role in 'fit for purpose' selection. The materials most widely used in cryogenics are copper, brass, stainless steel, glass, epoxy composites, Kapton™ and Teflon™ of various grades. In general the thermal conductivity depends on the purity (or less defect content) of the material. Higher purity metals will have higher thermal conductivity at cryogenic temperatures. In metals, purity is represented in terms of residual resistance ratio (RRR) from standard temperature and pressure (STP) to 4.2 K ($RRR \rightarrow R_{293\,K}/R_{4.2\,K}$). Generally, at room temperature the conductivity in metal is mostly due to thermal vibration of the lattice, but temperatures below 77 K are dominated by the effects of defects and impurities. In dry systems the challenges are to get the best of thermal conduction and electrical isolation (where the switches and joints sit in a dry environment and carry full transport current). The options are limited at present to using sapphire and/or beryllium oxide (beryllia) as a medium to isolate electrically from the cooler stage/heat sink. Low-thermal-conductivity components in the cryostat are usually austenitic stainless steel (AISI 304, 316, 321 of various composition) for cryostat structure and occasionally, for the support of heavy structures, together with Cu-Ni (0.1), Inconel 625, because in addition to having relatively low thermal conductivity at cryogenic temperatures, they are mechanically strong. For sensitive magnetic measurements titanium is preferred over stainless steel and AISI 321, because of its magnetic properties. Non-metallic materials are normally used for support where there is a need for low direct-heat leakage. They have the advantage of low thermal conductivity but have a long cool-down time compared with metal/metal alloys. In general, the materials used are fibre glass–epoxy composites (G10), Kapton™, Mylar™ and resins filled with glass, cloth or paper (Tufnol™).

- Thermal contraction – this parameter cannot be neglected for construction and needs to be matched closely to avoid the forces that develop between the elements that support the structure. The total thermal contraction in stainless steel, copper, brass, aluminium alloys and both filled and unfilled epoxy resin vary substantially when cooled from room temperature to 77 K but is generally minimal from 77 K to LHe temperatures. This is normally measured and defined in terms of $\partial l/l$ ($=L_{295\ \mathrm{K}} - L_T/L_{295\ \mathrm{K}}$) $\partial l/l$ (m/m) or percentage contraction[21]. Thermal contraction is normally classified in engineering terms as high (where $\partial l/l$ is >1%, seen in materials such as unfilled epoxy resin, silicone/vitone rubber), medium ($\partial l/l > 0.3\%$, are metals and alloys) and low ($\partial l/l < 0.3\%$ such as glass and ceramics for optical window-based applications).
- Heat capacity – This normally is classified in terms of specific heat, a measure of the amount of heat that can be removed by lowering the temperature by 1 K. In general heat capacity decreases substantially at cryogenic temperatures to about T^3 (lower operating temperature). Therefore at cryogenic temperatures some systems take longer to cool down and stabilise (thermal equilibrium) than others. Ceramics and epoxy/resin-filled composites have higher specific heat capacities of about an order of magnitude more than metals over the same range of temperature[22].

Mechanical properties

This is the prime concern for structural materials, and we do not use materials that embrittle on cooling. Hence this limits the use of most iron alloys such as carbon steel and Fe-Ni steels that are very good at room temperature. As discussed briefly above, FCC-structured materials such as copper, aluminium; brass and austenitic steel are used due to their excellent ductile properties at cryogenic temperatures. With careful design on thermal stress and strain, titanium hexagonal closed pack (HCP) and niobium for shielding body-centred cubic (BCC) can also be used. While designing the structural materials for room and cryogenic temperatures, the yield strength is considered at the operating temperature, along with its magnetic susceptibility properties.

Magnetic properties (magnetic susceptibility)

This is one of the factors that play an important role when selecting the materials for a magnetic environment. In high magnetic fields, the forces induced could be significant due to the choice of materials. Magnetic susceptibility of stainless steel is reasonably high when exposed to thermal cycles and cannot be neglected during the design stage. Details on the magnetic *mass* susceptibility and *volume* susceptibility are beyond the scope of this chapter so for further details on the properties of materials at low temperature readers are referred to reference 8. The materials preferred at low temperature are given in Table 6.2.

6.4.2 Vacuum technology

Vacuum (for heat load isolation) is needed at all times for containing cryogens within a cryostat to reduce the heat load (conduction and convection) from room temperature. Depending on the type of application, the level of vacuum requirement is defined as shown in Table 6.3. Most of the vacuum spaces are evacuated before cooling and subsequently sealed because at low temperatures the rate of desorption of gas trapped on the surface of materials is extremely low. Most of the cold surfaces act as a cryo-pumping surface, lowering the partial pressure of most gases in the system except helium, hydrogen and neon.

In addition to the selection of the right vacuum pump from Fig. 6.3, with correct vacuum ducting, another important factor in vacuum design is to limit the total

Table 6.2 Suggested materials in low-temperature environments

Material	Comments
SS 321	Not recommended for magnets
SS 304	Most commonly used (both for coil support and cryostat)
SS 304L	Commonly used (preferred for magnet/coil support and cryostat)
SS 316	Preferred for magnet vessel/cryostat
SS 316L	Preferred for magnet vessels
SS 316LN	Highly recommended for magnet vessel (high homogeneity systems)
Aluminium-1200(0)	Pure aluminum. Radiation shield and neutron beam applications (careful design for magnetic environment)
Aluminium-5083, 6082	Preferred supply except radiation shield – neutron beam
Brass-CW712R, CW507L	Most commonly used in magnet assembly joints, leads, structure
Copper- CW004A	Most commonly used
Copper- CW024A	Phosphorus de-oxidised
Copper-CW008A	OFHC (UHV application)
EP CC 301, 6F45, EP GC201	Standard sheet used for magnet, epoxy/cotton, epoxy/glass

Table 6.3 General classification of vacuum and its application[23]

Rough vacuum (Pa)	Medium vacuum (Pa)	High vacuum (Pa)	Very high vacuum (Pa)	Ultra high vacuum (Pa)	Extreme ultra high vacuum (Pa)
~99 000–3300	~3300–1 × 10^{-1}	~10^{-1}–10^{-4}	~10^{-4}–10^{-7}	~10^{-7}–10^{-10}	<10^{-10}
Vacuum forming	Sterilisation	Vacuum metalising		Plasma physics and fusion reactors	
Lifting	Freeze drying	Thin films		Surface physics and chemistry	
Fumigation	Distillation	Vacuum tubes, TV, X-ray tubes		Surface analytical instruments	
Dehydration	Vacuum casting and degassing	Low-density wind tunnels		Space simulation	
Vacuum chunks and tables	Vacuum coating	Chemical analysis		Special purpose vacuum tubes	
Vacuum packing	Lamp and flask manufacture	Cryogenic and electrical insulation, electron beams, vacuum metallurgy		Ultra pure gases	

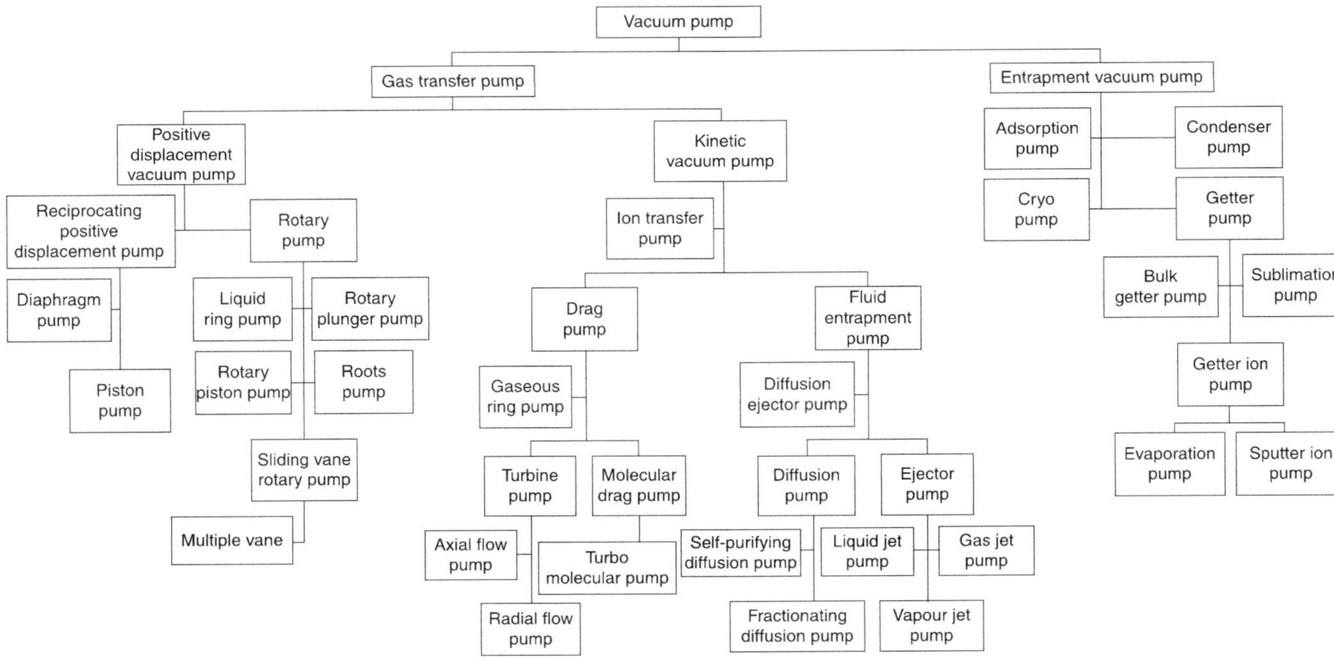

6.3 General overview of vacuum pump classification[16].

gas influx rate to a consistent level with the ultimate pressure required within the vacuum space. The sources of gas in vacuum system are given below.

- Vacuum vessel leak normally occurs at joints (permanent and/or temporary). For high vacuum (HV) and ultra-high vacuum (UHV) systems, materials selection is of prime importance. Materials are usually heated to a temperature >400 °C in order to remove any adsorbed gas within the vacuum chambers surface. Therefore, welding and hard soldering (brazing) are preferred in HV and UHV systems.
- Virtual leak is normally due to trapped gases within the vacuum system. This is critical because this limits the ultimate pressure in the vacuum space and prolongs the evacuation time. Therefore, it is always suggested wherever possible to have a continuous weld on the vacuum side.
- Degassing/out-gassing of materials is always of primary concern for any vacuum system. Most surfaces in the vacuum system contain adsorbed gases (this is a surface phenomena), e.g. moisture, gas molecules. This is not a major concern in rough- to medium-vacuum systems, but significantly affects HV and UHV systems. Therefore, materials with rough surfaces (which provide substantial surface areas), oxidised surfaces, insulated wires, greases, etc. are difficult to use in HV and UHV systems. In order to avoid out-gassing over time, vacuum chambers are degassed by baking the vacuum system while it is under vacuum[24]. It is proven that vacuum chambers need to be heated to remove the adsorbed gas molecules, but the question still remains as to what temperature and for how long? This is independent of the ultimate pressure that is desired after the bake out and cool down and more to do with the binding energies of the existing molecules and the materials of the vessel/cryostat. It has been shown that preheating between 100–150 °C and subsequently to about 400–450 °C for UHV systems gives good results in achieving the desired vacuum level. Vapour pressure in solids are complimentary to out-gassing where the vapour pressure of the materials/metals in vacuum systems must be selected for low vapour pressure, keeping the vapour pressure of the solids over the working temperature range below the desired ultimate pressure (vacuum).
- Permeation of gases is a prime concern for the cryogenic vacuum vessels that are usually evacuated at room temperature and sealed before cooling to have better cryo-pumping, reducing the pressure significantly. Details of the permeation of gases and its mechanism are therefore important for the detailed calculation of the vessel/container thickness[24].

6.4.3 Electrical wiring and connections

This section is huge and challenging. This will be an attempt to describe the basic requirement that needs to be addressed while designing and defining access into the operational environment (cold and/or vacuum). It is important to understand and plan wiring in general for any cryogenics system. This section will primarily highlight the various types of wiring in cryogenic engineering.

- DC and LF (<10 kHz) wiring. This wiring is normally tightly twisted pairs (voltage leads together and current leads together) in order to minimise voltage pick-up in a test environment. This is more effective than co-axial cables to reduce magnetically induced noise.
- AC HF (>10 kHz) wiring. In general, co-axial cables are used for measurement with special consideration towards providing thermal heat sinking.

Selection of the wiring for a cryostat is driven primarily by the following: ease of working with (soldering), heat load, thermo-electric drift and joule heating into the system (general guidelines given in Table 6.4).

The properties of the wiring materials considered are thermal conductivity, physical properties, residual resistance ratio (RRR) which defines the purity of conductors and insulation (based on the application and environment as in UHV). The wiring is a critical link from the cryogenic and vacuum environment to the outside world as shown in Fig. 6.4. Thus the design of vacuum and electrical feed-through both critical. These feed-throughs are primarily for instrumentation, diagnostics and electrical power and used both in cryogenic and non-cryogenic environments.

Table 6.4 Cryostat wiring recommendations

Application	Wire material
Voltage potential taps	Copper • For sensitive dc measurement, twisted paired wires • Preferred for continuous length
Heater	Constantan and phosphor bronze Phosphor bronze is preferred over constantan because it is easy to solder, use in magnet field and easy to handle but expensive
Thermometer leads	Phosphor-bronze and copper Phosphor bronze is preferred over copper Good thermal isolation (preferred when used in small lengths). For long length and routing, copper is preferred because of low resistivity compared to phosphor bronze
Low current leads	Copper and phosphor bronze Copper preferred because of low resistivity; phosphor bronze in magnetic field environment
Heater leads	Copper Low resistivity (less joule heating)

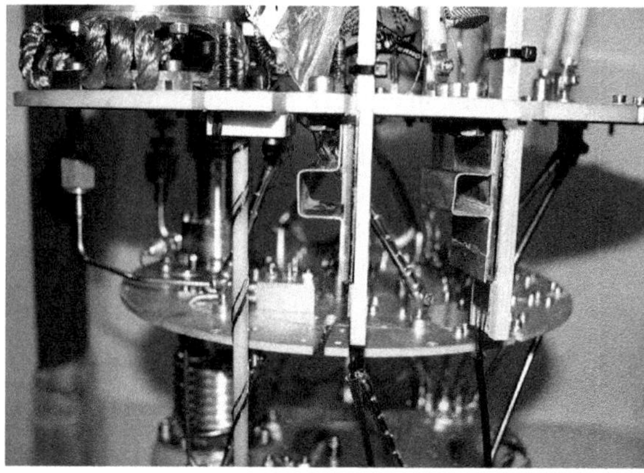

6.4 Typical wiring and support of an ultra low temperature insert in a cryogen-free environment <4.2 K (Source: Oxford Instruments Nanoscience, UK).

6.5 Cooler performance

In order to select a cryocooler, the cooler performance is evaluated:

- Cooler rating at different operating temperature and constant cooling power,
- Cooler rating at a combination of cooling power and operating temperature.

6.5.1 Cooling wet magnets

This is achieved by directly immersing the magnet in a cryogenic liquid. This method allows easy and efficient cooling independent of shape because applying direct cooling enhances the high heat transfer coefficient between the cooling medium and the heat removal surface. This type of cooling can be achieved in a liquid helium cryostat with a magnet sitting inside the Dewar flask and replenishing helium over time because of the normal boil off, maintaining a minimum level for magnet operation. The second method is by attaching a cryocooler to the cryostat, often called a recondensing Dewar where the magnet is situated inside the Dewar flask and cooled directly with LHe. The helium vessel is attached to the second stage of the cooler where the helium vapour due to normal boil-off is condensed and reused. The advantage over the normal helium cryostat is that this has a long liquid hold-time and does not require replenishing with LHc, though it does requires regular servicing of the cooler attached.

6.5.2 Cryo-cooling of HTS magnets

Cryogen-free (dry) and HTS magnets are currently cooled by a cryocooler, whose cold head region is attached to one end of the magnet assembly. Generally the first stage of the cooler is thermally attached to the radiation shield and the second stage of the cooler to the magnet assembly. The biggest challenge is to achieve and maintain a constant/minimum temperature gradient between the coldest region (second stage of the cooler) to the warmest point.

For simplicity, we see a typical scenario of an adiabatic volume where a magnet is cooled with the cryocooler (with no additional heat load). The mass of the magnet can be represented by a copper mass say (m_{Cu}) having a heat capacity (c_{Cu}) and copper density (ρ_{Cu}) in order to achieve a uniform temperature T_{Cu} over the entire magnet mass with second-stage cooler temperature T_{Cooler} by conduction cooling. The target is to achieve $T_{Cu} = T_{Cooler}$. From the first law of thermodynamics we can write for Q_{Cooler}

$$-Q_{Cooler} = \left(\frac{m_{Cu}}{\rho_{Cu}}\right).C_{Cu}.\frac{dT}{dt} \qquad [6.9]$$

while the cooling progress $dT/dt < 0$ ($-Q_{Cooler}$ represents the heat sink providing cooling). The primary concern is to select the right cooler, or the mass of the magnet that can be cooled within a reasonable time so that the magnet system can be designed. Equation 6.9 can be rewritten as

$$dt = \left(\frac{m_{Cu}}{\rho_{Cu}}\right).C_{Cu}.\frac{dT}{Q_{Cooler}}. \qquad [6.10]$$

The time constant τ_{Cu} can be calculated by integrating Eq. 6.10 between the operating temperature and target temperature. The time needed to cool the magnet is given in Eq. 6.11.

$$\tau_{Cu} = \frac{m_{Cu}}{\rho_{Cu}} \int_{T_{op}}^{T_F} \frac{C_{Cu}}{Q_{Cooler}}.dT = \frac{m_{Cu}}{\rho_{Cu}} \int_{T_{op}}^{T_F} \mathrm{K}.dT \qquad [6.11]$$

where $\mathrm{K} = C_{Cu}/Q_{Cooler}$, T_{op} is the operating temperature and T_F is the final temperature that needs to be achieved. This clearly shows that cooling time can be reduced primarily by a balance between the mass of the magnet and the cooling power available. A typical HTS magnet system for electron cyclotron resonance (ECR) ion source (with a pair of HTS coils using BSCCO-2223, operating at 23 K) within the customer's iron structure is shown in Fig. 6.5.

6.6 Thermometry

Temperature measurement and temperature sensors are essential when designing any cryostat system. Within an HTS magnet system, conductor temperature, associated cryogenics, protection from quench and thermal stability, temperature measurement and monitoring are important factors in

6.5 A typical HTS magnet system (Source: Mr Stephen Harrison, Scientific Magnetics, Abingdon, UK).

evaluation of its performance. This section in itself is very wide with books[8,25] and reviews[26] available.

In this section, we will discuss the commonly used temperature sensors in superconducting magnets (primarily between 4 K and 300 K). Generally the following sensors are used in superconducting magnets.

- Diodes: the junction voltage of the semiconductor device increases with decreasing temperature at a constant forward-biased current. The widely used sensors are based on Si and GaAlAs junctions.
- Resistors: resistance thermometers are both available with negative and positive temperature coefficient. For use in a magnetic field, negative temperature coefficient sensors based on germanium, carbon resistors and carbon-glass resistors are preferred over RuO. Positive temperature coefficient sensors, limited to low field applications only, use Pt-100, Pt-alloys and Rh-Fe. Presently Cernox™ (a zirconium oxy-nitride) thermometers are used for a wide range of temperature measurement between 4–300 K).
- Thermocouples: type E and Chromel–AuFe (0.07%) are the two primary thermocouples that have reasonably good signal level and sensitivity in high magnetic fields. These sensors are generally preferred for ac loss measurement employing the thermal/temperature rise method. The challenge of using thermocouples and temperature measurement under the influence of magnetic fields and in ac fields is its accuracy. Apart from the sensor sensitivity, the wiring associated with the sensor and the thermal dumps contributes to the signal-to-noise stability in an ac environment.

Type K thermocouples cover a wide range of temperature from 4–1500 K but have half the sensitivity of type E; type T (Cu–Constantan) are preferred in a vacuum environment, but the stability of type T in cryogenic environments is a challenge because it has high thermal conductivity from the extended copper arms of the sensor.

The challenges in cryogenic measurement are in accuracy/error of measurement. First is the basic sensor accuracy and stability, and second is the error seen due to wiring. In most cases, thermometer instrumentation and calibration are the primary reason for errors in measurement. The thermal connection between the sensor and sample/magnet to the measurement electronics needs detailed analysis before implementation. The following are good practice for any sensor wiring in a cryogenics environment.

- Thermally anchor the sensor to the sample/magnet through a high thermally conducting material interface.
- Ensure proper thermal grease is applied to thermally attach to the holder.
- For temporary sensor mounts (mostly flat sensors), beryllium copper springs are used along with thermal grease to the sensor holder and, in the case of permanent mounts, epoxy and In/In-3%Au solder is preferred.
- Thermal anchoring of the instrument leads serves as an intercept of the heat flow from elevated temperatures. For sensitive measurements, a number of thermal anchor points are advised, and the final heat sink should be close to the sample temperature to avoid thermo-electromotive force (emf) across the leads.
- In case of differential thermal contraction, the use of indium over Pb–Sn soft solder to accommodate the strain effect between sensor and holder is advised.

This section briefly discusses the placement and accuracy of the instrumentation used in any HTS system. Table 6.5 lists sensor types and accuracy that are essential to deciding the measurement schematic and performance monitoring.

For typical cryogenic system tests, the heat load to the system is usually measured based on pressure, temperature and mass flow rate of input and output. In most cryogenics systems, heat loads have been measured on large-scale systems, mostly to cover the issues surrounding the accuracy of measurement. The application of the first law of thermodynamics broadly provides the energy balance for the heat loads on different components. For a large system, the heat-load balance including the cryogenic is given by Eqs 6.12 and 6.13.

$$Q_{TOTAL} = Q_{HTS,System} + Q_{Cryogenics} = \dot{m}(h_1 - h_2) = \dot{m}c_p(T_1 - T_2) \qquad [6.12]$$

$$Q_{HTS,System} = (Q_{Conduction} + Q_{Radiation} + Q_{Joule-heating})_{CL\_Bus} + Q_{HTS} + Q_{Cryostat} \qquad [6.13]$$

Table 6.5 Instrument list and accuracy[8,26]

Type	Range (K)	Magnetic field application	Accuracy
Metallic resistance sensors (positive coefficient)			
Platinum	77–800	Yes, <0.1 % error when used >70 K	±0.6 K at 70 K, ±0.2 K at 300 K (uncalibrated) 20 mK at 77 K, 35 mK at 300 K (calibrated)
Rh-Fe	0.5–900	No, <80 K	10 mK at 4.2 K, 25 mK at 100 K, 35 mK at 300 K (calibrated)
Semi-conductor resistance sensors (negative coefficient)			
Cernox™	0.5–400	Yes, minimum error correction	5 mK at 4.2 K, 20 mK at 20 K, 50 mK at 100 K, 140 mK at 300 K (calibrated)
Carbon-glass	1–300	Yes, with correction	5 mK <10 K, 20 mK at 20 K, 55 mK at 50 K (calibrated)
Diode voltage sensor			
Si diode	1.5–475	No, <70 K	20 mK between 1.5 to 10 K, 55 mK >10 K (calibrated)
GaAlAs diode	1.5–325	Yes, error less than Si diode but >Cernox™	15 mK <20 K, 50 mK at 50 K, 110 mK at 300 K (calibrated)
Thermocouples			
Type E	3–1250	No	1.7 K from 75–273 K
Chromel-AuFe (0.07%)	3–325	No	–

6.7 Development and applications of cryogenics for HTS systems

In general, the demand for HTS systems is focused towards healthcare and energy systems. Most research in industries, national laboratories and universities across the world is primarily working towards electrical system applications of HTSs and has recognised the importance of cryogenics to enable the technology. Presently the research base is focused on the following applications with HTS technology:

- Industrial motor and generator (using Gifford McMahon (GM) single stage coolers and plans for using pulse tube refrigerator (PTR)),
- Transformer (GM coolers and LN2 sub-cooling),
- Cable (LN2 sub-cooling and reverse Brayton cycle cooling),
- Fault current limiter (GM single-stage coolers),
- SMES,
- Magnetic separation,
- MRI,
- Flywheel bearings (using GM single stage and/or two-stage coolers and/or PTR).

In short, principal objectives are focused on:

- Performance, cost and reliability of the cryogenic refrigeration systems that are expected and required for use in HTS power applications,
- Development of the present state of cryogenic refrigeration technology to match the expected needs of the HTS power applications.

6.8 Challenges and requirements

A concern for commercial equipment manufacturers is limited demand for the development of cryogenic systems to cater for individual systems through enhancing/expanding the existing proven technology. Therefore, the primary challenge is to get the large cryogenics system requirement under one umbrella and standardised. This will help in reducing the cost base and improve system reliability rather than individual design to support each HTS system requirement. Another objective is to reduce system heat loads (primarily the static heat load into the vessel/cryostat containing superconducting components) and engineering the dynamic load that arises due to ac loss in most of the HTS power application components.

The current need for research effort is to support the commercial applications of HTSs by improving existing refrigeration technology above 30% Carnot efficiency. This target will focus research and development upon improving compressor performance significantly, and system static heat loads and dynamic heat loads on a case-to-case basis. Coolers are evaluated in terms of the coefficient of performance (CoP). CoP for a cryocooler is normally expressed as the ratio of the work of the refrigeration output to the work of the compressor.

Equation 6.14 shows the theoretical CoP of a GM cryocooler, which is based on the ratio of the *P-V* work to the isothermal compression work.

$$COP_{GM} = \frac{W_{PV}}{W_{ISO}} = \frac{(P_{in} - P_{out}).V}{mRT.\log_e(P_{in}/P_{out})} \qquad [6.14]$$

where W_{PV} is the P-V work per cycle, W_{ISO} is the isothermal compressor work (per cycle), P_{in} is the intake pressure, P_{out} is the exhaust pressure, V is the expansion volume, m is the mass flow rate (per cycle), R is the gas constant for helium, and T is the compression temperature. Reducing the compression ratio of a real GM cryocooler improves CoP[27]. The running cost of a superconducting magnet is about 100 times less than the running cost for an equivalent copper magnet (low-field MRI system) shown in Table 6.6.

6.9 Typical guidelines for cryogen-free magnet systems

6.9.1 Cryogen-free magnets (LTS and HTS)

The cryogen-free magnet is cooled and maintained at its operating temperature by a closed-cycle cryocooler comprising a cold head, electrically driven compressor and connecting gas lines. Note that cryogen-free magnets are often designed with higher operating currents than equivalent wet magnets. The closed-cycle cooler may be of the Joule–Thomson or pulse tube design. Both cool the magnet by conduction cooling and *do not use liquid helium or nitrogen.*

6.9.2 Magnet design principles

A cryogen-free magnet design is generally bigger in size compared with magnets submerged in liquid helium. With wet magnets, heat generated within coils, joints and switches are accommodated within the enormous heat capacity that is available in LHe with little change in conductor temperature. But in cryogen-free

Table 6.6 Typical cost comparison of a copper magnet to superconducting magnet

Type of magnet system	Power (kW)	Energy required (continuous operation) (kW-hr)	Cost (£0.25/kW hr*)
Copper	2000†	17 520 000	£4.38 M
Superconducting	20	175 200	£43.8 K

* Typical cost assumed.
† Joule heating.

6.6 Oxford Instruments 16 T cryogen-free magnet assembly (Source: Oxford Instruments Nanoscience, UK).

magnets, heat generated can only be transferred out by conduction as shown in Fig. 6.6, and relies on a temperature gradient in the direction of heat flow that does not have the same capacity as a wet magnet. Overall heat being extracted in the second stage of the cryocooler results in temperature variation from the heat influx, resulting in significant changes in the operating temperature of the magnet.

The operating temperature challenge calls for the following:

- Systems designed for high operating temperature,
- Coils with a higher current margin, due to generation temperature,
- Higher permissible temperature at coils for better heat extraction (between coils to cooler),
- Coping with the high operating temperature imposed by the rest of the system (excluding the magnet).

6.9.3 Design for low operating temperature

Coils may be smaller and cheaper, and suggest the use of HTS, NbTi, Nb_3Sn (economics is a challenge)[3] as shown in Fig. 6.7. However, smaller coils and reduced coil margins (T_G) may give rise to quenching. The design of thermal extraction is very important to make low T_G designs in order to make this work.

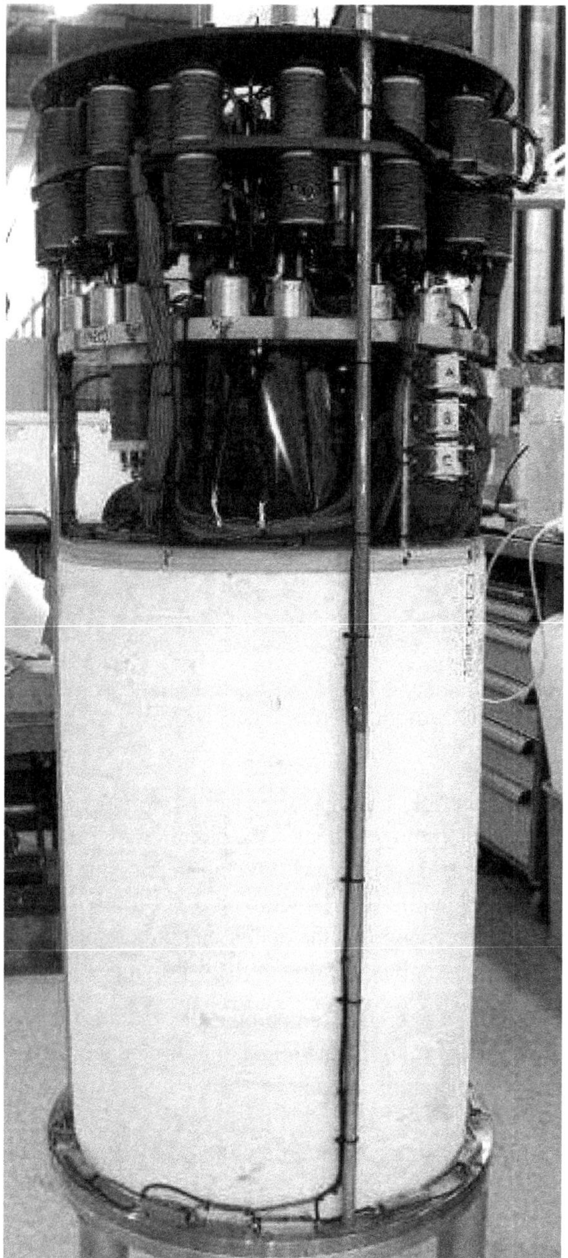

6.7 Oxford Instruments 22.5 T magnet. Hybrid NbTi+Nb_3Sn LTS and 2212-HTS (Source: Oxford Instruments Nanoscience, UK).

There will be a lot of heat generated during the early part of energisation (ac heat load as the magnet is swept through low fields). This heat must be removed at such a rate that the magnet recovers thermally in advance of the magnet reaching high fields.

The generation temperature is normalised for all magnet sections to ensure that no section has an engineering critical current density (J_E) too high, so as to keep the total heat generation (Q_{Gen}) down. Maximised cooling with minimised heat generation gives a low temperature rise during field change and allows a low T_G design resulting in smaller coils with smaller wire. There follow suggestions for maximising cooling and minimising heat generation.

- Longitudinal heat conduction gives direct route to end flanges with minimised eddy current heating,
- Aluminium tape applied longitudinally gives conduction cooling and improved emissivity,
- Material selection (high thermal conductivity),
- Selection of wire: short twist pitch, small filament diameter, matched critical current density,
- High superconductor to non-superconductor ratio,
- Within the structure break eddy current paths, materials,
- Other factors such as energisation rate (rapid energisation generates higher thermal load).

6.9.4 Switch (in case of persistence)

A cryogen-free switch is conduction cooled, opened using a pulse to the switch heater and the voltage of energisation to keep the resistive switch open. This can reduce thermal load rather than having the current through the switch heater continuously during the magnet ramp. Compared to a conventional wet switch, the cryogen-free switch cools from the inside diameter to the outside as the heat sink is on the inside. For HTS, persistence is still a challenge and development needs to address joint technology.

6.9.5 Current leads

The current leads are an important part of the overall system design and compete for space within the cryostat. This is a major connection between the magnet and the outside world carrying energisation current that introduces a lot of heat. Since there is no gas cooling, the method available to cool the leads is by conduction only.

Generally, current terminals at room temperature (RT) have resistive brass leads that are connected to HTS leads at just above the level of the cryocooler first stage (40 K). These brass leads are designed to give the best performance when carrying full magnet current, and the length and cross-section can be adjusted to optimise performance. Two heat loads must be considered: conducted heat due to the temperature difference between the RT vessel and cooler first stage, and joule heating due to current flowing through resistive brass. Brass is used in preference to high-purity copper because the resistivity changes less with change of operating temperature and gives less chance of thermal runaway.

The other ends of the HTS are soldered to conventional copper-stabilised niobium titanium leads that link across to the magnet start and end joints (4 K). Usually the 40 K point is a solid joint so that conducted heat from room temperature and ohmic heating from the current being carried through the brass can be efficiently intercepted. The situation is complicated because the leads need to be electrically isolated but in good thermal contact with the cryocooler at the 40 K and 4 K points, so direct-bonded copper interfaces are used.

The HTS sections of the leads between 40 K and 4 K are designed to carry all of the current in the superconductor so there is no joule heating. Conducted heat is limited by using Au-Ag substrate HTS material mounted on brass shunt protection. The NbTi link from the current leads to the joints of the magnet is kept as short as possible with sufficient cross-section so that the NbTi remains cold enough for it to be superconducting under operating conditions to take full magnet current, should the leads quench and the magnet not follow immediately.

6.9.6 Protection

The design should be made assuming that magnet and protection work together adiabatically during a quench and there is no cryogen to take away heat. Unless provided otherwise, protection circuits should be designed conservatively so that they do not overheat. Wire should be oversize to take the current and to aid cooling of the components by providing conduction paths (resistors and diodes are all mounted on insulators).

6.9.7 Design and manufacture of joints

Joints between coils and wire grades are likely centres of heat generation. Low resistance joints are preferred to minimise heating. Long-lap soldered tin joints and etched or spot welded NbTi joints can minimise heat generation in LTS but this is still a challenge for HTS Joint technology is not matured and the resistance is not comparable with LTS. HTS magnets are not persistent magnets, therefore they are run on the power supply and held across the power supply during operation. To conduct heat away from joints use heavy short lead runs and thermal braids from start and end joints to the thermal bus (electrically isolated). Thermal interception is particularly important if the magnet has a persistent mode switch (in LTS, heated when field is being changed). The use of joint cups with large thermal mass for the start and end joints may prove useful. These will form a thermal buffer against heating from energised current leads and heat conducted from an open switch.

6.9.8 Thermal shielding and anchoring

The radiation shield is the part of the thermal shield of the cryostat that is fixed to the first stage of the cryocooler and enables it to run at a temperature of about 40 K. In larger magnets a shield is normally applied around the magnet, joints, protection circuit and current leads and anchored to the second stage temperature. This shield protects the second stage components from radiation and with electrically insulating material (Kapton™) to stop short-circuiting coil terminations, current leads, and protection circuits to each other or ground.

6.10 Conclusions

The technology is generally proven, but enabling the highest cooling power demands is challenging. The primary challenge is to bring the cost down with the enhanced technology. The most distinctive feature of HTS over LTS is the operating temperature envelope. YBCO, BSSCO and MgB_2 have critical temperatures over 39 K. Therefore, a less intrusive cryogenic system above 4.2 K gives an added advantage working with HTS materials, because the running cost of these cryogenic systems decreases with increasing operating temperature. But in most applications targeted using HTS as discussed before, ac losses dominates cryogenic efficiency. This chapter only gives an overview of applied superconductivity and demands for cryogenics for beginners.

6.11 Sources of further information and advice

Practical Cryogenics, An Introduction to Laboratory Cryogenics (2000), Oxford Instruments Nanoscience, Abingdon (UK).

Reference Guide – Temperature measurement and control, Lake Shore Cryotonics, Inc (2002), Ohio (USA).

D A Cardwell and D S Ginley, *Handbook on Superconductors*, Vol. II, Inst. of Physics Press, Bristol (UK) (2003).

A Goyal, *Second-Generation HTS Conductors*, Kluwer Academic Publishers, Dordrecht (2004).

M N Wilson, *Superconducting Magnets*, Clarendon Press, Oxford (1983).

T S Sheahen, *Introduction to High Temperature Superconductivity*, Plenum Press, NY (1994).

Y Iwasa, *Case Studies in Superconducting Magnets – Design and Operational issues*, 2nd ed., Springer, NY (2009).

6.12 Acknowledgement

I would like to thank all my colleagues at Oxford Instruments NanoScience, particularly Dr Ziad Melhem, Mr Robin Brzakalik, Dr Steve Chappell, Mr Joe

Brown, Mr Andre Delieu, Mr Harish Agrawal, Dr Rod Bateman, Dr Tim Foster, Dr Sophie Walker and Cryogenics/ULT team for the open discussions we had in the past in understanding complication the hard way. Finally, I would like to thank my wife Dr Renuka Rajput-Ghoshal and my daughter Rudrani who made it possible for me to write at home with minimum household chore activities.

6.13 References

1. K Timmerhaus and T Flynn, *Cryogenic Process Engineering*, Plenum Press, NY (1989).
2. G K White and P J Meeson, *Experimental Techniques in Low-Temperature Physics*, 4th ed, Oxford University Press, Oxford, UK (2002).
3. *Superconductor Week*, 11, p2 (September 2008).
4. J W Ekin, S E Russek, C C Clickner, and B Jeanneret, '*In situ* noble metal YBa2Cu3O7 thin-film contacts' *Applied Physics Letter*, **62**, p369 (1993).
5. Y Hu, *et al.*, *Advances in Cryogenic Engineering (Materials)*, **44**, p381.
6. Y S Cha, M T Lanagan, K E Gray, V Z Jankus, Y Fang, 'Analysis and interpretation of critical current experiments for bismuth-based high-temperature superconductors made by power-in-tube processing', *Appl. Supercond*, **2**, p47 (1994).
7. J W Ekin, T M Larson, N F Bergren, A J Nelson, A B Swartzlander *et al.*, 'High T_c Superconductor/noble-metal contacts with surface resistivities in the 10–10 Ωcm^2 range', *Applied Physics Letter*, **52**, p1819 (1988).
8. J W Ekin, *Experimental Techniques for Low Temperature Measurements*, Oxford University Press, NY (2006).
9. P K Ghoshal, T A Coombs and A M Campbell, 'Calorimetric method of ac loss measurement in a rotating magnetic field', *Review of Scientific Instruments*, **81**, 074702 (2010).
10. J W Ekin, 'Offset criterion for determining superconductor critical current', *Applied Physics Letter*, **55**, p905 (1989).
11. A Godeke, H J G Krooshoop, B Haken Ten, H H J Kate Ten, 'Experimental verification of the temperature and strain dependence of the critical properties in Nb_3Sn wires' *IEEE Trans. Appl. Supercond.*, **11**, p1526 (2001).
12. D P Hampshire, H Jones and E W J Mitchell, 'An in-depth characterization of $(Nb,Ta)_3$ Sn filamentary superconductor', *IEEE Trans. Magnetics*, MAG21, p289 (1985).
13. B Ten Haken, 'Strain effects on the critical properties of high-field superconductors', *PhD thesis*, University of Twente, Enschede, The Netherlands (1994).
14. F Kreith and M S Bohn, *Principles of Heat Transfer*, 6th ed., Brooks/Cole, Pacific Grove, CA (2001).
15. R F Barron, *Cryogenic Systems*, 2nd ed., Oxford University Press, NY (1985).
16. A Roth, *Vacuum Technology*, Elsevier Science B.V., Amsterdam (1990).
17. J R Hull, *Cryogenics*, **29**, p1116 (1989).
18. A Ballarino, 'High temperature superconducting current leads for the large hadron collider', *IEEE Trans. Appl. Supercond.*, **9**, p523 (1999).
19. M. Lakrimi, J Brown, P Cetnik, M Wilkinson, D Clapton *et al.*, 'Low boil-off HTS current leads', *IEEE Trans. Appl. Supercond.*, **17**, p2270 (2007).
20. R P Reed and A F Clark, *Materials at Low Temperatures*, American Society for Metals, Metals Park, Ohio (1983).

21. R Radebaugh, E D Marquardt, J Gary, A O'Gallagher, 'Regenerator Behavior with Heat Input or Removal at Intermediate Temperatures' *11th International Cryocooler Conf.*, Keystone, CO, 20–22 June 2000, *http://www.cryogenics.nist.gov*, (2001).
22. The Goodall Chart, APT Division, UKAEA Culhum Laboratory, Abingdon, England (2009) (British Cryogenic Council, www.bcryo.org).
23. *The Vacuum Technology Book*, Pfeiffer vacuum – V1, Germany (www.pfeiffer-vacuum.net).
24. G F Weston, *Ultra High Vacuum Practice*, Butterworths, London (1985).
25. F Pobell, *Matter and Methods at Low Temperature*, 2nd ed., Springer-Verlag, Berlin (1996).
26. L G Rubin, 'Cryogenic thermometry: A review of progress since 1982', *Cryogenics*, **37**, p341 (1997).
27. H Nakagome, T Okamura and T Usami, 'Research on improvement in the efficiency of the GM refrigerator', *Cryocoolers* **14**, International Cryocooler Conference Press, Boulder, CO, p187 (2007).

7

Electromagnetic modeling of high temperature superconductor (HTS) materials and applications

F. GÖMÖRY, Slovak Academy of Sciences, Slovak Republic

Abstract: The modeling of electromagnetic phenomena is essential in the design of electric power devices. For the applications of high temperature superconductors (HTS) the critical state model used for low temperature superconductors (LTS) provides a valuable starting point. Consistent with the more complex behavior of HTS is the incorporation of such features as the magnetic anisotropy, non-uniformity, thermal activation and close contact with ferromagnetic material. Specific numerical techniques have been developed to cope with these tasks. This chapter provides an overview of these methods and shows their use in practical examples.

Key words: finite element calculations in superconductors, critical state model, AC losses.

7.1 Introduction

The main reason for using superconductors in electrical engineering is their particular ability to carry large electrical currents. This feature is linked to their unique electromagnetic property – the pinning of magnetic flux in the volume of superconducting material. Superconductors exhibiting this property are called hard superconductors. This term has been introduced (in analogy to magnetic materials) to underline the fact that flux pinning leads to a hysteresis of macroscopic magnetization with respect to applied magnetic field changing in time. The physical mechanism behind this is now quite well understood; however it is not easy to link the interaction between one flux line and involved pinning centres with the quantitative prediction of macroscopic properties. In the investigation of low temperature superconductors (LTS) such as NbTi or Nb_3Sn this problem was resolved by the phenomenological description introduced by Bean (Bean 1962) commonly known as the critical state model.

Modeling the electromagnetic behavior of HTSs is based on the knowledge developed during the investigation of LTS. Therefore we start with the basic formulation of the critical state model: locally, in any part of a hard superconductor one can find either no electrical current, or the current with density J equal to the so called critical current density, J_c. Its value is the property of material and is used either fully or not at all:

$$|J| = \begin{cases} 0 \text{ in the places that never experienced electrical field} \\ J_c \text{ elsewhere} \end{cases} \quad [7.1]$$

The first applications of hard superconductors appeared in the 1960s, when analytical models were used in many fields of engineering. Within such an approach, the methods for designing of DC superconducting magnets (Wilson 1983) have been developed, principal formulas for the estimation of AC dissipation in superconducting devices derived (Norris 1970, Campbell 1982, Carr 2001) and first numerical techniques proposed (Ashkin 1979). It is incredible how well these analytical models served the engineers in research and development of large-scale applications of superconductors. With the rapid development of computer technology in both computing power and price affordability in the 1990s, numerical methods started to appear in the field of hard superconductors addressing several issues:

- Electromagnetic properties of superconductors are non-linear, hence analytical methods provide solutions that could be approximations only.
- Distribution of electromagnetic quantities in superconductors depends critically on its shape, but analytical methods resolve only few simplified geometries.
- In comparison with LTSs, the critical current density J_c in HTSs is often such a complex function of electrical field, magnitude and direction of magnetic field, position, etc. that the use of a constant value for J_c is too a crude simplification.

Current activity in numerical modeling of superconductors is quite intensive. At the first international workshop on numerical modeling of HTS in Lausanne, Switzerland, 5–7 May 2010, about 50 participants decided to form an international network that would organize regular scientific meetings, exchange knowledge and develop benchmarks and standards.

This chapter gives an overview of the methods used to model the interaction of magnetic flux with hard superconductors. These methods are for the prediction of properties of materials, elements, devices, development of an optimal design and interpretation of experimentally observed behavior. The intention is to provide the reader who is not an expert in the field of numerical simulations in superconductors with the knowledge necessary to formulate the problem and correctly estimate the expertise that is necessary to resolve it. In many cases there exists a simplified approach that could be easily implemented using a commercial software tool. Selected examples of such calculations helping to understand the basic features of the investigated problem will be shown.

The shapes of HTSs are either bulks consisting exclusively of HTS material or composites containing other materials. In the case of multifilamentary wires it is sometimes desirable to see HTS filaments separately (Oota *et al.* 1995, Johnston *et al.* 1997, Fabbricatore *et al.* 2000, Stavrev *et al.* 2003, Inada *et al.* 2004, Dutiot *et al.* 2005, Inada *et al.* 2007). Alternatively one can treat the whole filamentary zone as one superconducting material with an effective critical current density (Amemiya *et al.* 1998a)

$$J_c = \lambda J_{c,SC} \qquad [7.2]$$

where $J_{c,SC}$ is the critical current density of the superconducting material and λ is the fraction that the superconductor occupies in the filamentary zone. Such a transformation is illustrated in Fig. 7.1. This approach is fully justified when the filaments are straight (not twisted) and the redistribution of current is easy because of a highly conductive matrix. In the case when an external magnetic field varying in time is applied, the importance of currents forming the electromagnetic response of the filamentary zone as a whole can be considered by determining the coupling length, l_c (Wilson 1983, p. 175). This is the distance at which the transversal current connecting individual filaments across metallic matrix reached the longitudinal transport capacity of superconducting portion:

$$l_c = \sqrt{\frac{8J_{c,SC}d_f\rho_m}{dB/dt}} \qquad [7.3]$$

where d_f is the typical transversal dimension of a single filament (in the case of a round filament this is the filament diameter), ρ_m is the electrical resistivity of the matrix and dB/dt is the rate of change of applied magnetic field. To see the electromagnetic response of individual filaments, either the wire should be much shorter than the coupling length or the filaments must be twisted with the pitch $l_p << l_c$.

Another simplification that is common in numerical calculations of HTS electromagnetic properties is the use of two-dimensional formulation (2D) instead of three-dimensional representation (3D). This means that when we calculate the critical current or AC loss in an HTS tape (which is always of finite length) the problem of electrical current transfer from the terminations is neglected. Generally one assumes that the tape is of infinite length, and the distribution of electrical current and magnetic field resolved in the cross-section is repeated everywhere. This is schematically shown in Fig. 7.2. This geometry will be used as the reference case in this chapter unless mentioned otherwise.

7.2 Overview of modeling approaches for HTS materials

The first samples of HTS were prepared by sintering the powders of precursors to obtain a pellet of ceramic material with random orientation of grains and poor

7.1 Cross-section of HTS tape containing $Bi_2Sr_2Ca_2Cu_3O_x$ filaments in silver matrix (left) and the replacement of its filamentary zone by an elliptic core with the same critical current (right).

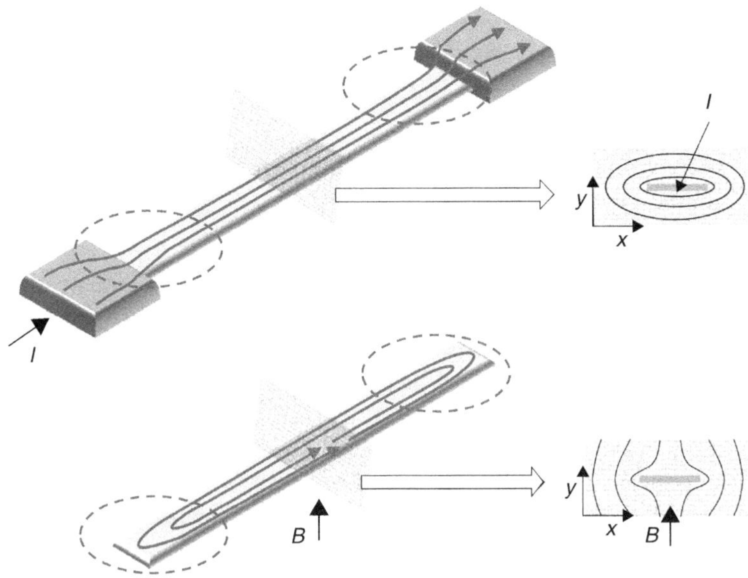

7.2 Two-dimensional approximation of HTS tape transporting electrical current *I*, or exposed to magnetic field *B*. The existence of end effects is neglected.

electrical contacts between them. Now, inter-grain currents achieve a level allowing the existence of intragrain currents to be neglected (Amemiya *et al.* 1998b). The possibility to use Bean's critical state model for HTS materials was early recognized (Clem 1988, Gömöry and Lobotka 1988). The formula shown in Eq. 7.1 is the first approximation of their properties. This also means that the extensive studies of magnetic field interaction with LTSs provided a valid base for further improvements necessary to cope with the properties of HTSs. One such significant achievement from the LTS period is the modification assuming that the value of J_c in Eq. 7.1 is not constant because of dependence of critical current density on the magnetic field B. Among the empirical models the formula suggested by Kim *et al.* (Kim *et al.* 1962) acquired the widest acceptance:

$$J_c = \frac{J_{cB0}}{1+\dfrac{B}{B_0}} \qquad [7.4]$$

This expression contains two parameters: J_{cB0} is the asymptotic value that J_c would reach at zero magnetic field ($B = 0$) and B_0 is the magnetic field below which the critical current density smoothly saturates (Jiang *et al.* 2006). Equation 7.4 allows analytical results to be obtained for very specific geometries or in the hyperbolic approximation $B >> B_0$ (e.g. Babaei Brojeny and Clem 2005).

It is now commonly recognized that, in the case when the original model with J_c = constant does not provide valid predictions, the following modifications should be considered:

- Incorporation of thermal activation,
- More complex expression for field dependence,
- Recognition of anisotropy,
- Non-uniformity of superconductor properties.

The critical state – when the gradients of magnetic field persist – is thermodynamically metastable, with a tendency towards uniform distribution of vortices. At liquid helium temperatures the thermal energy providing the individual vortices' fluctuating energy component is too low compared to the height of barriers confining the positions of vortices. However, at 77 K the thermal energy is ~20 times higher and the thermal activation is appreciable, leading to the phenomenon that acquired the name 'giant flux creep' (Yeshurun and Malozemoff 1990). The most practical way of incorporating the thermal activation for the purpose of numerical modeling is the use of an electrical field-dependent critical current density instead of a constant value:

$$J_c = J_{c0E}\left(\frac{E}{E_c}\right)^n. \qquad [7.5]$$

In this expression, J_{c0E} is the critical current density determined at the electrical field E_c and n is the exponent characterizing the importance of thermal activation. It roughly corresponds to the proportion between the height of the energy barrier opposing the movement of vortices, U_b, and the energy of thermal activation k_BT (where k_B is the Boltzman constant 1.38×10^{-23} J/K), $n \approx U_e/k_BT$. Conventionally in the determination of critical current density the values of $E_c = 10^{-4}$ or 10^{-5} V/m are used. At reaching this level of electric field it is assumed that all of the cross-section of HTS wire is filled with critical current. Equation 7.5 has been shown to be representative of commercial HTS tapes, and this has been confirmed by many experiments (Rhyner 1993, Bossavit 1994, Amemiya *et al.* 1997). The value for n could significantly drop with magnetic field: for multifilamentary tapes containing $Bi_2Sr_2Ca_2Cu_3O_{10}$ in Ag matrix (this material is often referred to as Bi-2223/Ag) the reduction from n = 18 determined at zero magnetic field down to n = 4 at 0.2 T applied in a perpendicular direction with respect to the tape's wide face was observed (Amemiya *et al.* 1998b).

The critical currents in the first sintered HTS materials sharply decreased in a superimposed DC magnetic field. With the improvement of inter-grain contacts the sensitivity to magnetic field has also been reduced. The behavior of a wide range of samples could be covered by adding one more parameter to Kim's model:

$$J_c = \frac{J_{cB0}}{\left(1+\dfrac{B}{B_0}\right)^{\beta}} \qquad [7.6]$$

with Kim's model corresponding to $\beta = 1$, and the original field-independent Bean model is recovered using $\beta = 0$.

Anisotropy of superconducting properties is not surprising in HTS materials when one takes into account the highly anisotropic shape of the elementary crystal cell. As a consequence of this anisotropy, the pinning force depends on the orientation of vortices. This is observed in the determination of the critical current density's dependence on applied magnetic field for the difference between critical current density when magnetic field is applied parallel to the Cu-O planes and that measured with the magnetic field oriented perpendicular to the Cu-O planes (Stavrev *et al.* 2002a). Moreover, different pinning centers of anisotropic shape (e.g. columnar or planar defects) are often incorporated. As a result of all these factors, the angle dependence of pinning force could be rather complex. Depending on the type of numerical modeling, two approaches are then possible: when the field dependence is an important factor, e.g. in the case when no additional DC magnetic field is applied, or when the material anisotropy is extremely high; it is recommended that $J_c(B,\alpha)$ is inserted in a tabular form so that the numerical calculation could be used for the interpolation in every point. This approach could be time-consuming, and the convergence of calculations not secured; therefore it is recommended to start with a rough description of the anisotropy by the elliptical formula

$$B_{eff} = \sqrt{B_{\parallel}^2 + k_a^2 B_{\perp}^2} \qquad [7.7]$$

with the definition of directions (parallel, perpendicular) with respect to Cu-O planes and k_a the parameter characterizing the anisotropy. Merging the expressions Eqs 7.6 and 7.7 into one single formula leads to representation of J_c containing four parameters

$$J_c = \frac{J_{cB0}}{\left(1 + \frac{\sqrt{B_{\parallel}^2 + k_a^2 B_{\perp}^2}}{B_0}\right)^{\beta}} \qquad [7.8]$$

An important achievement of using these numerical methods is the possibility of accounting for a spatial variation of the superconductor properties (Stavrev *et al.* 2002b, Tsukamoto 2005, Nishioka *et al.* 2005, Kimura and Amemiya 2007, Inada *et al.* 2008, Nguyen *et al.* 2009). The simplest way is to include a modifying function $K_h(x,y)$ in the characterization of the critical current density

$$J_c = K_h(x, y) J_{c00} \qquad [7.9]$$

For example, in Gömöry *et al.* (2007) it has been found that the agreement between experimental and calculated critical currents for Bi-2223/Ag tape could be improved by taking into account the non-uniformity of superconducting filaments by an empirical expression

$$J_c = \left(1 + \frac{k_x^2}{a_{SC}^2} + \frac{k_y^2}{b_{SC}^2}\right) J_{c00} \qquad [7.10]$$

where J_{c00} is the value of critical current density in the centre of the core and the increase in two main directions is characterized by two independent parameters k_x and k_y.

To summarise the modifications of the original critical state model with constant J_c, the complete expression for critical current density in HTS materials acquired the following form

$$J_c = J_{c0} K_B(\vec{\boldsymbol{B}}) K_E(E) K_h(\vec{\boldsymbol{r}}) \qquad [7.11]$$

where the modifying function for the magnetic field dependence is

$$K_B(\vec{\boldsymbol{B}}) = \frac{1}{\left(1 + \frac{\sqrt{B_{\parallel}^2 + k_a^2 B_{\perp}^2}}{B_{é}}\right)^{\beta}} \qquad [7.12]$$

Thermal activation is taken into account by introducing the term

$$K_E(E) = \left(\frac{E}{E_c}\right)^{\frac{1}{n}} \qquad [7.13]$$

and the non-uniformity of current-carrying capability of the superconducting material is accounted for by the function of space coordinates $K_h(\vec{\boldsymbol{r}}) = f(x, y, z)$ describing the variations of the local critical current density in the superconducting volume.

The essential step in formulation of the appropriate model for a given problem regarding HTS materials is the proper choice of complexity in expressing the critical current density. The use of Eq. 7.11 should in principle cover all expected phenomena, but very probably would lead to excessive computing time and demands on computing resources. As the first step it is therefore recommended to estimate which of the modifications is expected to play the most significant role in the modeled process. For example, when the range of actual magnetic fields inside the HTS material is significantly suppressed by the application of a strong background DC field, one can use the approximation $K_B(\vec{\boldsymbol{B}}) = 1$. Unless there is a plausible reason to expect that the superconducting properties vary in the HTS volume, the hypothesis of material uniformity $K_h(\vec{\boldsymbol{r}}) = 1$ is a good starting point. Also, in the author's experience, the flux-creep effects do not bring significant modification to the prediction of AC losses in commercial HTS materials at 50 Hz, and the models assuming $K_E(E) = 1$ give valid predictions. On the other hand, at frequencies in kilohertz or megahertz range the use of the electric-field-dependent critical current density in HTS materials would be necessary. In the following sections of this chapter, examples of modeling problems that always use a simplification of the general expression Eq. 7.11 will be shown.

7.3 Modeling methods for DC conditions

The situation when an HTS carries DC is generally a metastable state, because the distribution of the trapped magnetic field depends also on the history. The calculation that would take this into account will be dealt with in the next section. This section shows how to model the only situation in the DC regime that is not ambiguous: reaching the critical current.

The reason why the critical current is a single-valued quantity is easily seen in the original Bean formulation (Eq. 7.1). Assume an HTS wire with the superconductor cross-section S_{SC} perpendicular to the current flow. When a DC current is supplied to the wire, its absolute value could be either lower than the critical current I_c, or equal to I_c. In the first case, not all of the cross-section of the wire is used to transport the electrical current, and the voltage between the wire ends is zero. At reaching the critical current – noticed by an appearance of voltage at the wire terminals – all of the superconducting portion of the wire section is transporting the critical current density, thus

$$I_c = \int_{S_{SC}} J_c \, dS \qquad [7.14]$$

In the simplest approach that assumes J_c = constant,

$$I_c = J_c S_{SC} \qquad [7.15]$$

This expression is widely used in the experimental determination of critical current density. Because it assumes a constant J_c one should check if the local magnetic field produced by the transport current is negligible compared to the external applied field. Often such a value is a characteristic that would not necessarily correspond to any portion of the investigated sample.

With the development of commercial finite-element method-based software, the incorporation of critical current dependence on magnetic field or electrical field is possible. This will be shown in three practical examples.

Example 1 is low-field critical currents in Bi-2223 tape. Let us analyze the situation when an HTS wire carrying DC current is placed in a static magnetic field B_a, as shown schematically in Fig. 7.3. When the value of such a field is comparable or even lower than the field produced by the critical current, the distribution of local magnetic fields is rather complex (Majoros *et al.* 2001, Stavrev *et al.* 2002a). The most important factor that should be included in the model is the field-sensitive critical current density such as, for example, given by Eq. 7.8. This means we start using

$$K_B(\vec{B}) = \frac{1}{\left(1 + \frac{\sqrt{B_{\parallel}^2 + k_a^2 B_{\perp}^2}}{B_{é}}\right)^{\beta}} \qquad [7.16]$$

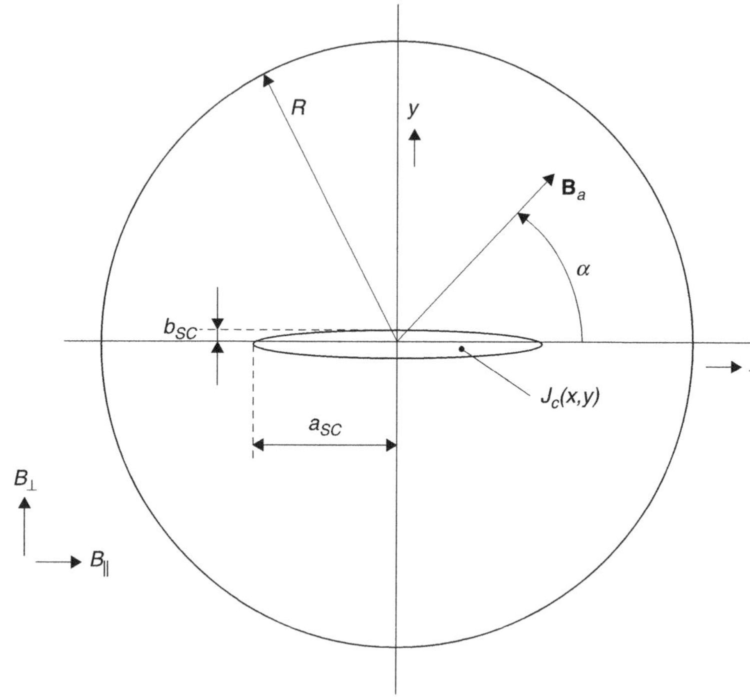

7.3 Typical geometry used in this chapter. The superconductor (in this case of elliptic cross-section with semi-axes a_{SC} and b_{SC}, respectively) is placed coaxially in the calculation box of radius R.

where $K_h(\vec{r}) = 1$; $K_E(E) = 1$. The finite-elements must allow the dependence of material properties on the magnetic field components. In this case the filamentary zone is replaced by an elliptical core limited by two main semi-axes a_{el}, b_{el}, and with the equivalent critical current density given by Eq. 7.2.

The uniform magnetic field with absolute value B applied at the angle α with respect to the horizontal direction could be set, for example, by the boundary condition for the vector potential of the magnetic field at the boundary of the calculation box, A_b:

$$A_b = B(y\cos\alpha - x\sin\alpha) \qquad [7.17]$$

The use of vector potential **A** defined as the rotor of the magnetic flux density **B** by the expression $\vec{B} = \nabla \times \vec{A}$ is favored in two-dimensional formulations because then it contains only one component in the direction perpendicular to the drawing plane in Fig. 7.3. In the cross-section of the elliptical core the current density J_c is imposed as the function of the local magnetic field according to the combination of Eqs 7.6, 7.8 and 7.11, thus it is not uniform in the core. The finite element software resolves the magnetostatic approximation of the Maxwell equation

$$\nabla \times \left(\frac{\vec{B}}{\mu} \right) = \nabla \times \left(\frac{1}{\mu} \nabla \times \vec{A} \right) = \vec{J} \qquad [7.18]$$

where μ is the magnetic permeability and J the density of electrical current. In the case of a hard superconductor the Meissner currents and the existence of the lower critical magnetic field B_{c1} can be neglected and the permeability of vacuum, $\mu_0 = 4\pi \times 10^{-7}$, is inserted for μ. An example of calculated distribution of magnetic field and current density is shown in Fig. 7.4.

From the calculated distribution of electrical current, J_{calc}, the critical current is determined with the help of Eq. 7.14, taking into account that the calculation has been done for the effective critical current density in the whole filamentary zone:

$$I_c = \int_{S_{zone}} J_{calc}\, dS_{zone} \qquad [7.19]$$

From the series of calculations performed for a set of applied magnetic fields at different orientations the field-dependence of critical current can be calculated as shown in Fig. 7.5, and compared with experimental data.

Example 2 is the distribution of current density in porous superconducting film. When we are interested in seeing the effect of pores, a practical geometry for the model is that shown in Fig. 7.6. Similar to the previous case, the real three-dimensional problem has been reduced to two dimensions assuming, instead of a film with finite thickness in the direction perpendicular to the drawing plane, a rectangle with perfect translational symmetry in the z-direction. The porosity of film has been inserted by the series of squares. Except for the holes the superconducting material itself is uniform, and in the first approximation we neglect the influence of the magnetic field produced by the current. Therefore we use

$$K_B(\vec{B}) = 1; \quad K_h(\vec{r}) = 1; \quad K_E(E) = \left(\frac{E}{E_c} \right)^{\frac{1}{n}} \qquad [7.20]$$

The governing equation for the flow of electrical current links the local current density with the gradient of electrostatic potential, V, through the resistivity ρ as

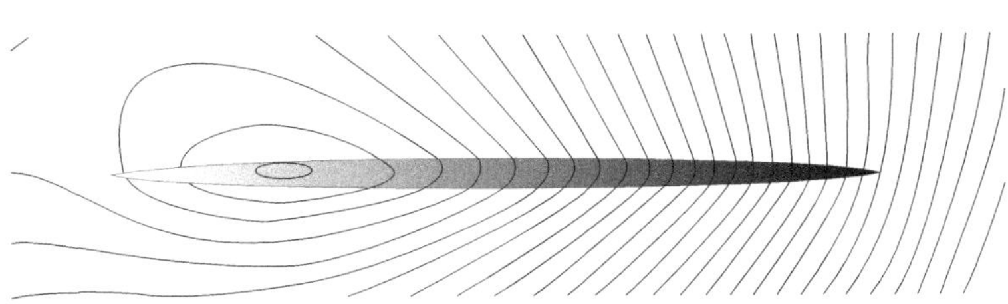

7.4 Distribution of critical current density in the elliptic core of HTS tape (grey scale) and the lines of constant vector potential of magnetic field in for the wire transporting critical current in magnetic field applied at the angle of 30° with respect to the horizontal plane.

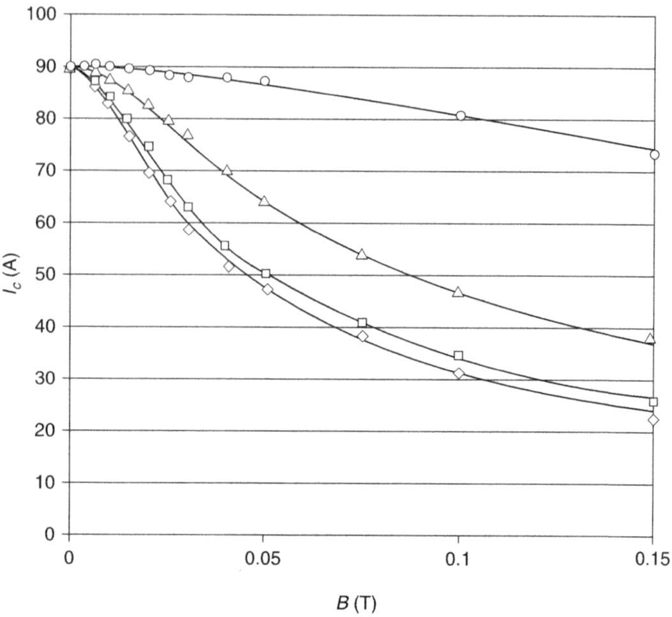

7.5 Dependence of critical current on magnetic field determined from the current distributions such as that shown in Fig. 7.4 calculated for different values of applied magnetic field, *B*, applied at the angles 0°, 30°, 60°, 90° (from top to bottom) with respect to the longer dimension of the elliptic core.

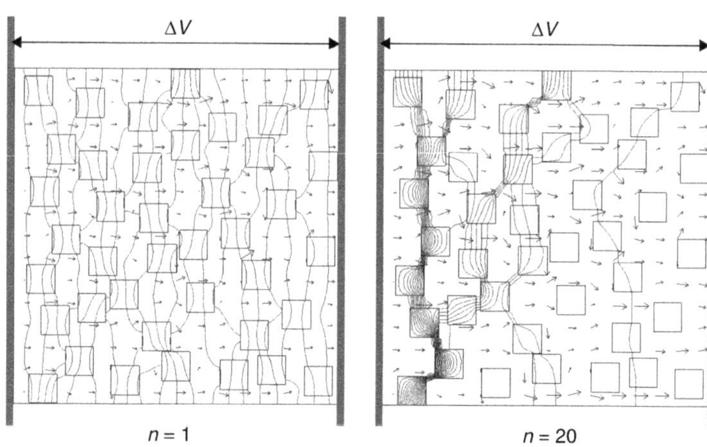

7.6 Distribution of electric field (equipotential lines) and current density (arrows) in a thin film containing holes of quadratic shape. On the left-hand side the normal metal with constant resistivity exhibits essentially uniform distribution of electrical field. In the case of a superconductor with the power-law current–voltage curve shown on the right-hand side one can see the concentration of electrical field in so-called flux channels.

$$\vec{J}(x, y) = \frac{\nabla V(x, y)}{\rho}. \qquad [7.21]$$

The resistivity of the superconductor can be derived from Eq. 7.5 as

$$\rho_{SC} = \frac{E}{J} = \frac{E_c}{J^n_{c0E}} J^{n-1}. \qquad [7.22]$$

The holes have resistivity $\rho_{hole} \rightarrow \infty$. At the two opposite faces of the sample is applied a difference in electrical potential, ΔV. The finite element calculation (Bartolome *et al.* 2007) for the distribution of electrical current density performed by resolving Eq. 7.21 will give different results for different values of n, as shown in Fig. 7.6. In the case of $n = 1$ corresponding to a normal metallic material, the distribution of electric potentials is roughly uniform. In the case of a superconductor with $n = 20$ the same porosity pattern results in a concentration of the electric field causing a so-called 'flux jet' (Friesen and Gurevich 2001) close to the electrode on the left.

Example 3 is the calculation of the critical current of a superconducting coil. It is assumed that all the turns in a coil are connected in series, and thus carry the same current. The distribution of magnetic field can then be calculated for any current, allowing identification of the place in the winding experiencing the highest magnetic field, B_{max}. The 'load curve' represents the dependence of the magnetic field in that locality upon current. Intersection of the load curve with the $I_c(B)$ curve of the conductor used to wind the coil, as shown in Fig. 7.7, determines

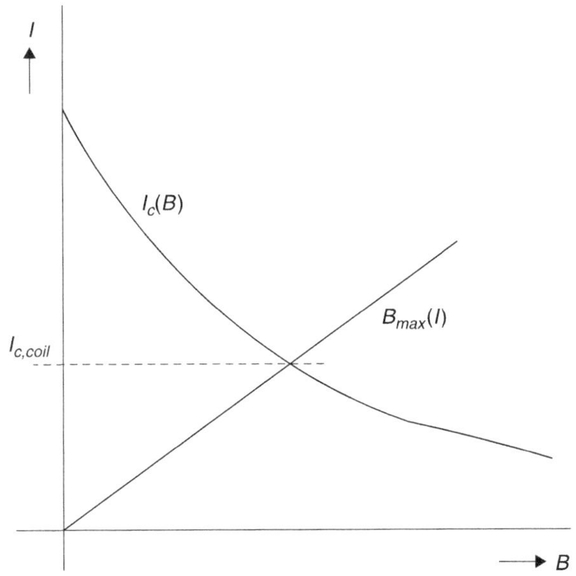

7.7 The intersection of $I_c(B)$ dependence with the load curve of a superconducting coil $B_{max}(I)$ provides the estimation of maximum current in the coil.

the maximum current before the appearance of a resistive voltage, sometimes called the critical current of the coil. The corresponding value on the B axis indicates the upper limit of magnetic field found in the coil winding.

In the case of HTSs, because of strong dependence of critical current on the magnetic field direction, the place with maximum absolute value of magnetic field will not indicate automatically the first place where the critical current is reached (Pitel *et al.* 2002). One can cope with this problem in the following way. After defining the geometry of the winding, its cross-section is considered as filled with uniform current density J_1. The conductor is characterized by the (anisotropic) field-dependent critical current, e.g. according to the approximation

$$I_c(\vec{\boldsymbol{B}}) = \frac{I_0}{\left(1 + \frac{\sqrt{B_{\parallel}^2 + k_a^2 B_{\perp}^2}}{B_{\acute{e}}}\right)^{\beta}}. \quad [7.23]$$

For the cross-section S_{turn} that one turn of the coil occupies in the winding, the engineering critical current density will be given as

$$J_{c,eng} = I_c / S_{turn} \quad [7.24]$$

and compared with the current density J_1 assumed in the calculation of magnetic field distribution. In the case $J_{c,eng} < J_1$ in any part of the winding, the probing current density J_1 is below the limit that the coil could carry. In the next step the same analysis is performed with the probing current density $J_2 > J_1$, and vice versa; when observing $J_{c,eng} > J_1$ in a significant portion of the winding, the probing current density should be reduced. This is repeated until the current density in some part of the winding overcomes the critical value (Eq. 7.24). The choice for the portion of winding with $J_{c,eng} > J_1$ that should stop the procedure for determining the critical current of the superconducting magnet is not critical: in the case of a simple solenoidal magnet the critical current found at the requirement that 0.0001% (or 10^{-6}) of the winding carries the current beyond the critical value was only 0.2% less than that found when using a criterion of 0.01%. In Fig. 7.8 are shown the results of such a calculation for a simple solenoid. The portion of winding where the critical current density has been superseded is indicated in solid black. Figure 7.8a is the result found for a coil made from coated conductor tape with weak anisotropy of critical current density, ($k_a = 0.75$). The overcritical zone develops rapidly from the median plane of the solenoid, inner turns in the first order. In contrast, when assuming the same coil to be wound from a generic Bi-2223 tape with much stronger anisotropy characterized by $k_a = 0.1$, the first part to reach the critical current are the upper and lower extremities, as shown in Fig. 7.8b.

7.8 Regions transporting higher than critical current (black) in the cross-section of HTS coil at the currents of 56, 58, and 60 A (from left to right). Only one half of the cylindrical winding is shown; the symmetry axis is vertical. Superconducting turns are circular, larger dimension of the tapes in vertical direction: (a) coil from HTS material with negligible anisotropy like YBCO-coated conductor tapes, (b) coil made from Bi-2223 tape.

7.4 Modeling methods for AC conditions

The critical state model (Eq. 7.1) was developed to describe the behavior of a hard superconductor in a magnetic field varying in time. Therefore in the modeling of AC conditions the superconductor is characterized by its critical current density that, in the simplest approach, is a constant. However, in the analysis taking into account the key factors influencing the electromagnetic behavior of HTSs the generalized expression Eq. 7.10 is to be used.

AC excitation of an HTS – by AC current or AC field or both simultaneously – is a complex process resulting in the cyclic variation of magnetic field distribution in the volume of superconductor, but also around it. The quantity of interest in many applications is the heat dissipated because of flux pinning in the superconductor volume, commonly called the AC loss. The loss behavior, i.e. the dependence of energy dissipated during one AC cycle on the excitation current or field, has two uses. Firstly, its value is used to calculate the required cooling power, and is essential for the engineer designing a superconducting device. Secondly, the analysis of loss behavior is used as a diagnostic tool, allowing understanding of the magnetic field interaction with the superconductor.

The modeling method calculates the distributions of physical quantities (magnetic field, current density, electric field and vector potential of magnetic field) for a limited number of instants during the AC cycle. We will call the snapshot taken at the time t_s as *the solution* for time t_s. AC loss is not the only quantity that could be evaluated using these solutions. For example, a mapping of the magnetic field around the conductor could give important information about the distribution of electrical currents inside the superconducting volume (Dutoit *et al.* 2005). On the other hand, the analysis of the voltage signal detected on a superconducting wire carrying AC current could be used to verify the proposed models of magnetic flux dynamics (Klinčok *et al.* 2005).

Let us first summarize the points that the methods presented below have in common. Because of the dependence of the actual solution upon the history, there is always a need to use the solution obtained for the previous time instant, t_p, in the calculation for actual instant t_s. In some procedures it is not the time interval that is set, but the difference in transported current, $I_s - I_p$, or the applied magnetic field, $B_s - B_p$. In these expressions, 's' means 'actual' and 'p' is 'previous'. The choice of the number of instants in the AC cycle for which the solution is calculated is a trade-off between the wish for detailed knowledge and limited computing capacity. In the author's experience, when using less than 20 solutions in one AC cycle, one risks neglecting some important feature of the process.

In order to reduce the computation time, symmetries in space and in time should be utilized. Examples of reducing the area in which the solution is found are illustrated in Fig. 7.9. The dashed lines in reduced geometries signify the condition imposed on the direction of magnetic flux density: $\vec{n} \times \vec{B} = 0$, i.e. only the component normal to the boundary survives.

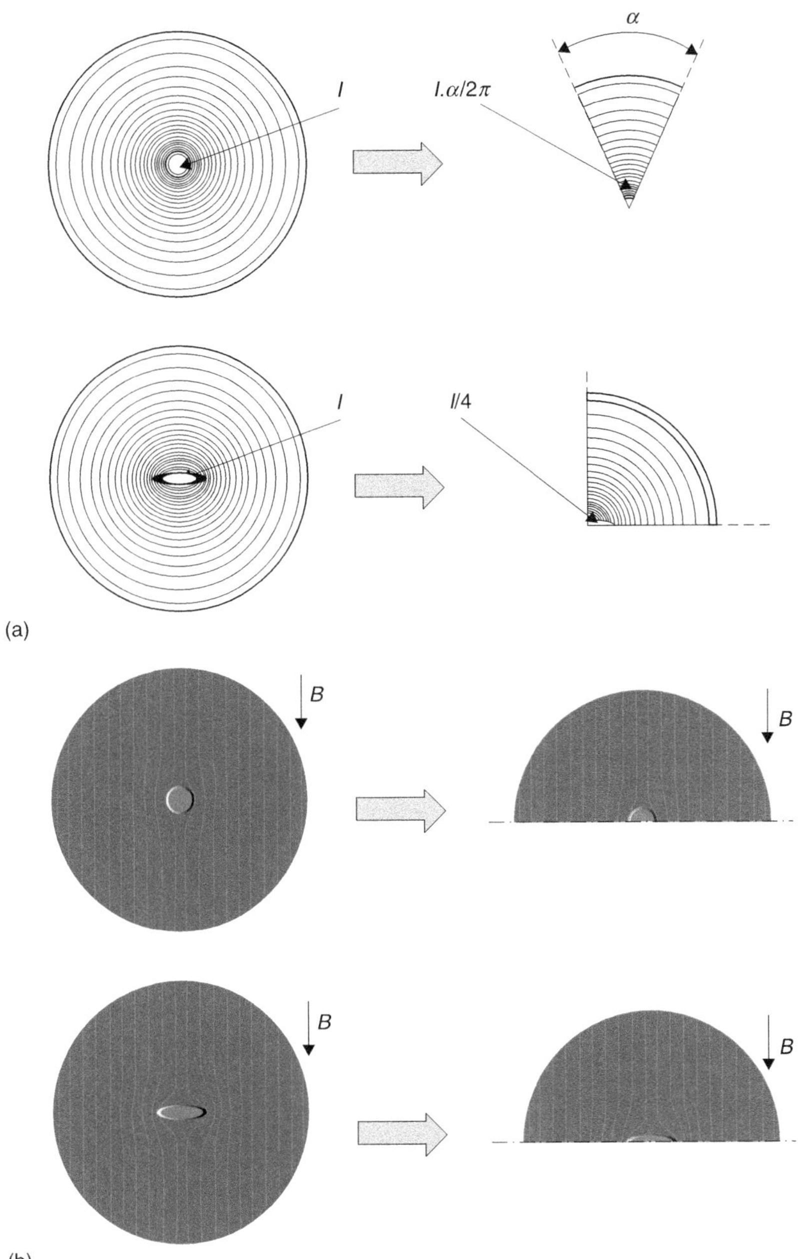

7.9 Some examples of using symmetries in the calculation for (a) current transport, (b) applied magnetic field.

Other kinds of symmetries are those with respect to time. They could be used to find solutions without carrying out the calculations for all time instants. For example, in the case of a harmonic AC current with period T the solution for $I(T/2 + t_s)$ has the opposite sign with respect to that found at $I(t_s)$. However this rule does not hold for the initial transient period of switching on the current.

7.4.1 The $J(\Delta A)$ method

From the numerical methods presented in this chapter, this one method requires the minimum knowledge of mathematics behind numerical techniques. It could be easily utilized in a commercial finite element program to solve the problems in two-dimensional geometry shown in Fig. 7.3. The current density in a superconductor is perpendicular to the local magnetic field lying in the drawing plane, $\vec{B} = (B_x, B_y, 0)$ and the electric field as well as the vector potential of magnetic field exhibit only the z-component (perpendicular to the drawing plane), E_z and A_z, respectively.

The original idea (Campbell 2007) to relate the current density in a superconductor, J_s, to the local value of vector potential of the magnetic field, A_z, was further elaborated (Gömöry *et al.* 2009) following the scheme illustrated in Fig. 7.10. The critical state model (Eq. 7.1) replaced the real current–voltage curve (dashed) in the left panel by the rectangular one (continuous line). Thus the current density in the superconductor is

$$J_s = J_c \text{sign}(E_z) \qquad [7.25]$$

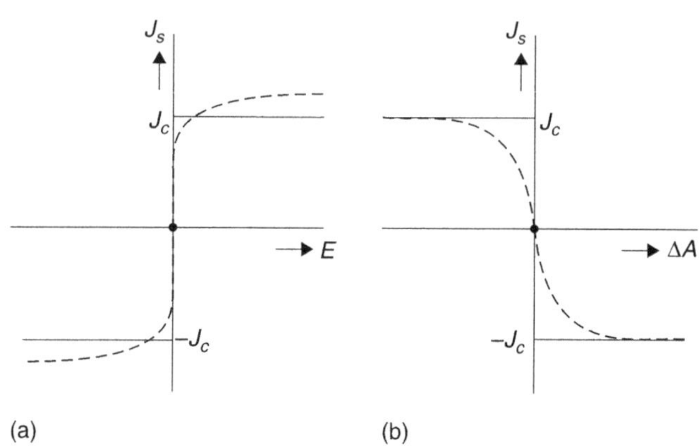

7.10 (a) Replacement of a smooth current–voltage curve (dashed line) by the rectangular one (continuous line). Right: replacement of the step-like dependence of current density in superconductor, J_s, on the change of vector potential of magnetic field (continuous line) by a curve with finite derivatives (dashed line).

where sign(x) is the signum function. The electrical field stems from the change of magnetic field – expressed through its vector potential A_z – and the gradient in electrostatic potential

$$E_z = -\frac{\partial A_z}{\partial t} - \nabla\varphi. \quad [7.26]$$

Let us first consider the case of one single superconductor exposed to AC excitation. The value of φ is the same across the cross-section thus it cannot influence the distribution of current density. Therefore we can simplify Eq. 7.26 to

$$E_z = -\frac{\partial A_z}{\partial t} \quad [7.27]$$

with a proper choice of boundary conditions for A_z as will be shown later. This means the definition of rectangular current voltage curve can be written as

$$J_s = J_c \operatorname{sign}(E_z) = J_c \operatorname{sign}(-\frac{\partial A_z}{\partial t}) = J_c \operatorname{sign}(-\Delta A_z) \quad [7.28]$$

where the last step is allowed because $\partial t > 0$. Equation 7.28 allows transformation of Eq. 7.18 into the governing equation for A_z in a superconductor in two-dimensional geometry as

$$-\frac{1}{\mu_0}\left(\frac{\partial^2 A_z}{\partial x^2} + \frac{\partial^2 A_z}{\partial y^2}\right) = -J_c \operatorname{sign}(\Delta A_z) \quad [7.29]$$

where

$$\Delta A_z(x, y) = A_s(x, y) - A_p(x, y) \quad [7.30]$$

is the difference between the actual value of vector potential component, A_s, and the value in the previous time step, A_p. To help the numerical calculation it is appropriate to replace the stepwise sign(x) by a smoother function $f(x)$ with similar properties, i.e. $f(x \to -\infty) = -1$, $f(x \to 0) = 0$, $f(x \to \infty) = 1$. The author's preference is $\tanh(x/x_0)$, where the step sharpness could be controlled with the choice of x_0. This replacement is shown in the right hand side of Fig. 7.10. Then the distribution of the z-component of the vector potential in the superconductor cross-section is found by resolving the equation

$$\frac{1}{\mu_0}\left(\frac{\partial^2 A_s(x,y)}{\partial x^2} + \frac{\partial^2 A_s(x,y)}{\partial y^2}\right) = J_c \tanh\left(\frac{A_s(x,y) - A_p(x,y)}{A_0}\right). \quad [7.31]$$

The remaining physical quantities directly derive from the solution for A_z, e.g. the current density in a superconductor is

$$J_s(x,y) = J_c \tanh\left(\frac{A_p(x,y) - A_s(x,y)}{A_0}\right) \quad [7.32]$$

and the magnetic field in all the space where the calculation is performed is obtained as

$$B_x(x,y)=\frac{\partial A_s(x,y)}{\partial y},\quad B_y(x,y)=-\frac{\partial A_s(x,y)}{\partial x}. \qquad [7.33]$$

One important point in using the $J(\Delta A)$ method is the choice of boundary conditions that would secure the existence of the so-called *neutral zone*. This is the portion of superconductor with zero electrical field, thus zero current density. For example, in the wire transporting AC current with amplitude smaller than the critical current, the neutral zone is the central part not used by the (critical) current density. In the case of a slab magnetized by a strong magnetic field the neutral zone shrinks to a thin layer between the parts filled with current densities of opposite sign. The way of using this method will be illustrated in the following examples.

Example 4 is of AC transport in a wire with an elliptical cross-section. The geometry of the problem is that used in Section 7.3.1 in the critical current calculation (Fig. 7.3) with one important modification. Instead of assuming that the critical current density J_c is filling all of the cross-section of the wire, the flow through the superconductor is the history-dependent current density J_s given by Eq. 7.32.

During the period of an AC cycle, T, we need to find the solutions for time instants t_s. The AC current $I = I_a \sin\omega t$ is defined by its angular frequency $\omega = 2\pi/T$ and amplitude I_a. A convenient way of controlling the actual value of the transport current is with the condition of the vector potential, A_b, on the boundary $R(x, y) = \sqrt{x^2 - y^2}$. The current I_{set} that would be consistent with the set value of A_b can be estimated from the known analytical expression for the round wire with radius R_w calibrated to reach a zero value in the wire centre:

$$A_b = -\frac{\mu_0 I_{set}}{4\pi}\left(2\ln\left(\frac{R}{R_w}\right)+1\right). \qquad [7.34]$$

For our wire with an elliptical cross-section as shown in Fig. 7.3 we could have an equivalent radius $R_w = \sqrt{a_{SC}^2 + b_{SC}^2}$. It is not guaranteed that the result of the calculation

$$I = \int_{S_{SC}} J_s \, dS_{SC} \qquad [7.35]$$

will give the transport current equal to I_{set}; however this does not pose any problem in AC loss evaluation.

We start from an initial state $t = 0$, when the electromagnetic properties are zero everywhere. Then we perform a series of calculations, increasing the value of vector potential at the boundary, A_b, given by Eq. 7.34. This is the initial part of the AC cycle. For each solution we determine the transport current with the help of Eq. 7.35. When this value has reached the amplitude I_a, we revert the sign of increments in A_b and calculate the series of solutions for decreasing current (the return part of the AC cycle), until the evaluated transport current is less than $-I_a$. Then we again change to positive increment in A_b to complete the AC cycle until $I = 0$. This procedure is schematically shown in the left side of Fig. 7.11. The index i is introduced to underline the fact that the calculation is carried out for a limited number of time instants. In this example 37 solutions have been assumed during the complete AC cycle.

It is now interesting to analyze the relation between the value of A_b, which according to Eq. 7.34 is defined by I_{set}, and the total current transported by a superconductor, I. The values for I_i are quite similar to $I_{set,i}$, but not identical. The hysteresis of magnetic flux interaction with the superconductor is reflected by this difference. To see it better, we introduce an arbitrary correction to A_b:

$$A_c = A_b - \mu_0 kI \quad [7.36]$$

where k is a number chosen to see clearly the difference between the increasing and decreasing portion of the plot showing A_b versus I. In Fig. 7.12 the results for two cycles with different amplitude are shown using the transformation Eq. 7.36. The value of k was chosen so that at the beginning of the return the slope approaches zero. One can clearly see that the initial part of the cycle (the first increase of current) is not repeated, and the data from the return part are representative for all the remaining cycle(s). This feature could be used to reduce the number of solutions required to characterize the AC cycle as schematically shown in the right-hand side of Fig. 7.11, where the additional modification constantly increments current. Because of the rectangular current–voltage curve such a change of the current waveform does not change the hysteresis loop shown in Fig. 7.12. In the first step, the change from zero to the amplitude value is performed. Then the solutions in the return part of the cycle are calculated. The remaining solutions are found using the symmetry

$$I_{\uparrow}(A_b) = I_{\downarrow}(-A_b) \quad [7.37]$$

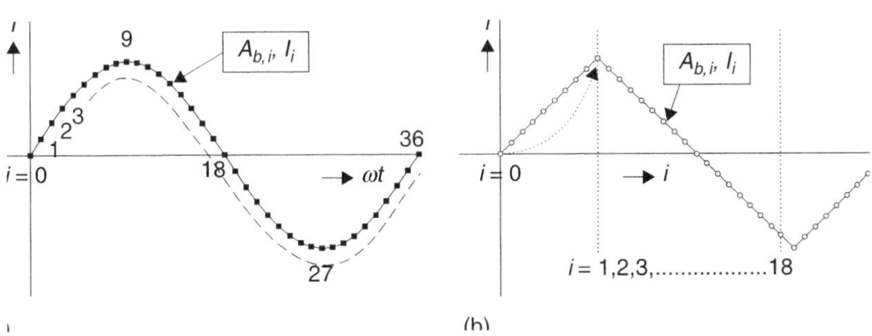

7.11 Calculation scheme for the AC cycle. Starting from the virgin state with no current, the solutions are found for a certain number of time instants during the cycle. (a) Calculation preformed for all the cycle, (b) calculation performed only for one half of the cycle, using a constant step in AC current instead of that one in time.

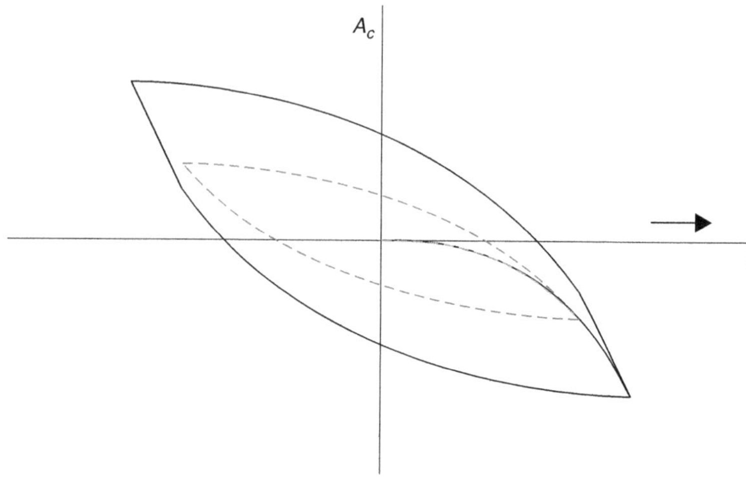

7.12 Hysteresis loops for two cycles of AC current of different amplitude. The quantity plotted on vertical axis is the vector potential set at calculation boundary (8.33) corrected according to (8.35) in the way conserving the areas of loops.

where $I_{\uparrow}$ are the currents at the increasing part of the cycle and $I_{\downarrow}$ those at the decreasing part.

Determination of transport AC loss can be performed in different ways; the most practical can be understood from Fig. 7.13. On the left-hand side is the geometry used in the measurement of AC loss (Clem *et al.* 1996) based on extracting from the measured voltage, U, the fundamental harmonic component in phase with current, U_{loss}. From the Maxwell equation $U = -\partial\Phi/\partial t$ one finds for the two-dimensional representation on the right-hand side of Fig. 7.13 that

$$E_z = \frac{U}{l} = \frac{\partial A_b}{\partial t} \qquad [7.38]$$

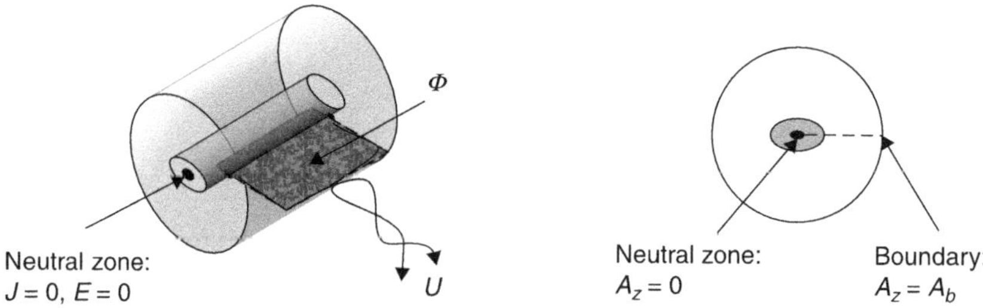

7.13 Experimental arrangement of transport AC loss determination (left) and its two-dimensional representation (right).

because the magnetic flux, Φ, in the loop of length l in the z-direction is $\Phi = l(A_z(0,0) - A_b) = -lA_b$ and the neutral zone persists at every current lower than the critical current. Then the AC loss per cycle and unit length of the wire will be

$$Q = \int_0^{2T} E_z I \,\mathrm{d}t = \int_0^{2T} \frac{\partial A_b}{\partial t} I \,\mathrm{d}t = \oint I \,\mathrm{d}A_b = \oint I \,\mathrm{d}A_c \qquad [7.39]$$

The last equality holds because the linear term in the transformation Eq. 7.36 does not contribute to the integral (Eq. 7.39) calculated over the entire cycle. The values of Q are actually the areas of hysteresis loops plotted in the A_c versus I plane like those shown in Fig. 7.12.

From the calculations performed for various amplitudes I_a the loss dependence on AC current can be constructed. A theoretical prediction can be derived assuming constant J_c (Norris 1970). Numerical calculations could take into account the $J_c(B)$ characteristic of the superconductor or its non-uniformity in the wire cross-section. The influence of these refinements is illustrated in Fig. 7.14.

Example 5 is of a coated conductor (CC) tape with magnetic substrate in a perpendicular AC magnetic field. CC tapes contain thin superconducting layers deposited on a metallic substrate (Fig. 7.15). The substrate, separated from the HTS material by a buffer layer, is often magnetic. Its permeability will influence the distribution of magnetic field in the superconducting layer and thus must be taken into consideration in the model. Here we use the data for Ni-5%W substrate (Claassen and Thieme 2008), shown in Fig. 7.16. Therefore in the substrate the equation

$$\frac{1}{\mu_0 \mu_r}\left(\frac{\partial^2 A_z}{\partial x^2} + \frac{\partial^2 A_z}{\partial y^2}\right) = 0 \qquad [7.40]$$

will be solved with magnetic-field dependent μ_r. The superconducting layer is represented through its critical current density J_c, and the actual current density in the superconductor is calculated by applying Eq. 7.32. Different with respect to the previous example is the boundary condition: for an arbitrary orientation of an applied magnetic field the general expression Eq. 7.17 should be used. Here we

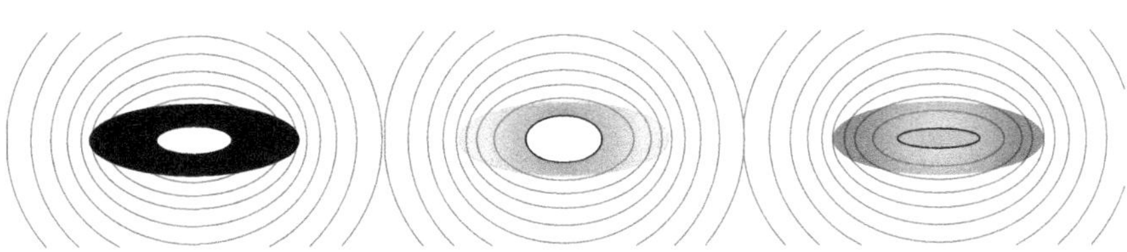

(a) (b) (c)

7.14 Distribution of current density at the amplitude of AC cycle with current below the critical current, calculated assuming (a) constant J_c, (b) $J_c(B)$ dependence (middle), (c) better J_c at the tape edges. Darker color means higher current density.

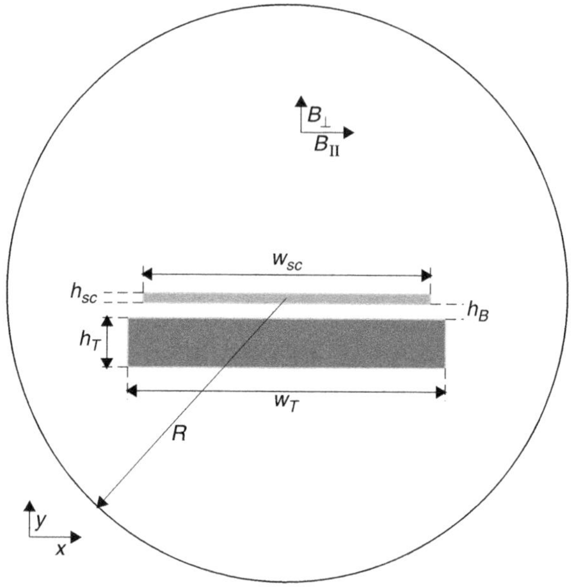

7.15 Cross-section of CC tape consisting of the HTS layer (width w_{SC}, thickness h_{SC}) separated from the metallic substrate (width w_T, thickness h_T) by the distance h_B corresponding to the thickness of buffer layer.

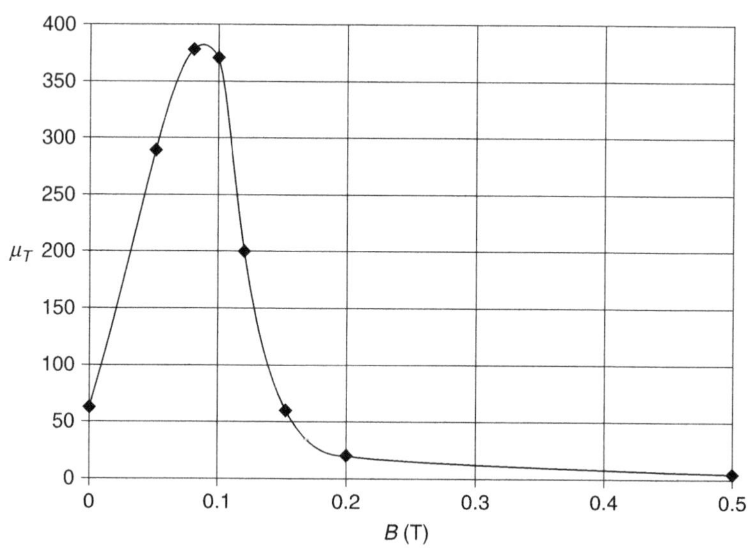

7.16 Ni-5%W substrate magnetic permeability used in the model.

assume the AC field $B_{ac} = B_a \sin\omega t$ in a perpendicular direction with respect to the layer, therefore

$$A_b = -xB_{ac} \qquad [7.41]$$

The calculation procedure is analogous to that presented in the previous example for AC transport. An example of the calculated distribution is shown in Fig. 7.17. One can clearly see that the shape of the neutral zone is asymmetric in the vertical direction because the substrate 'sucks' the flux lines. The important quantity determined for each solution is the magnetization of the superconducting layer (Brandt and Indenbom 1993)

$$M(B_{ac}, B_a) = \frac{1}{S_{sc}} \int_{sc} -xJ(x, y)\, dS_{sc} \qquad [7.42]$$

that is used to calculate the AC loss with the help of the expression for the volume density of loss in a body magnetized by an external field with amplitude B_a:

$$Q_{SC}(B_a) = -\oint M(B_{ac}, B_a)\, dB_{ac}. \qquad [7.43]$$

The index 'SC' has been used to stress that this is only the loss in the HTS layer. On top of this dissipation there will also be the magnetization loss in the substrate. This contribution cannot be calculated exactly unless the domain structure is considered. In various calculations (Nguyen *et al.* 2010, Miyagi *et al.* 2007) the simplified approach is used assuming that, during the AC cycle, parts of the substrate undergo magnetization cycles defined by the amplitude of local magnetic field at the maximum applied external field. This means that only once in the AC

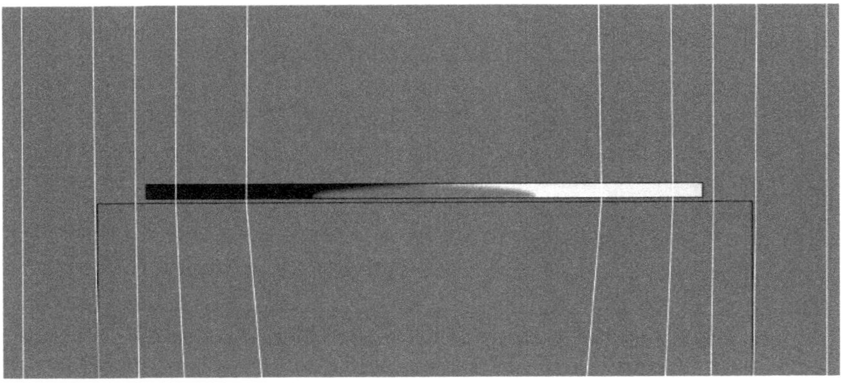

7.17 Distribution of current density (grey scale) in superconducting layer of CC tape with ferromagnetic substrate. The vertical scale is ×100 enlarged. White lines are the contours of constant magnetic vector potential.

cycle, at maximum applied field, the hysteresis loss density Q_h in the substrate cross-section should be integrated. A practical approximation for Q_h that contains two parameters is

$$Q_h = Q_s \begin{cases} \left(\dfrac{B}{B_s}\right)^2 & B \leq B_s \\ 1 & B > B_s \end{cases}. \qquad [7.44]$$

In the case of Ni-5%W substrate we have found that $Q_s = 100$ J/m^3 and $B_s = 0.15$ T.

From the calculations performed for a series of AC field amplitudes, B_a, the loss dependence shown in Fig. 7.18 has been determined (Gömöry *et al.* 2010). One can see that at low fields the hysteresis loss in the substrate prevails; on the other hand the re-shaping of flux lines led to the reduction of loss in the superconducting layer. The agreement with experiment is not perfect, however the main features have been correctly predicted.

Example 6 has two parallel wires in an AC magnetic field. This is the first step towards more complex structures such as cables or coils. For simplicity we assume two round wires of an HTS with constant critical current density, J_c, placed at a certain distance in the horizontal plane $y = 0$. When the magnetic field B_{ac} is

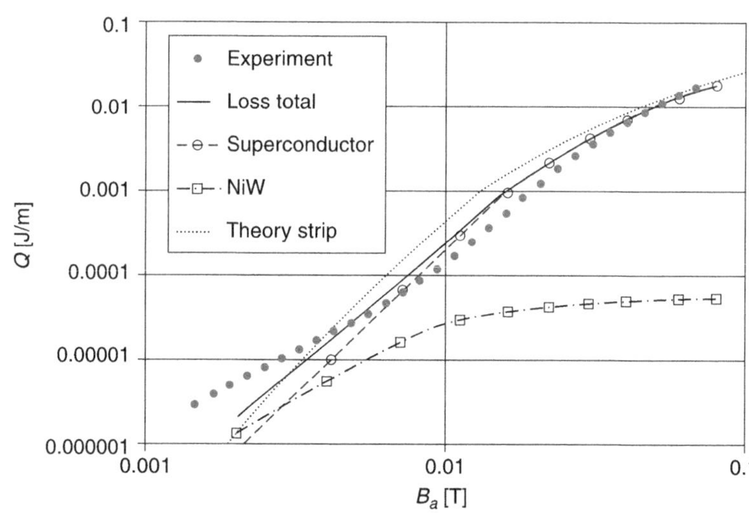

7.18 Result of the evaluation of AC loss in CC tape on magnetic substrate when exposed to AC magnetic field perpendicular to the tape wide face. The loss in superconductor (open circles) is slightly lower than the analytic prediction for a thin strip (dotted line). Adding the hysteresis in magnetic substrate (open squares) results in estimation for total magnetization loss in the tape (full line) that can be compared with experimental data (full circles).

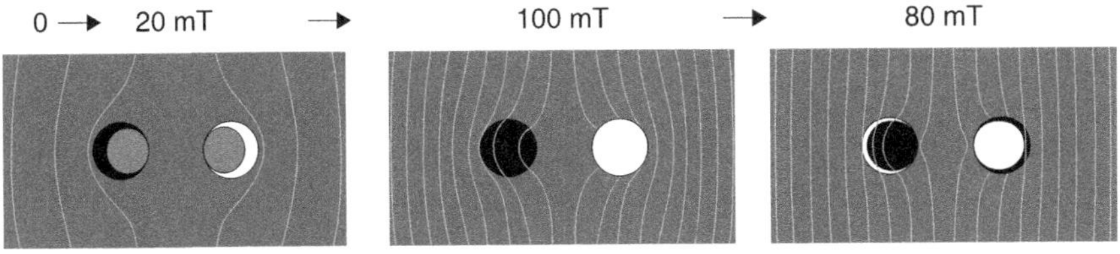

7.19 Distribution of current density (grey scale) in a pair of round wires with perfect coupling exposed to magnetic field in vertical direction. The solutions for three instants of the magnetization process are shown.

applied in the y-direction using the boundary condition (Eq. 7.40) and as the governing equation for A_z in superconductor is used (Eq. 7.32) the distributions shown in Fig. 7.19 are found. Obviously there is a net current flowing in each of the wires, and one can interpret these distributions as forming a loop in which the current flowing in the left-hand wire returns through the right-hand one and vice versa. Such two wires are 'coupled'. When the current induced by the external magnetic field returns through the same wire – in such a case the wires will be 'uncoupled' – the full expression for vector potential (Eq. 7.26) should be used instead of the simplification (Eq. 7.27) that works for single wires and some special cases of stacked wires. This requires the imposition of a difference in electrostatic potential φ between two wires that would control the total currents. The practical way is to introduce a dimensionless parameter ε that will modify Eq. 7.32 for current density in the left-hand wire

$$J_s(x,y) = J_c \tanh\left(\frac{A_p(x,y) - A_s(x,y)}{A_0} + \varepsilon\right) \qquad [7.45]$$

while in the right-hand wire

$$J_s(x,y) = J_c \tanh\left(\frac{A_p(x,y) - A_s(x,y)}{A_0} - \varepsilon\right). \qquad [7.46]$$

With the choice of ε the strength of coupling between wires is controlled. In Fig. 7.20 are shown the solutions obtained when a value of ε has been established

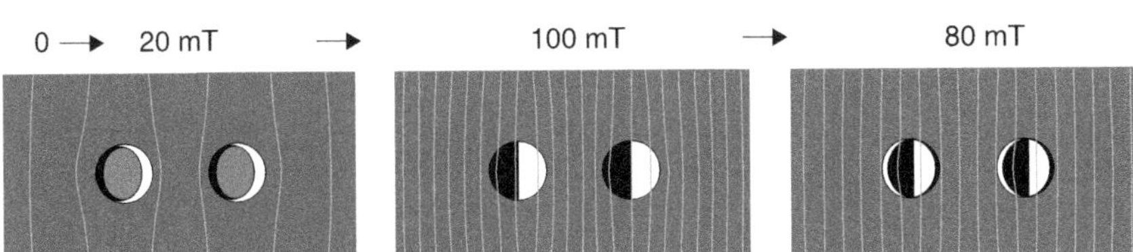

7.20 Distribution of current density (grey scale) in a pair of round wires that are not coupled at all. When exposed to magnetic field in vertical direction, the current returns through the same wire.

such that the net current in each wire is zero. In a similar way, the distribution of net currents in a set of parallel conductors can be controlled by the proper choice of ε in each of them.

The total magnetic moment can be evaluated from the solution found at any instant of the applied magnetic field $B_{ac} = B_a \sin \omega t$ as

$$m(B_{ac}, B_a) = \int -xJ(x,y)dS \qquad [7.47]$$

where the non-zero contribution to the integral is expected only in the cross-sections of two wires. Because the distributions are symmetrical, the result does not depend on the choice of wire position with respect to $x = 0$.

The magnetic moment may be used to evaluate the AC loss. An intermediate step is the calculation of magnetization

$$M(B_{ac}, B_a) = \frac{m(B_{ac}, B_a)}{S_{2w}} \qquad [7.48]$$

where S_{2w} is the cross-section to which the magnetization is attributed. Its choice is relevant for the determination of the volume density of AC loss given by Eq. 7.42 but falls out when the total loss is calculated by multiplying the loss density with the cross-section.

Figure 7.21 shows the magnetization loops for the two cases treated above (coupled vs uncoupled) evaluated using the same total volume S_{2w}. The AC loss is

$$Q_{2w}(B_a) = -S_{2w} \oint M(B_{ac}, B_a)dB_{ac} = \oint m(B_{ac}, B_a)dB_{ac} \qquad [7.49]$$

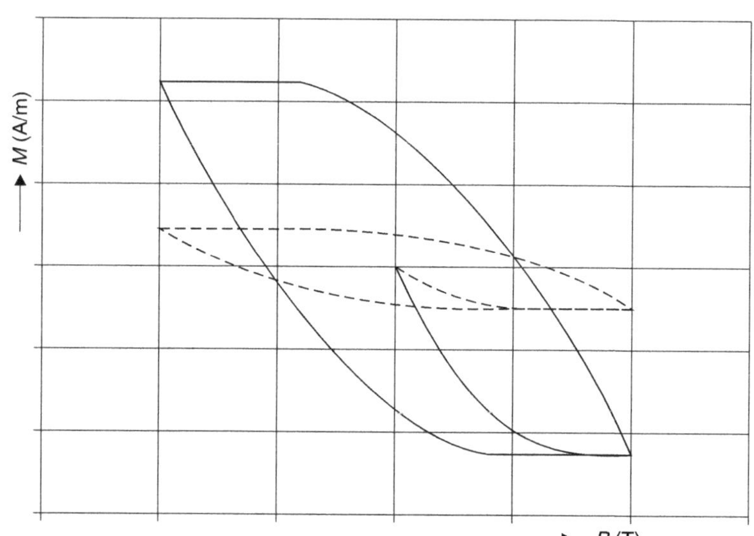

7.21 Magnetization curves determined for the pair of coupled wires (full line) and the same for uncoupled wires (dashed line).

With this example the part dedicated to the $J(\Delta A)$ method has finished. The following section will briefly explain the principles of other modeling methods that are well documented in the literature.

7.4.2 The minimum magnetic energy variation (MMEV) method

Similar to the $J(\Delta A)$ method, this method works with the rectangular current–voltage curve (continuous line on the left-hand side plot in Fig. 7.10). The cross-section of the superconductor is divided into elements that in the initial part of the cycle are in stepwise manner 'switched on' to carry the critical current density. Figure 7.22 shows the superconductor of rectangular shape (width w, height h) assumed to transport current that reached 31% of critical current (10 of 32 identical elements already filled with J_c). After a further increase of transport current, one of the elements that is actually current-free should start to carry J_c. Which one will it be? One intuitively feels that in the example shown in Fig. 7.22 it would be the one marked in grey. To resolve this problem Sanchez and Navau (Sanchez and Navau 2001a, 2001b) suggested searching for the minimum in magnetic energy (MME)

$$W_m = \frac{1}{2\mu_0} \int_{S_A} B^2 \, \mathrm{d}S_A \qquad [7.50]$$

and they also provided a fast non-standard algorithm for the purpose of finding the solution. The first models were for cylinders in a magnetic field with cylindrical symmetry, in particular, a uniformly applied field or a levitating cylinder on a circular magnet (Navau and Sanchez 2001). Later they extend the MME to tapes with transport current, together with Pardo (Pardo *et al.* 2004a) or applied field (Pardo *et al.* 2004b) and multiple tapes (Pardo *et al.* 2003, Pardo *et al.* 2005). The general situation of any combination of transport current and applied magnetic field in any direction, including rotating fields, requires an essential generalization.

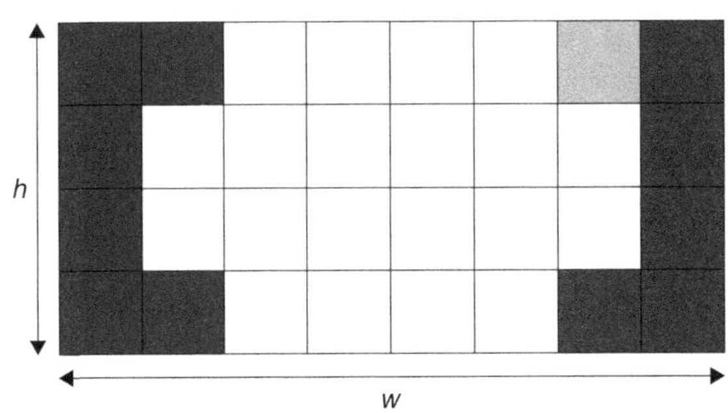

7.22 Rectangular cross-section of superconducting wire transporting electrical current that filled the portion marked in black. After a further increase of current the next element (marked in grey) should be filled.

It is necessary to minimize the function proposed by Prigozhin, who deduced it from first principles and the sharp $E(J)$ relation of the critical state model (Prigozhin 1996, 1997). This function is the energy of the current variation between two time instants or, in other words, the magnetic energy variation

$$\delta W_m = \frac{1}{2}\int_{SC} J(x,y)A_{s,J}(x,y)\,dS - \int_{SC} J(x,y)A_{p,J}(x,y)\,dS - \int_{SC} J(x,y)\,(A_{s,a}(x,y)-A_{p,a}(x,y))\,dS \qquad [7.51]$$

where the vector potential of the applied external field is $A_{s,a}$, that created by the currents in a superconductor is $A_{J,a}$, and the analog quantities in the previous solution are indexed by p instead of s. The advantage of expressing the magnetic energy in terms of $J.A$ is that the integration is performed only in the cross-section of the superconductor(s). Proper gauge for the vector potential is

$$\nabla \cdot A = 0 \qquad [7.52]$$

and the constraints that are applied at finding the minimum of δW_m are

$$I = \int_{SC} J(x,y)\,dS, \quad |J_s| \le J_c \qquad [7.53]$$

The resulting minimum magnetic energy variation (MMEV) method was developed by Pardo (Pardo *et al.* 2007) for simultaneous in-phase alternating applied field and current; subsequently, it was also used for out-of-phase applied field and current (Vojenčiak *et al.* 2006), levitation with translational symmetry (Del-Valle *et al.* 2007, Del-Valle *et al.* 2008), pancake coils (Pardo 2008, Souc *et al.* 2009, Pardo *et al.* 2009) and Roebel cables (Terzieva *et al.* 2010, Grilli and Pardo 2010). The final development is to predict the interaction of linear magnetic materials with arbitrary susceptibility (Pardo *et al.* 2009, Grilli *et al.* 2010). Most published MMEV results are for constant critical current density but it can also simulate a field-dependent critical current density. This was first done for MME and then for the full MMEV. Another kind of MME is for two-dimensional thin films of finite length (Navau *et al.* 2008).

7.4.3 Brandt's method

In contrast to the two methods described in the previous part of this Chapter, the numerical procedure developed by Brandt (1996) uses an E-J relation to calculate the current distribution in the superconductor. This involves $J(B)$ dependence (Brandt and Mikitik 2005) or non-uniform J_c in the calculations and takes into account the London currents (Brandt 2001a, Brandt 2001b). Similarly to the MMEV method, the cross-section is divided into small rectangular areas with identical surface (Fig. 7.22). The low-frequency approximation of Maxwell's equation (Eq. 7.18) is now formulated as

$$-\frac{1}{\mu_0}\left(\frac{\partial^2 A_z}{\partial x^2}+\frac{\partial^2 A_z}{\partial y^2}\right)=J_s \qquad [7.54]$$

where the current density J_s in the superconductor now depends on the electrical field that is given by Eq. 7.25. A significant achievement of the method is that the calculation of distributions is carried out only in the cross-section of the superconductor. For this purpose the rearrangement of expressions for the vector potential is introduced.

Assume that on the HTS wire transporting AC current the AC magnetic field, B_{ac}, is also applied in the y-direction. Then the vector potential is the superposition of two terms: A_J is the component generated by the transport current and $-xB_{ac}$ is caused by the AC field:

$$A_z = A_J - xB_{ac} \qquad [7.55]$$

and the solution of Eq. 7.53 can be expressed as

$$A_z(x,y,t)+xB_{ac}(t)=-\mu_0\int_{SC} Q(x,y,x',y')J(x,y,t)\,dS \qquad [7.56]$$

where the integral kernel

$$Q(x,y,x',y')=\frac{1}{2\pi}\ln\left(\frac{\sqrt{(x-x')^2+(y-y')^2}}{r_0}\right) \qquad [7.57]$$

contains an arbitrary constant r_0 (Yazawa *et al.* 1998). The distribution of current can be found by inverting Eq. 7.55 as

$$J(x,y,t)=-\frac{1}{\mu_0}\int_{SC} Q^{-1}(x,y,x',y')\left[A_z(x',y',t)+x'B_{ac}(t)\right]dS \qquad [7.58]$$

where the inverse kernel is defined by

$$\int_{SC} Q^{-1}(x,y,x',y')Q(x',y',x'',y'')\,dS=\delta\left(\sqrt{(x-x'')^2+(y-y'')^2}\right) \qquad [7.59]$$

Because of the smooth current–voltage curve the process of flux penetration is now frequency dependent and the solution comprises also the distribution of electrical field in superconductor. In the case of power-law dependence (Eq. 7.5) the governing equation is

$$E_{c0E}\left(\frac{J(x,y,t)}{J_c}\right)^n+\nabla\varphi-x\frac{\partial B_{ac}(t)}{\partial t}$$

$$=\mu_0\int_{SC} Q(x,y,x',y')\frac{\partial j(x',y',t)}{\partial t}\,dS \qquad [7.60]$$

In the case where the superconductor cross-section can be divided into rectangular elements with dimensions Δx and Δy, the discretized form of Eq. 7.57 that can be used for numerical calculation is

$$J_k(t+\Delta t) = J_k(t) + \frac{\Delta t}{\mu_0 \Delta x \Delta y} \sum_{l=1}^{N} Q_{kl}^{-1} \left[E_{c0E} \left(\frac{J_l(t)}{J_c} \right)^n + \nabla\varphi(t) - x_l \frac{\partial B_{ac}}{\partial t} \right] \quad [7.61]$$

and from Eq. 7.59 it follows that

$$\nabla\varphi(t) = \frac{\dfrac{\mu_0 \Delta I}{\Delta t} - \sum_{k=1}^{N} \sum_{l=1}^{N} Q_{kl}^{-1} \left[E_{c0E} \left(\frac{J_l(t)}{J_c} \right)^n - x_l \frac{\partial B_{ac}}{\partial t} \right]}{\sum_{k=1}^{N} \sum_{l=1}^{N} Q_{kl}^{-1}} \quad [7.62]$$

with

$$Q_{kl} = \frac{1}{4\pi} \ln \frac{(x_k - x_l)^2 + (y_k - y_l)^2}{r_0^2} \quad [7.63]$$

The expression for the integral kernel will be different for shapes with axial symmetry (Brandt 1998a, Brandt 1998b).

A significant feature of this method is that the kernel Q_{kl} as well as its inverse has to be evaluated only once, at the beginning of the calculation. Examples of distributions calculated for the self-field case (no applied magnetic field, thus the terms with B_{ac} are left out in Eqs 7.60 and 7.61) are shown in Fig. 7.23. The power law exponent, n, has significant influence on the sharpness of the boundary between the portion filled with current density and the neutral zone (Klinčok 2007).

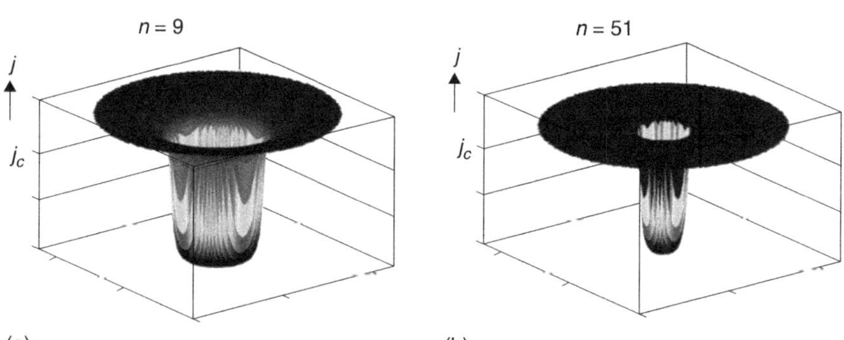

7.23 Distributions of current density in the cross-section of a round HTS wire calculated assuming the power law exponent: (a) $n = 9$, (b) $n = 51$.

7.5 Future challenges and trends

This chapter gives basic information about some modeling methods for electromagnetic behavior of HTS materials, and examples of well-established procedures have been presented. It is not by chance that many issues of practical importance have not been covered. At the present level of knowledge we are not able to give a rigorous recipe for solving the following important tasks.

7.5.1 Coupling currents and coupling loss

First attempts in this direction (Grilli *et al.* 2003b) revealed that the value added to traditional analytical models (e.g. Kwasnitza and Clerc 1994, Campbell 1982, Banno and Amemiya 1999) is hardly adequate for the required effort. Recently, a combination of analytical approximation with finite element simulation of coupling current provided surprisingly good predictions for LTS wires (Fabbricatore *et al.* 2009). The problem is crucial when CC tapes are considered for AC coils for transformers, motors and pulse magnets. Recent success in these directions (Kimura and Amemiya 2007, Grilli *et al.* 2009, Grilli and Pardo 2010) indicates rapid progress in this topic.

7.5.2 Critical state in three dimensions

This is not as much the problem of modeling methods as missing a clear physical guide. Several approaches have been proposed to define the critical state in the situation where current density is not perpendicular to magnetic field, thus the Lorentz force also has a component along the flux line (Clem and Perez-Gonzales 1984, Navarro and Campbell 1991, Badia and Lopez 2002, Ruiz and Badia-Majos 2010). Numerical techniques to cope with three-dimensional problems are developed with a lot of effort and diligence (Vanderbemden *et al.* 2007, Lousberg *et al.* 2009), however the confirmation of results by experimental data is far from being complete (Stavrev *et al.* 2002c, Grilli *et al.* 2005).

7.5.3 AC loss in power devices

The main purpose of modeling HTS electromagnetic properties is to develop methods for predicting the behavior of power devices, where the HTS wire transporting AC current feels the magnetic field generated by other wires (Kajikawa *et al.* 2001). Currently, the problem is well understood for power transmission cables (Grilli *et al.* 2003a, Costa-Bouzo *et al.* 2004, Grilli and Sjöström 2004, Jiang *et al.* 2006, 2008, Amemiya *et al.* 2008, Grilli *et al.* 2009, Amemiya *et al.* 2010). Significant understanding has been reached in the anti-parallel configuration suitable for fault current limiters (Majoros *et al.* 2007,

Li *et al.* 2008, Nguyen *et al.* 2009) where the results can also be confronted with analytical prediction (Clem 2008). Only a few results have been published on modeling the current distribution and AC loss in HTS stacks (Oomen *et al.* 2003, Grilli *et al.* 2006, Yuan *et al.* 2009) and AC coils (Pardo 2008, Pardo *et al.* 2009, Šouc *et al.* 2009). It is only a question of time before more extensive work on simulating HTS coils and comparison with experimental results will take place.

7.5.4 CC tapes – wires with extreme aspect ratio

The shape of the superconducting layer in the most promising HTS wire – the coated conductor tape – poses an additional problem to numerical simulation because the aspect ratio is greater than 1000 (Daumling 1998). Methods replacing the two-dimensional penetration of magnetic flux into the HTS layer by a one-dimensional process are manageable (Sirois and Grilli 2008, Brambilla *et al.* 2009, Nguyen *et al.* 2010) but the validity of results should always be considered carefully. Another possibility is the use of a variable meshing that would provide enough elements in the direction of the expected movement of the flux front (Solovyov and Gömöry 2011).

7.5.5 HTS combined with ferromagnetic materials

It is not clear whether in the future that low-cost CC tapes could be prepared on a non-magnetic substrate. In some cases the ferromagnetic component in the wire can improve its properties (Majoros *et al.* 2000, Nakahata and Amemiya 2008, Genenko *et al.* 2009, Safran *et al.* 2010). Therefore it is important to develop a realistic model of a superconductor-ferromagnetic composite (SCFMC). Currently a non-constant permeability of ferromagnetic material can be inserted, however it could not be hysteretic. Therefore the hysteresis loss in ferromagnetic material should be added by a separate calculation. The methods used for estimation of hysteresis loss in SCFMC provide surprisingly good predictions, but neglect the existence of a domain structure in the ferromagnetic part (Farinon *et al.* 2005, Miyagi *et al.* 2007, Gömöry *et al.* 2009, Amemiya *et al.* 2010).

7.5.6 Coupling electromagnetic and thermal problems

This topic is very important in the design of fault current limiters. Fortunately, commercial finite element software is available to couple different physical models. The published results indicate that a significant improvement in the device design can be achieved when proper models for both of the coupled problems have been developed (Duron *et al.* 2007, Roy *et al.* 2008, Fujishiro and Naito 2010).

7.6 Conclusions

In this chapter an attempt was made to provide the basic understanding about possibilities that the current techniques and software tools offer in numerical modeling of HTS. The ways to determine two key properties – critical current and AC loss – in HTS wires have been presented, and possibilities to extend these procedures to more complex objects such as HTS devices have been discussed. Examples have been incorporated to encourage the readers to perform the modeling on their own.

The extent of this chapter does not allow coverage of the field in full. Topics not explained are: techniques based on current vector potential and the scalar magnetic potential, so-called T-Ω method (Amemiya *et al.* 1998a, Amemiya *et al.* 2010), the use of magnetic field formulation with H as an independent variable (Hong *et al.* 2006, Hong *et al.* 2007, Grilli *et al.* 2009), tracking of the flux front (Coombs *et al.* 2001) and other power-law representations for the $E(J)$ curve (Bossavit 1994, Maslouh *et al.* 1997). These original ideas can be found in the references. For the same reason the choice of a suitable software tool for the finite-element calculations presented has not been discussed. This is a specific topic bringing interesting new inspirations (Gu and Han 2005, Farinon *et al.* 2010, Stenvall and Tarhasaari 2010).

Before starting to build up a numerical model it is wise to check for analytical solutions that have been worked out for different arrangements of HTS wires (Vellego and Metra 1995, Mawatari 1996, Magnusson *et al.* 1999, Babaei Brojeny and Clem 2005, Mawatari 2008, Mawatari and Kajikawa 2008, Clem 2008, Mawatari *et al.* 2010) or use an estimate based, for example, on HTSs in states of complete shielding (Fabbricatore *et al.* 2000) or complete saturation (Rhyner 2002).

As mentioned already in the introduction, the numerical modeling of HTS is generally understood as one of the key elements in the development of applications in electrical power. Because it is an interdisciplinary field, further progress depends on tight interaction between industry engineers and research scientists from the field of superconductivity applications and, at the same time, on the developments in various fields outside superconductivity such as computer hardware, software tools and mathematical models for systems with non-linear and hysteretic properties; also numerical simulations able to assess, for example, the effect of gaps between tapes on AC loss in power transmission cable (Amemiya *et al.* 2008) or to optimize the basic element of a superconducting resistive fault current limiter (Nguyen *et al.* 2009).

Systematic activity on a global scale is carried out by a recently established workgroup that organizes topical workshops with a regular schedule. The reader interested in following this development can consult http://supra.epfl.ch/modelinggroup/Workshop.asp and http://www-g.eng.cam.ac.uk/epec/HTSModelingWorkgroup/program/index.php which provide information about the first two workshops.

7.7 Acknowledgment

I would like to thank E. Pardo, M. Vojenčiak, E. Seiler, B. Klinčok, V. Zermeno and M. Solovyov for useful discussions and J. Šouc for inspirational experiments. Financial support from Slovak grant agency VEGA under contract 2/0172/09 is acknowledged.

7.8 References

Amemiya N, Miyamoto K, Banno N and Tsukamoto O (1997), 'Numerical analysis of AC losses in high-T superconductors based on E-J characteristics represented with n-value', *IEEE Trans. Appl. Supercond.*, **7**, 2110–2113.

Amemiya N, Muruwawa S, Banno N and Miyamoto K (1998a), 'Numerical modelings of superconducting wires for AC loss calculations', *Physica C*, **310**, 16–29.

Amemiya N, Miyamoto K, Muruwawa S, Banno N and Mukai H (1998b), 'Finite element analysis of AC loss in non-twisted Bi-2223 tape carrying AC transport current and/or exposed to DC or AC external magnetic field', *Physica C*, **310**, 30–35.

Amemiya N, Jiang Z, Li Z, Nakahata M, Kato T, *et al.* (2008), 'Transport losses in single and assembled coated conductors with textured-metal substrate with reduced magnetism', *Physica C*, **468**, 1718–1722.

Amemiya N, Nakahata M, Fujiwara N and Shiohara Y (2010), 'AC losses in two-layer superconducting power transmission cables consisting of coated conductors with a magnetic substrate', *Supercond. Sci. Technol.*, **23**, 014022.

Ashkin M (1979), 'Flux distribution and hysteresis loss in a round superconducting wire for the complete range of flux penetration', *J. Appl. Phys.*, **50**, 7060–7066.

Babaei Brojeny A A and Clem J R (2005), 'Self-field effects upon the critical current density of flat superconducting strips', *Supercond. Sci. Technol.*, **18**, 888–895.

Badia A and Lopez C (2002), 'Vector magnetic hysteresis of hard superconductors', *Phys. Rev. B.*, **65**, 104514.

Banno N and Amemiya N (1999), 'Analytical formulae of coupling loss and hysteresis loss in HTS tape', *Cryogenics*, **39**, 99–106.

Bartolome E, Gömöry F, Granados X, Puig T and Obradors X (2007), 'Universal correlation between critical current density and normal-state resistivity in porous YBa2Cu3O7-x thin films', *Supercond. Sci. Technol.*, **20**, 895–899.

Bean C P (1962), 'Magnetization of hard superconductors', *Phys. Rev. Letters*, **8**, 250–253.

Bossavit A (1994), 'Numerical modeling of superconductors in three dimensions: A model and a finite element method', *IEEE Trans. Magn.*, **30**, 3363–3366.

Brambilla R, Grilli F, Nguyen D N, Martini L and Sirois F (2009), 'AC losses in thin superconductors: the integral equation method applied to stacks and windings', *Supercond. Sci. Technol.*, **22**, 075018.

Brandt E H (1996), 'Superconductors of finite thickness in a perpendicular magnetic field: Strips and slabs', *Phys. Rev. B.*, **54**, 4246–4264.

Brandt E H (1998a), 'Superconductor disks and cylinders in an axial magnetic field. I. Flux penetration and magnetization curves', *Phys. Rev. B.*, **58**, 6506–6522.

Brandt E H (1998b), 'Superconductor disks and cylinders in an axial magnetic field. I. Nonlinear and linear ac susceptibilities', *Phys. Rev. B.*, **58**, 6523–6533.

Brandt E H (2001a), 'Type II superconductors with applied current and magnetic field: effects of geometry and London depth', *Supercond. Sci. Technol.*, **14**,704–707.

Brandt E H (2001b), 'Theory of type-II superconductors with finite London penetration depth', *Phys. Rev. B.*, **64**, 024505.

Brandt E H, and Indenbom M (1993), 'Type II superconductor strip with current in perpendicular magnetic field, *Phys. Rev. B.*, **48**, 12893–12906.

Brandt E H and Mikitik G P (2005), 'Anisotropic superconducting strip in an oblique magnetic field', *Phys. Rev. B.*, **72**, 024516.

Campbell A M (1982), 'A general treatment of losses in multifilamentary superconductors', *Cryogenics*, **22**, 3–16.

Campbell A M (2007), 'A new method of determining the critical state in superconductors', *Supercond. Sci. Technol.*, **20**, 292–295.

Carr W J jr (2001), *AC Loss and Macroscopic Theory of Superconductors*, 2nd edition, Taylor and Francis, London.

Claassen J H and Thieme C L (2008), 'Magnetic properties of Ni-based substrates for HTS tape', *Supercond. Sci. Technol.*, **21**, 105003.

Clem J R and Perez-Gonzalez A (1984), 'Flux-line-cutting and flux-pinning losses in type-II superconductors in rotating magnetic fields', *Phys. Rev. B.*, **30**, 5041–5047.

Clem J R (1998), 'Granular and superconducting-glass properties of the high temperature superconductors', *Physica C.*, **153–155**, 50–55.

Clem J R, Benkraouda M, Pe T and Mcdonald J (1996), 'Penetration of magnetic flux and electrical current density into superconducting strips and disks', *Chinese Journal of Physics*, **34**, 284–290.

Clem J R (2008), 'Field and current distributions and ac losses in a bifilar stack of superconducting strips', *Physical Review B.*, **77**, 134506.

Coombs T A, Campbell A M, Murphy A and Emmens M (2001), 'A fast algorithm for calculating the critical state in superconductors', *Compel – The International Journal for Computation and Mathematics in Electrical Engineering*, **20**, 240–252.

Costa-Bouzo M, Grilli F and Yang Y (2004), 'Modeling of coupling between superconductors of finite length using an integral formulation', *Supercond. Sci. Technol.*, **17**, 1103–1112.

Daumling M (1998), 'AC power loss for superconducting strips of arbitrary thickness in the critical state carrying a transport current', *Supercond. Sci. Technol.*, **11**, 590–593.

Del-Valle N, Sanchez A, Navau C and Chen D X (2007), 'Optimizing levitation force and stability in superconducting levitation with translational symmetry', *Appl. Phys. Lett.*, **90**, 042503.

Del-Valle N, Sanchez A, Navau C and Chen D X (2008), 'Lateral-displacement influence on the levitation force in a superconducting system with translational symmetry', *Appl. Phys. Lett.*, **92**, 042505.

Duron J, Grilli F, Antognazza L, Decroux M, Dutoit B *et al.* (2007), 'Finite-element modeling of YBCO fault current limiter with temperature dependent parameters', *Supercond. Sci. Technol.*, **20**, 338–344.

Dutoit B, Duron J, Stavrev S and Grilli F (2005), 'Dynamic field mapping for obtaining the current distribution in high-temperature superconducting tapes', *IEEE Trans. Appl. Supercond.*, **15**, 3644–3647.

Fabbricatore P, Farinon S, Innocenti S and Gömöry F (2000), 'Magnetic flux shielding in superconducting strip arrays', *Phys. Rev. B.*, **61**, 6413–6421.

Fabbricatore P, Farinon S, Incardone S, Gambardella U, Saggese A *et al.* (2009), 'The transverse resistivity in S/C multifilament wires studied through ac susceptibility measurements', *J. Appl. Phys.*, **106**, 083905.

Farinon S, Fabbricatore P, Gömöry F, Greco M and Seiler E (2005), 'Modeling of current density distribution in critical state by commercial FE codes', *IEEE Trans. Appl. Supercond.*, **15**, 2867–2870.

Farinon S, Fabbricatore P and Gömöry F (2010), 'Critical state and magnetization loss in multifilamentary superconducting wire solved through the commercial finite element code ANSYS', *Supercond. Sci. Technol.*, **23**, 115004.

Friesen M and Gurevich A (2001), 'Nonlinear current flow in superconductors with restricted geometries', *Phys. Rev. B.*, **63**, 064521.

Fujishiro H and Naito T (2010), 'Simulation of temperature and magnetic field distribution in superconducting bulk during pulsed field magnetization', *Supercond. Sci. Technol.*, **23**, 105021.

Genenko Y A, Rauh H, Kruger P and Narayanan N (2009), 'Finite-element simulations of overcritical states of a magnetically shielded superconductor strip', *Supercond. Sci. Technol.*, **22**, 055001.

Gömöry F and Lobotka P (1988), 'Determination of shielding current density in bulk cylindrical samples of high Tc superconductors from AC susceptibility measurement', *Solid State Comm.*, **66**, 645–648.

Gömöry F, Pardo E, Vojenčiak, M and Šouc J (2009), 'Magnetic flux penetration and transport AC loss in superconductor coated conductor on ferromagnetic substrate', *IEEE Trans. Applied Supercond.* **19** 3102–3105.

Gömöry F, Šouc J, Vojenčiak M and Klinčok B (2007), 'Phenomenological description of flux pinning in non-uniform high-temperature superconductors in magnetic fields lower than self-field', *Supercond. Sci. Technol.*, **20.**, S271–S277.

Gömöry F, Vojenčiak M, Pardo E, Solovyov M and Šouc, J (2010), 'AC losses in coated conductors', *Supercond. Sci. Technol.*, **23**, 034012.

Grilli F, Stavrev S, Dutoit B and Spreafico S (2003a), 'Numerical modeling of a HTS cable', *IEEE Trans. Applied Supercond.*, **13**, 1886–1889.

Grilli F, Costa-Bouzo M, Yang Y, Beduz C and Dutoit B (2003b), 'Finite element method analysis of the coupling effect between superconducting filaments of different aspect ratio', *Supercond. Sci. Technol.*, **16**, 1228–1234.

Grilli F and Sjöström M (2004), 'Prediction of resistive and hysteretic losses in a multi-layer high-Tc superconducting cable', *Supercond. Sci. Technol.*, **17**, 409–416.

Grilli F, Stavrev S, Le Floch Y, Costa-Bouzo M, Vinot E *et al.* (2005), 'Finite-element method modeling of superconductors: from 2-D to 3-D', *IEEE Trans. Applied Supercond.*, **15**, 17–25.

Grilli F, Ashworth S P and Stavrev S (2006), 'Magnetization AC losses of stacks of YBCO coated conductors', *Physica C.*, **434**, 185–190

Grilli F, Brambilla R, Martini L, Sirois F, Nguyen D N *et al.* (2009), 'Current density distribution in multiple YBCO coated conductors by coupled integral equations', *IEEE Trans. Applied Supercond.*, **19**, 2859–2862.

Grilli F and Pardo E (2010), 'Simulation of ac loss in roebel coated conductor cables', *Supercond. Sci. Technol.*, **23**, 115018.

Gu C and Han Z (2005), 'Calculation of AC losses in HTS tape with FEA program ANSYS', *IEEE Trans. Appl. Supercond.*, **15**, 2859–2862.

Hong Z, Campbell A M and Coombs T A (2006), 'Numerical solution of critical state in superconductivity by finite element software', *Supercond. Sci. Technol.*, **19**, 1246–1252.

Hong Z, Campbell A M and Coombs T A (2007), 'Computer modeling of magnetisation in high temperature bulk superconductors', *IEEE Trans. Appl. Supercond.*, **17**, 3761–2764.

Inada R, Nakamura Y, Oota A, Fukunaga T and Zhang P X (2004), 'Effect of wire arrangements on AC transport losses in cylindrical conductors composed of Ag-sheathed Bi2223 tapes', *Physica C.*, **412**, 1148–1153

Inada R, Tateyama K, Nakamura Y, Oota A, Li C S *et al.* (2007), 'Total AC loss of Ag-sheathed Bi2223 tapes with various filament arrangements carrying AC transport current in AC parallel transverse magnetic field', *Supercond. Sci. Technol.*, **20**, 138–146.

Inada R, Nakamura Y and Oota A (2008), 'Numerical analysis for AC losses in single-layer cables composed of rectangular superconducting strips with various lateral J(c) distributions', *Journal of Physics Conference Series*, **97**, 012324.

Jiang Z, Amemiya N, Nakahata M, Iijima Y, Kakimoto K *et al.* (2006), 'AC loss characteristics of YBCO coated conductors with varying magnitude of critical current', *IEEE Trans. Applied Supercond.*, **16**, 85–88.

Jiang Z, Amemiya N, Kakimoto K, Iijima Y, Saitoh T *et al.* (2008), 'Numerical calculation of AC losses in multi-layer superconducting cables composed of coated conductors', *Supercond. Sci. Technol.*, **21**, 015020.

Johnston M D, Everett J, Dhalle M, Caplin A D, Friend C M *et al.* (1997), 'Current and field distribution within multifilamentary Bi2223/Ag tapes', *IEEE Trans. Applied Supercond.*, **7**, 1339–1342.

Kajikawa K, Takenaka A, Kawasaki K, Iwakuma M and Funaki K (2001), 'Numerical simulation for AC losses of HTS tapes in combined alternating transport current and external AC magnetic field with phase shift', *IEEE Trans. Applied Supercond.*, **11**, 2240–2243.

Kim Y B, Hempstead C F and Strnad A R (1962), 'Critical persistent currents in hard superconductors', *Phys. Rev. Lett.*, **9**, 306–309.

Kimura F and Amemiya N (2007), 'Numerical analysis of AC losses in multifilamentary coated conductors with non-uniform electric properties', *IEEE Trans. Applied Supercond.*, **17**, 2402–2405.

Klinčok B, Gömöry F and Pardo E (2005), 'The voltage signal on a superconducting wire in AC transport', *Supercond. Sci. Technol.*, **18**, 694–700.

Klinčok B (2007) 'Study of the high temperature superconductor tape carrying AC transport current', PhD Thesis, Institute of Electrical Enginnering, Slovak Academy of Sciences, Bratislava.

Kwasnitza K and Clerc S (1994), 'AC losses of superconducting high Tc multifilament Bi-2223/Ag sheathed tapes in perpendicular magnetic fields', *Physica C.*, **233**, 423–435.

Li Z Y, Ma Y H, Ryu K, Choi S, Park K B and Oh I S (2008), 'Influence of the conductor's arrangement and current on the AC loss characteristics of a fault current limiting coil', *Physica*, **468**, 2063–2066

Lousberg G P, Ausloos M, Geuzaine C, Dular P, Vanderbemden P *et al.* (2009), 'Numerical simulation of the magentization of high-temperature superconductors: a 3D finite element method using a single time-step iteration', *Supercond. Sci. Technol.*, **22**, 055005.

Magnusson N, Mugala G, Schönborg and Hörnfeldt S (1999), 'Critical currents of high-temperature superconductors in homogeneous magnetic fields', *J. Low. Temp. Phys.* **117**, 1519.

Majoros M, Glowacki B A and Campbell A M (2000), 'Transport ac losses and screening properties of Bi-2223 multifilamentary tapes covered with magnetic materials', *Physica C.*, **338**, 251–262

Majoros M, Glowacki B A and Campbell A M (2001), 'Critical current anisotropy in $Ag/(Pb,Bi)_2Sr_2Ca_2Cu_3O_{10+x}$ multifilamentary tapes: influence of self-magnetic field', *Supercond. Sci Technol.*, **14**, 353–362.

Majoros M, Ye L, Campbell A M, Coombs T A, Sumption M D and Collings E W (2007), 'Modeling of transport ac losses in superconducting arrays carrying anti-parallel currents', *IEEE Trans. Applied Supercond.*, **17**, 1803–1806.

Maslouh M, Bouillault F, Bossavit A and Verite J C (1997), 'From Bean's model to H–M characteristic of a superconductor: Some numerical experiments,' *IEEE Trans. Applied. Superc.*, 7, 3797–3801.

Mawatari Y (1996), 'Critical state of periodically arranged superconducting-strip lines in perpendicular fields', *Phys. Rev. B.*, 54, 13215–13221.

Mawatari Y (2008), 'Magnetic field distributions around superconducting strips on ferromagnetic substrates,' *Phys. Rev. B.*, **77**, 104505.

Mawatari Y and Kajikawa K (2008), 'Hysteretic ac loss of polygonally arranged superconducting strips carrying ac transport current', *Appl. Phys. Lett.*, **92**, 012504.

Mawatari Y, Malozemoff A P, Izumi T, Tanabe K, Fujiwara N *et al.* (2010) 'Hysteretic ac losses in power transmission cables with superconducting tapes: effect of tape shape', *Supercond. Sci. Technol.*, **23**, 025031.

Miyagi D, Amadutsumi Y, Takahashi N and Tsukamoto O (2007), 'FEM analysis of effect of magnetism of substrate on AC transport current loss of HTS conductor with ferromagnetic substrate', *IEEE Trans. Applied Supercond.*, **17**, 3167–3170.

Navarro R and Campbell L J (1991), 'Magnetic-flux profiles of high-Tc superconducting granules: Three-dimensional critical-state-model approximation', *Phys. Rev. B.*, **44**, 10146–10157.

Nakahata M and Amemiya N (2008), 'Electromagnetic field analyses of two-layer power transmission cables consisting of coated conductors with magnetic and non-magnetic substrates and AC losses in their superconductor layers', *Supercond. Sci. Technol.*, **21**, 015007.

Navau C, Sanchez A, Del-Valle N and Chen D X (2008), 'Alternating current susceptibility calculations for thin-film superconductors with regions of different critical-current densities', *J. Appl. Phys.*, **103**, 113907.

Navau C and Sanchez A (2001), 'Magnetic properties of finite superconducting cylinders. II. Nonuniform applied field and levitation force', *Physical Review B.*, **64**, 214507.

Nguyen D N, Grilli F, Ashworth S P and Willis J O (2009), 'AC loss study of antiparallel connected YBCO coated conductors', *Supercond. Sci. Technol.*, **22**, 055014.

Nguyen D N, Ashworth S P, Willis J O, Sirois F and Grilli F (2010), 'A new finite-element method simulation model for computing AC loss in roll assisted biaxially textured substrate YBCO tapes', *Supercond. Sci. Technol.*, **23**, 025001.

Nishioka T, Amemiya N, Enomoto N, Jiang Z, Yamada Y *et al.* (2005), 'AC loss of YBCO coated conductors fabricated by IBAD/PLD method', *IEEE Trans. Applied Supercond.*, **15**, 2843–2846.

Norris W T (1970), 'Calculation of hysteresis losses in hard superconductors carrying ac: isolated conductors and edges of thin sheets', *J. Phys. D.*, **3**, 489–507.

Oomen M P, Nanke R and Leghissa M (2003) 'Modeling and measurement of ac loss in BSCCO/Ag tape windings', *Supercond. Sci. Technol.*, **16**, 339–354.

Oota A, Fukunaga T, Matsui M, Yuhya S and Hiraoka M (1995), 'AC losses of Ag-sheathed (Bi,Pb),Sr,Ca,Cu,O, monofilamentary and multifilamentary tapes,' *Physica C.*, **249**, 157–165.

Pardo E, Sanchez A and Navau C (2003), 'Magnetic properties of arrays of superconducting strips in a perpendicular field', *Phys. Rev. B.*, **67**, 104517.

Pardo E, Chen D X, Sanchez A and Navau C (2004a), 'Alternating current loss in rectangular superconducting bars with a constant critical-current density', *Supercond. Sci. Technol.*, **17**, 83–87.

Pardo E, Chen D X, Sanchez A and Navau C (2004b), 'The transverse critical-state susceptibility of rectangular bars', *Supercond. Sci. Technol.*, **17**, 537.

Pardo E, Sanchez A, Chen D X and Navau C (2005), 'Theoretical analysis of the transport critical-state ac loss in arrays of superconducting rectangular strips', *Phys. Rev. B.*, **71**, 134517.

Pardo E, Gömöry F, Šouc J and Ceballos J M (2007), 'Current distribution and ac loss for a superconducting rectangular strip with in-phase alternating current and applied field', *Supercond. Sci. Technol.*, **20**, 351–364.

Pardo E (2008), 'Modeling of coated conductor pancake coils with a large number of turns', *Supercond. Sci. Technol.*, **21**, 065014.

Pardo E, Šouc J and Vojenčiak M (2009) 'Ac loss measurement and simulation of a coated conductor pancake coil with ferromagnetic parts', *Supercond. Sci. Technol.*, **22**, 075007.

Pitel J, Korpela A, Lehtonen J and Kováč P (2002), 'Mathematical model of voltage-current characteristics of Bi(2223)/Ag magnets under an external magnetic field', *Supercond. Sci. Technol.*, **15**, 1499–1506.

Prigozhin L (1996), 'The Bean model in superconductivity: Variational formulation and numerical solution', *J. Comp. Phys.*, **129**, 190–200.

Prigozhin L (1997), 'Analysis of critical-state problems in type-II superconductivity', *IEEE Trans. Applied Supercond.*, **7**, 3866–3873.

Roy F, Dutoit B, Grilli F and Sirois F (2008), 'Magneto-thermal modeling of second-generation HTS for resistive fault current limiter design purposes', *IEEE Trans. Applied Supercond.*, **18**, 29–35.

Rhyner J (1993), 'Magnetic properties and AC losses of superconductors with power law current-voltage characteristics', *Physica C.*, **212**, 292–300.

Rhyner J (2002), 'Vector potential theory of ac losses in superconductors', *Physica C.*, **377**, 56–66.

Ruiz H S and Badia-Majos A (2010), 'Smooth double critical state theory for type-II superconductors', *Supercond. Sci. Technol.*, **23**, 105007.

Safran S, Vojenčiak M, Gencer A and Gömöry F (2010) 'Critical current and AC loss of DI-BSCCO tape modified by the deposition of ferromagnetic layer on edges', *IEEE Trans. Applied Supercond.*, **20**, 2294–2300.

Sanchez A and Navau C (2001a), 'Magnetic properties of finite superconducting cylinders. i. uniform applied field', *Physical Review B.*, **64**, 214506.

Sanchez A and Navau C (2001b), 'Magnetic properties of finite superconducting cylinders. ii nonuniform applied field and levitation force', *Physical Review B.*, **64**, 214507.

Sirios F and Grilli F (2008), 'Numerical considerations about using finite-element methods to compute ac losses in HTS', *IEEE Trans. Applied Supercond.*, **18**, 1733–1742.

Solovyov M and Gömöry F (2011), 'Numerical simulation of magnetic flux penetration and AC loss in HTSC coated conductor tapes', *J. Supercond. Novel Magn.* **24** 69–74.

Šouc J, Pardo E, Vojenčiak M and Gömöry F (2009), 'Theoretical and experimental study of ac loss in high temperature superconductor single pancake coils', *Supercond. Sci. Technol.*, **22**, 015006.

Stavrev S, Dutoit B and Nibbio N (2002a), 'Geometry considerations for use of Bi-2223/Ag tapes and wires with different models of Jc(B)', *IEEE Trans. Applied Supercond.*, **12**, 1857–1865.

Stavrev S, Grilli F, Dutoit B, Klutsch I, Vinot E *et al.* (2002b), 'Numerical modeling of Bi-2223 multifilamentary tapes with position-dependent Jc', *Physica C.*, **372–376**, 1800–1805.

Stavrev S, Grilli F, Dutoit B, Nibbio N, Vinot E *et al.* (2002c), 'Comparison of numerical methods for modeling of superconductors', *IEEE Trans. Magnetics*, **38**, 849–852.

Stavrev S, Dutoit B and Lombard P (2003), 'AC losses of multifilamentary Bi-2223/Ag conductors with different geometry and filament arrangement', *IEEE Trans. Applied Supercond.*, **13**, 3561–3565.

Stenvall A and Tarhasaari T (2010), 'Programming finite element method based hysteresis loop computation software using non-linear superconductor resistivity and T-φ formulation', *Supercond. Sci. Technol.*, **23**, 075010.

Terzieva S, Vojenčiak M, Pardo E, Grilli F, Drechsler A *et al.* (2010), 'Transport and magnetization ac losses of ROEBEL assembled coated conductor cables: measurements and calculations', *Supercond. Sci. Technol.*, **23**, 014023.

Tsukamoto O (2005), 'AC losses in a type II superconductor strip with inhomogeneous critical current distribution', *Supercond. Sci Technol.*, **18**, 596–605.

Vanderbemden P, Hong Z, Coombs T A, Denis S, Ausloos M *et al.* (2007), 'Behavior of bulk high-temperature superconductors of finite thickness subjected to crossed magnetic fields: Experiment and model', *Phys. Rev. B.*, **75** 174515.

Vellego G and Metra P (1995), 'An analysis of the trasnport losses measured on HTSC single-phase conductor prototypes', *Supercond. Sci. Technol.*, **8**, 476–483.

Vojenčiak M, Šouc J, Ceballos J M, Gömöry F, Klinčok B *et al.* (2006), 'Study of ac loss in Bi-2223/Ag tape under the simultaneous action of ac transport current and ac magnetic field shifted in phase', *Supercond. Sci. Technol.*, **19**, 397.

Wilson M N (1983), *Superconducting Magnets*. Oxford, Clarendon.

Yazawa T, Rabbers J J, Ten Haken B, Ten Kate H H J and Yamada Y (1998), 'Numerical calculation of current density distributions in high temperature superconducting tapes with finite thickness in self field and external field', *Physica C.*, **310**, 36–41.

Yeshurun Y and Malozemoff A (1988), 'Giant flux creep and irreversibility in an Y-Ba-Cu-O crystal – an alternative to the superconducting-glass model, *Phys. Rev. Lett.*, **60**, 2202–2205.

Yuan W, Campbell A M and Coombs T A (2009), 'A model for calculating the AC losses of second-generation high temperature superconductor pancake coils', *Supercond. Sci. Technol.*, **22**, 075028.

Zhang G, Dai S, Song N, Guo W, Zhang J *et al.*, (2011) 'The construction progress of a high temperature superconducting power substation in China', Applied Superconductivity Conference, Washington, August 2010.

Part II

Application of high temperature superconductor (HTS) materials and components in energy systems

8
Superconducting fault current limiters and power cables

W. HASSENZAHL, Advanced Energy Analysis, USA

Abstract: This chapter addresses the technology of superconducting power cables and superconducting fault current limiters. The first part of the chapter is a general summary of the electric power grid. This sets the stage for a discussion of the historical development of superconducting AC and DC cables, which begins with low temperature superconductor-based systems. It then describes the present state of superconducting cable technology and gives a summary of some of the major programs around the world. Whereas power cables may be thought of as a replacement for conventional power system components, fault current limiters and fault current controllers have no conventional equivalent in the power grid. Several conceptually different fault current limiter designs are described. Their state of development and installation on the power grid are described.

Key words: electric power, power grid, power cable, AC power cable, DC power cable, fault current limiter, fault current controller, high temperature superconductor (HTS), low temperature superconductor (LTS).

8.1 Introduction

8.1.1 Background

Shortly after superconductivity was discovered in 1911, many in the budding utility industry thought that superconductivity might provide a boon to the electric power industry, just as the development of undersea cables had increased worldwide communications in the late-nineteenth century. That did not happen. In fact, the first half-century after the discovery of superconductivity saw little progress in material characteristics that would allow them to compete with conventional materials. In particular, they did not have the ability to carry sufficient current to allow the design and construction of effective electrical power devices. That changed in the early 1960s with the discovery of Nb-Ti and Nb_3Sn, and the subsequent development of conductors based on these materials. The discovery in 1986 of the high-temperature superconductors (HTS) caused a boom in interest in a variety of superconductor-based applications. After a period of hype, researchers settled down to a marathon of activities on superconducting materials and a variety of applications.

Today, some superconductors have current densities as great as 100 times that of copper and have zero resistance in DC operation. These capabilities provide superconducting power applications with several potential advantages over conventional devices, including higher efficiency, smaller size, and reduced weight. In addition, superconductors have the unique capability of changing between the superconducting state and the normal state under controlled conditions. This phase change, in which their resistivity can change by several orders of magnitude, provides an additional possibility for applications that are beyond the realm of conventional materials. As a result, there is today an ever-increasing promise of real power applications based on superconducting materials.

8.1.2 The electric power grid

Although the electric power grid is seemingly familiar to many, few are aware of the level of complexity built into it and the critical need for stability and reliability. It is a very complicated interactive system with a variety of controls, checks, and balances that generally allow it to operate smoothly. A great deal more information is needed to understand why, 'we, the users, can expect to flip a switch and see the lights go on' (Hassenzahl 2005). An in-depth study of the electric power grid is a life-long challenge. Here we look at a few grid characteristics before moving on to HTS power applications.

It has been claimed that the North American power grid is the largest machine in the world. Power grids, and other large machines, function effectively because of controls, which are, in general, based on maintaining several parameters within small tolerances. One of the most critical parameters in the power system is the nominal frequency of transmitted electric power, which is usually 50 or 60 Hz. Figure 8.1 is a graphic modified from Eto *et al.* (2010) of how various actions are taken to assure the proper frequency of the grid. Various mechanisms are in place to maintain that frequency over several temporal regimes. In the short term, seconds to minutes–and for small frequency variations – the power output of generators is adjusted. For greater frequency changes, a variety of approaches are used. One novel approach is to use rapid-response flywheel storage energy systems. Because of the possibility of loss of a generator or a line from a generator to the grid, government power regulating organizations require a certain fraction of the power generation that is online at any given time to be spinning, but not delivering power. This is referred to as spinning reserve, and companies that can supply spinning reserve are paid a premium for their maintaining that capability. For extreme frequency adjustment, the utilities shed load via prearranged contracts with major users.

On a longer term, minutes to hours, the frequency is controlled by ramping of generation up or down to accommodate the forecast load. This latter control is often referred to as dispatch or, somewhat more correctly, economic dispatch. The term 'economic' is used because the utility prefers to use the most economical – and usually the most efficient – form of generation that is available at any given time. This dispatch mechanism requires some anticipation of the expected load. For example, in anticipation of the morning power demand, power generators that

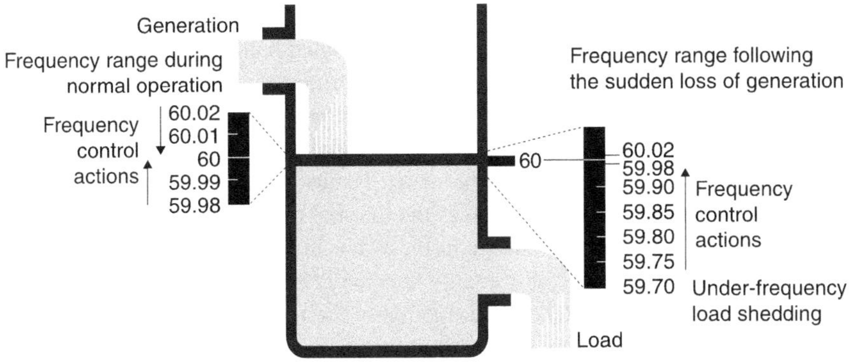

8.1 Simplified concept of frequency regulation and what controls are used to maintain frequency at the nominal value. The level of the 'water' represents the frequency. Normal load and generation variations are such that the nominal frequency of 60 Hz seldom varies by more than ±0.02 Hz. Generation is controlled for small adjustments. Should greater frequency variation occur, usually when generation is lost and the frequency decreases even more, additional controls are needed. These include using the spinning reserve, adding additional generation or using a fast response energy storage system. Ultimately, if the frequency drops below a prescribed level, the utility sheds load. (Source: Lawrence Berkeley National Laboratory.)

are not needed during evening and nighttime hours, and which require some time to reach full power, are carefully ramped up. Dispatch does not require a rapid response and, for many years was accomplished manually by engineers in a central control facility at each local utility. As computer controls became more effective, economic dispatch was taken over by electronic components, though often the generation and load mix remained the same. While mentioning dispatch, it is important to comment on renewables such as wind and solar. Neither is dispatchable in the traditional sense, and wind in particular can cause considerable upset to the normal operation of the grid where there is high penetration, i.e. where they produce a significant fraction of the power. Finally, on a daily basis, usually at a period of low load near midnight, the power output of a significant fraction of controllable generators is adjusted to achieve a small frequency error so that the total number of cycles during each day is precisely 216 000 for a 60 Hz system and 180 000 for a 50 Hz system.

Here we only address large power systems, but utility grids do not exist in many areas. Remote communities operate power systems without ties to an extended power grid (this is often referred to as an island, and the process is referred to as islanding). In addition, some large ships today require the equivalent of a small power grid for their operation. In some cases, even the propulsion system consists of an electric motor directly coupled to the propellers.

It is one thing to know about some characteristic of the power grid, but that does not provide us with a true understanding of what happens in the conversion of potential, chemical, solar, wind, wave, tidal, or nuclear energy into the power that is delivered to us for use on an almost instantaneous basis. To explore more deeply, we must follow the energy from the original source to the ultimate user.

Electrical energy is not available in nature. It must be produced by conversion from some other energy source, such as the thermal energy from the combustion of a fossil fuel, the potential energy in the water in a dam, etc. The method of electricity production from most energy sources is by converting heat to rotary mechanical motion. However, only a fraction of the initial energy is converted to rotary motion. The efficiency of that conversion is determined by various thermodynamic processes. In particular, most of our electricity comes from fossil and nuclear sources, which produce a high temperature. The efficiency of the production of electricity is determined by the upper temperature and the lower temperature in a turbine. This relationship is referred to as the Carnot efficiency (Moran and Shapiro 2003). Generally, the efficiency of gas and steam turbines is in the range 30–50%. Hydroelectric power generation is 85–90% efficient. The United States, Europe, and Japan use almost all the water available to produce electricity. Brazil and Paraguay share the 12.6 GW hydroelectric facility at the Itaipu dam on the Parana River. Brazil uses even more than their share of the electricity generated by purchasing some of Paraguay's share. China has recently developed the Three Gorges facility with a capacity of 18.2 GW. Each of these has several times the power capacity of the largest hydroelectric facilities in the US, such as the Boulder (or Hoover) Dam's 2.08 GW.

Additional sources of energy are wind and solar. They will eventually become important sources of electricity but, in the near term, they continue to have their own sets of barriers to large-scale implementation. They are both relatively inefficient. Commercial solar cells can convert 10–15% of the normally incident light into electricity. The filling factor of most solar configurations reduces that to between 8–10%. Note that some experimental cells have up to 40% conversion efficiencies. However, there is no clear path to their commercial production at a reasonable cost. Under optimum conditions, wind generators can convert about 35% of the energy that passes through them into electricity. However, they do not intercept a great fraction of the wind that is passing by, and the wind is intermittent.

In most cases, electricity is created in a rotating generator, and a transformer steps the voltage up to a level appropriate for long-distance transmission. Near the end of its journey to the user, the voltage is transformed to a level that is appropriate for local distribution systems, and is carried to the site of use by a distribution line or an underground cable. Along the energy flow path, the controls and feedback systems mentioned above ensure the near-continuous operation of the power grid. That is true even under many upset or fault conditions. For example, a large generator may fail or need to be taken out of service unexpectedly, or a significant load may drop from the grid. The former would tend to cause the frequency to decrease and the latter would cause it to increase. In addition, lightning may strike a power line or a transformer, trees may fall on power lines, overheated lines may sag during overload conditions and arc to ground, vandals may shoot the insulators on transmission lines, etc. These latter situations cause not only frequency variations, but significant disruption of power flow and service to small, and occasionally to large areas.

Most generators are over 99% efficient in the conversion of rotational energy into electricity. The high efficiency is, in part, a result of their compact size, which is possible because they use iron and can operate with small separation between the stationary and rotating parts. One result of this design is that generators work best at modest voltages in the 10 to 50 kV range. Transformers have losses that are typically less than 0.5% of the energy that passes through them. Because of these two facts, for most of the twentieth century, AC was essentially the only widespread form of electricity. Today it is straightforward to convert AC electricity to DC electricity and vice versa. Thus the utility and the user have a choice of form of electricity that can be used in any situation and in any part of the grid. High voltage DC power lines are used in many situations to transmit large quantities of power over long distances.

Nearly 10% of the electrical energy generated is dissipated in the path from generator to user. Though this may seem to be a small fraction of the total, it is an enormous quantity of energy. Future technical improvements, including superconductivity, can hope to save only a portion of this 10%.

8.1.3 Superconductors in the power grid

In principle, superconductors could replace conventional conductors in the existing components. However, conventional power technology has become highly sophisticated over the past century so that existing power system elements are effective, efficient, economical and, most importantly, simple and easy to use. These attributes, combined with large-scale production, deliver economical functionality and present a significant challenge to the introduction of any new technology (Haldar and Abetti 2011).

After type II superconductors became a practical reality in the 1960s, some early programs explored power applications such as generators, cables, transformers and motors. In the same period, new concepts with no conventional equivalent, such as flux pumps for converting power and charging high-current magnets, including superconducting magnets for energy storage, and fault current limiters (FCL) were suggested. Designing superconducting devices with the same capabilities as existing conventional systems is a tremendous challenge. In addition, only in the last decade have HTS materials become available that can operate at the temperature of liquid nitrogen. The requirement is that they support high current densities and have the ability to operate in the magnetic fields found in power applications such as generators, motors, and superconducting magnetic energy storage (SMES).

Several other criteria must be met by superconductors to make them an effective part of most power applications. In particular, several mechanisms in the superconductors themselves and in the combined structure of a superconductor and normal metal convert electrical energy to heat. These include simple resistance, hysteresis losses, coupling currents and eddy currents (Wilson 1983). Whereas losses in conventional conductors are removed at near ambient temperatures by conduction or convection, losses in superconductors occur at cryogenic temperatures and must be removed by a refrigerator. Just as the Carnot efficiency determines the effectiveness of converting heat into rotational energy, it also determines the thermodynamic efficiency of removing heat from a cryogenic environment (Hassenzahl 2006). In addition, refrigeration systems only function at a fraction of ideal. As a result, depending on the temperature at which the heat is generated and the total amount of heat, the room temperature electrical energy required to remove the heat may be 10–50 times as large as the thermal energy itself. The implication is that if efficiency is an important factor, the superconducting system must be 15–50 times as efficient as the conventional technology that it is replacing (Gouge *et al.* 2002).

Nevertheless, a great deal of development and demonstration efforts are in the works today on a broad range of superconducting power technologies. Most of the programs have considerable government support, though there are some exceptions, such as a recent FCL installed by Nexans in Germany. The Nexans technology will be discussed in the section on FCLs. As electrification increases, industrial, urban and commercial areas require more and more power. The impact on the power grid is mainly seen as the need for greater and greater power flow along a particular corridor, greater power transfer in individual substations, etc. These factors are independent of any increase in overall power use. Superconducting power technologies such as transformers, cables, and fault current limiters will likely find use in niche situations where conventional solutions are inadequate. The fact that their enhanced performance can in some cases offset initial development costs will accelerate their use.

We can separate superconducting power applications into two groups. One includes those in which conventional materials are replaced by superconductors. The other includes novel technologies that are based on unique characteristics of superconductors. Cables, motors, generators, transformers, and flywheels form the first group and SMES, FCLs and fault current controllers (FCCs) are in the second. Additional applications of superconductivity are a superconducting substation (Zhang *et al.* 2011, Hassenzahl 2000) and integrated power transmission systems in which liquid hydrogen is transferred as a fuel and cools a superconducting cable (Bartlit *et al.* 1972).

The IEEE has been instrumental in publications that address superconducting applications in various fields, including power. Two issues of the *IEEE Proceedings* focused on superconductivity and had articles that show some of the historical developments and give snapshots of the status in August 1989 and October 2004

(Kirtley and Edeskuty 1989, Hassenzahl *et al.* 2004, Kalsi *et al.* 2004). In addition several articles on power applications of superconductors were published in the May, June and August 2000 issues of the *Power Engineering Review* (PER), at the time the official journal of the IEEE Power Engineering Society. The July 1997 issue of *IEEE Spectrum* was dedicated to developments of HTS applications for electric power systems.

8.1.4 What may the future hold?

Superconductivity power applications have great promise, but it seems that the challenges today are exceptional and will slow development for at least five reasons:

- The cost of superconductor, which is several times that of copper.
- The previously described issue of removing heat from a cryogenic environment.
- The existing high level of development of conventional power components.
- The need for large-scale production to achieve competitive costs. Both large quantities of superconductor and multiple instances of individual devices will be needed as steps along the way to commercialization.
- The economic environment today is such that governments around the world are reducing their support for all but the most applied research.

Having heard the bad news, there is some hope because developments of superconducting materials continue to advance the technology, and there will be specific niche cases where the capabilities of the superconducting applications will overcome both the development costs for few-of-a-kind devices and the higher cost of materials. Thus, we can expect to see a few, perhaps even dozens, of superconducting FCLs, power cables, and transformers installed at utilities over the next decade. In addition, it is likely that some small generators will be installed in the nacelles of offshore wind-power generators. The other power technologies will attend future breakthroughs.

8.2 Fault current limiters

8.2.1 Introduction

Electric transmission and distribution systems are designed to operate in a variety of conditions. Normal operation has a fixed voltage at every location in the system and a nominal current that is usually in the range of a few hundred to a few thousand amperes. A variety of upset conditions can cause the current to increase to higher levels. For example, loss of a transformer in a substation typically requires one or more transformers in the same substation or nearby substations to operate at higher power levels. Transformers are designed conservatively and have overload ratings for operation at higher power levels for a certain period of time with little effect on operation or their functional life. During some emergency conditions, however, transformers are loaded to even higher levels, which cause the temperature to increase. Insulation life is reduced if excessive temperatures occur for long periods. Similarly, other components of the system are designed to accommodate the various off-normal conditions.

One upset condition that can have serious consequences is a fault, or short circuit. Faults can occur almost anywhere in the transmission and distribution system. Because they can be 10–50 times the rated current, they can cause major damage to various components. The damage is typically a result of increased forces, which are proportional to the square of the current, $F \propto I^2$, or, extreme heating when the fault continues for a long enough period. The temperature reached on any conductor in the system during a fault is given by:

$$T(t_f) = \int_0^{t_f} \frac{J^2 \rho}{C} dt \qquad [8.1]$$

where T is the temperature, t is time, t_f is the time at which the fault is cleared, J is the current density in the conductor, ρ is its resistivity, and C its specific heat. The temperature can rise above the melting point of aluminum or even copper in less than a second during some fault conditions. Note that this same equation is used in the design of fuses and in determining the maximum quench temperature of a superconductor.

Faults may occur between phases of a three-phase system, maybe between one phase and ground, may be associated with a lightning strike, an induced short, etc. Fault duration, if unchecked, varies from a few cycles to much longer periods. No matter the source of the fault or the resulting current, each component of the grid is designed to withstand the mechanical and thermal loads for a few cycles. Early in the development of power systems, engineers came up with four simple lines of defense against faults. The first is to have a fusing element in the circuit that simply melts or vaporizes before other system components are damaged. The second is a circuit breaker, a device that is forcibly opened when the current exceeds a preset limit. The third is an explosive-driven current interrupter. Fuses, circuit breakers, and explosive interrupters require a few cycles to operate. The fourth is the use of inductors or reactors that are permanently installed at critical locations in the grid.

As electrification expands, the number of power stations, other grid components, and interconnections on the existing system increase. The result *vis-à-vis* fault current is that it continues to increase in almost every transmission and distribution system. Technically, the reasons for the increase are twofold: first, the increased number of interconnections reduces the effective system impedance; and second, the total power available in any area generally increases over time. Since the voltage is fixed, the lower impedance creates larger fault currents. It is not reasonable (nor economical) to replace older network elements just to deal with higher fault currents; thus, some method of limiting them is needed. In principle, an increase of the internal impedance of each element in the grid might accomplish this end. There are, however, two undesirable side effects of this type of solution. First, higher impedances cause increased energy loss; and second, fluctuating loads would require stronger control systems to avoid reduced overall grid stability.

Today, the electric utilities most frequently control the effects of fault currents, by either inserting reactors (inductors) in critical locations, or by interrupting the fault current with special switchgear. An example of reactors that are used is shown in Fig. 8.2. These reactors are passive devices that provide a voltage drop that is proportional to the current. At operating current that voltage is usually between 0.5–3.0% of the local system voltage. When a fault occurs and the current increases, the voltage across the reactor increases linearly with the current and may exceed half of the local system voltage; i.e. during a fault, the impedance of the reactors is a large fraction of the impedance of the entire grid. The increased voltage drop causes the overall fault current to be reduced considerably. Steady-state reactors have two disadvantages: firstly, they consume power on a continuous basis and thus reduce the efficiency of the power system; secondly, the increased reactance decreases the responsiveness of the grid.

The second solution to controlling fault current is to install high-voltage circuit breakers that open when the current exceeds some predetermined value. Figure 8.3 shows some of the switchgear in use today. The interested reader may find further information on circuit breakers in Dam (2009). Though the existing

8.2 Three reactors in elevated tanks (one is hidden from view) at the TIDD substation on the AEP power grid. The tanks for these reactors are live, i.e. they are tied to one terminal of the reactor. The site has been chosen for a demonstration of a superconducting FCL, which will eventually replace the reactors. (Courtesy of Zenergy Power and AEP.)

8.3 Six circuit breakers at a high-voltage substation in California. They are dead-tank breakers, i.e. the tanks are at ground potential. These circuit breakers are oil-filled and operate independently. They open during a fault and reclose after a fixed number of cycles. They then reopen if the current remains high, i.e. if the fault has not cleared.

switchgear is effective today, the technology is near the theoretical limit of current interrupting capability. One issue is that the circuit must be broken in such a way that arcs, which are created as the switches are opened, can be quenched at the zero current point of the cycle. Often a combination of reactors and circuit breakers are used together. Note that circuit breakers alone have no effect on maximum fault current during the first cycle, thus, without a fault current limiter, all grid components must be designed to accommodate the forces produced by the maximum fault current.

A device having a negligible impedance during normal operation and the ability to develop a considerably higher impedance within a very short time (less than a quarter of an AC cycle) after a fault occurs could limit the short-circuit current to a reasonable value. Ideally, such an FCL would operate passively and would return to its normal operational mode immediately after the fault clears. Note that electric power literature prior to the 1970s sometimes referred to reactors and circuit breakers as fault current limiters. Electric power utilities became interested in this issue as power levels were increasing in the 1970s. The first conference on the topic, which included superconductors as a possible component, was hosted by EPRI in 1976 (EPRI 1977). The rest of this chapter describes some of the history of superconducting FCLs and the present status of their development. It is of note that in addition to superconducting FCLs, there has been considerable effort on the use of silicon-based electronics to control current and power flow in various parts of the grid. There is a wide set of applications that use power electronics for this type of control. The general term is flexible alternating current transmission systems (FACTS) (Hingorani and Gyugyi 1999). The technology evolved from the power converters that are used for the connection of high voltage direct current (HVDC) power cables and power lines to the AC grid. One specialized aspect of the technology is the development of silicon-based devices for the control of fault currents. This technology is often referred to as fault current controllers (FCCs). One of the superconducting technologies discussed below is a FCC.

8.2.2 Superconducting fault current limiters and controllers (SFCLs and SFCCs)

The first article with a proposal for the use of the phase change characteristics of practical superconductors for controlling fault currents in a power system was by Laquer (1968). Earlier, he provided the first description of superconducting switches and flux pumps for a variety of applications including charging superconducting magnets (Laquer 1962a, 1962b, Laquer *et al.* 1966). Since that seminal effort, hundreds of patents and thousands of papers have been written on superconducting FCLs. It is beyond the scope of this summary chapter to discuss a significant fraction of the concepts and the intricacies of the various designs. However, it is instructive to comment on some other early work and to include in the references some review papers that are specific to FCLs or that include FCLs as a part of a general review of superconducting power applications.

Early developments of superconducting FCLs occurred long before the perovskite superconductors were discovered in 1986. In fact, most of the concepts that are the basis of ongoing developments were conceived prior to that discovery. One of the earliest approaches (Rogers *et al.* 1983, Boenig and Paice 1983) was to use a superconducting coil that was switched into the circuit when the current exceeded a predetermined level. Several electronic switches were installed so that the coil could be in the circuit twice during each cycle. The advantage of using a DC coil is that AC losses are avoided during normal operation and cryogenic refrigeration requirements are small. The first FCC constructed and tested in a utility environment was based on this concept (Leung *et al.* 1997). Development began in the early 1990s and two devices were fabricated using HTS BSCCO–2223. The first, a 2.2 kV, 2.4 kA_{rms} device was completed in 1995 and was successfully tested at the Southern California Edison (SCE) Center Test Substation in Norwalk, CA. A second, larger device, specified at 26 MVA and designed to operate at 12.5 kV, 1.2 kA_{rms} and was tested at the same facility in 1999. This device was designed so that it would be transportable, as shown in Fig. 8.4. Arcing occurred in the connections between one of the three superconducting coils and the connection to a bushing during full voltage tests. The device was subsequently

8.4 A transportable fault current controller at the substation in Los Alamos. The picture was taken during testing that occurred after electrical arcing during the initial tests at SCE were repaired. (Source: Los Alamos National Laboratory.)

repaired by the staff at the Los Alamos National Laboratory and one phase was successfully tested at a Los Alamos substation where the active voltage was 13.7 kV (Waynert *et al.* 2003). Other work on this concept and several variations thereof have been carried out by various researchers; Hoshino *et al.* (2001) is one example. However, no other large devices have been constructed. This technology's loss of favor is mainly due to the physical size of the device, the amount of superconducting material needed, and the fact that it is not passive, i.e., it requires a feedback and control system to operate.

Today, the functionality of all commercial and pre-commercial superconducting FCLs (SFCL) depends on a phase change in either a superconductor or in a magnetic material. Perhaps the most significant factor in this choice is that relying on a phase change to accomplish current control allows the devices to operate passively. Over the past decade, several general papers on FCLs describe how they function, discuss the dozen or so ways in which they can be used in the grid, and give a snapshot of the status of the technology (Noe and Oswald 1999, Nagata *et al.* 2001, Hongesombut *et al.* 2003, Hassenzahl *et al.* 2004, Noe and Steurer 2007, Young and Hassenzahl 2009, Hassenzahl and Young 2010). Here we focus on three FCL technologies and present some information on existing and planned installations.

Three conceptually different types of SFCLs are under development in various laboratories and industries around the world. These are best described as resistive, shielded core, and saturated (or saturable) core FCLs. The latter two types, which rely on a ferromagnetic core, are sometimes referred to collectively as inductive type FCLs. However, because the way they operate is so different, they are treated

Table 8.1 General characteristics of three superconducting fault current limiter concepts

Technology	Losses	Triggering	Recovery	Size/weight	Distortion
Resistive SFCL	Hysteretic (depends on HTS materials).	Passive	HTS conductor must be re-cooled.	Potential to be small, because HTS performs limiting action.	Only during first cycle.
Shielded-core SFCL	Hysteretic (amount depends on HTS materials).	Passive	Faster than resistive, but re-cooling required.	Large and heavy due to iron core and windings.	Only during first cycle.
Saturable-core SFCL	Continuous power for DC coil that saturates the iron core, and joule heating in copper coils.	Passive	Immediate.	Large and heavy due to iron core and conventional windings.	Some; caused by nonlinear characteristic of iron core.

separately here. The three types are described briefly in Table 8.1, which is adapted from Xiao (2009).

8.2.3 Resistive SFCLs

Figure 8.5 shows the principle of the resistive SFCL. The current limiting element consists of a superconductor, which can be wire, tape, or bulk material that provides a path R_{SC} for the current. A parallel resistor R_P may be in the form of a sheet of normal material in contact with the superconductor, or may be a separate element that has no physical contact with the superconductor. During normal operation, the resistance of the superconductor is zero; it carries the entire operating current and the voltage drop is zero. In the event of a fault in which the current increases over some threshold, the superconductor must rapidly transition to the normal state over its entire volume in a fraction of a cycle. Completely quenching the superconductor avoids hot spots and assures sharing, particularly if multiple parallel superconducting elements are used. Once the superconductor quenches, its resistance increases by several orders of magnitude and the normal material begins to carry the majority of the current. Even though the overall current is greater than the normal operating current, the superconductor current $I_{SC}(t)$ decreases sufficiently so that the superconductor is not damaged:

$$I_{SC}(t) = I_{ac}(t)\frac{R_P}{R_P + R_{SC}(t)} \tag{8.2}$$

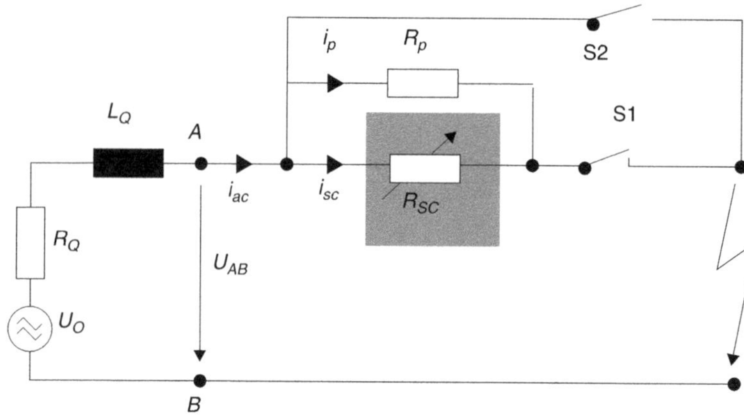

8.5 Schematic of a resistive SFCL showing the superconductor as a variable resistance. There are several variations on this design that have been used by various manufacturers.

The normally closed switch, S1, may be included as part of the FCL or it may already be a part of the grid. Its operation will depend on the overall capabilities of the FCL. If the FCL can recover under load, then S1 will remain closed after the fault. If the FCL requires some time to recover, then switch S2 can be closed and switch S1 opened until the FCL is ready to operate again.

The passive transition of the superconductor and the high normal resistance causes the actual current to be considerably less than the prospective fault current. The value of the limited fault current depends on R_p and the impedance of the load. The ratio of the prospective fault current to the resulting fault current can be as much as 10, and depends on the design of the FCL. The action protects the power grid and decreases the maximum current that is seen by the circuit breakers.

Several groups have developed variations on the theme, shown in Fig. 8.5. Termed hybrid resistive FCLs, they may include a separate, fast switch in series with the superconducting element. This switch quickly isolates the superconductor after most of the current has transitioned to the shunt element, thereby allowing the superconducting element to begin the recovery cycle while the limiting action is sustained by the shunt. Other possibilities are to install a reactor in parallel with the FCL. The current automatically transfers to the reactor when the superconductor quenches.

8.2.4 Shielded core SFCLs

The shielded core FCL combines the phase change of a superconductor with the transformer effect of a ferromagnetic core. In a sense, it may be considered a variation of the resistive FCL with the HTS electrically and cryogenically isolated from the high-voltage grid, as shown in Fig. 8.6. An electromagnetic connection

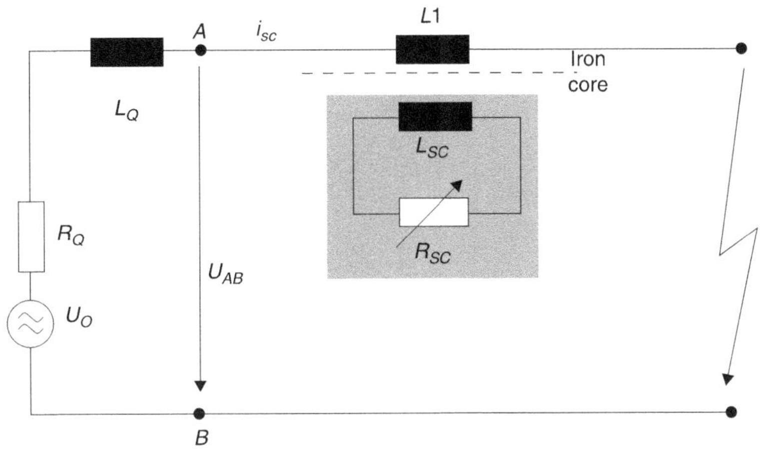

8.6 Schematic of a shielded core SFCL showing that the high voltage circuit is isolated from the cryogenic environment.

is made between the line and the HTS element through mutual coupling of a high-voltage AC coil and a closed-circuit superconducting coil. In essence, this FCL is a transformer in which the secondary is a shunted HTS element. During a fault, the increasing current on the superconductor causes the HTS element to quench. In the resistive state the shunted transformer effect is lost, and the iron core and the high-voltage AC coil form a reactor that reduces the fault current. Following a quench only a small current flows in the superconducting coil. Thus, recovery can be quite rapid once a fault has cleared.

8.2.5 Saturable core SFCLs

The saturable core SFCL is conceptually quite different from other types of SFCLs. Rather than using a phase transition in the superconductive material; it relies on a phase change in an iron core. The behavior of the magnetic properties of iron is dynamic in that it can change from the saturated state to the unsaturated state or vice versa almost instantaneously. The exact time for a transition depends on a variety of factors including material characteristics and the format of the core, which is usually made of many thin sheets of a specialty steel. Because of the extensive development of transformers over the past century, the transition time can be measured in microseconds. An example of the concept for one phase of a three-phase AC system is shown in Fig. 8.7. There are two independent iron cores and two AC windings for each phase. The AC windings use conventional conductors that are wrapped around the core to form an inductance in series with the AC line. Both coil and core assemblies are positioned inside a superconducting coil that saturates the iron. As shown in the figure, the two conventional coils are in series and are wrapped with opposite helicities. Because the current alternates

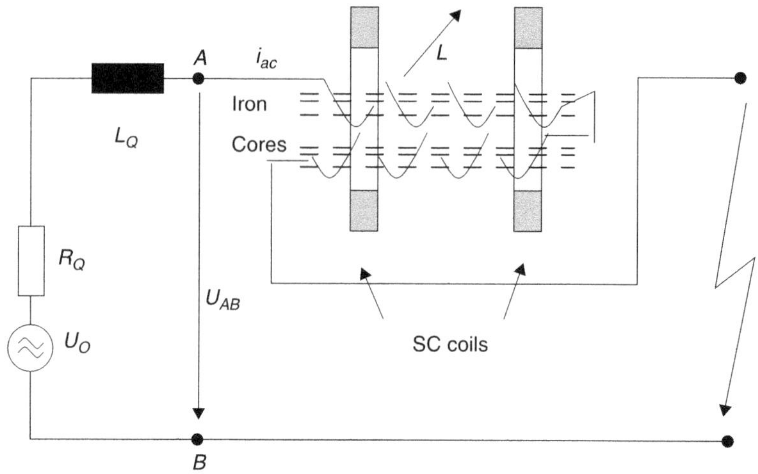

8.7 Schematic of one phase of a saturated core SFCL showing the two high voltage coils that are connected to the power grid and superconducting solenoids that saturate the iron cores. There are several variations on this design that have been used by various manufacturers.

in the coils, during one half of the cycle the field in one of the coils bucks the superconducting field and the other boosts the superconducting field. During the next half cycle, the condition reverses. The geometry shown in Fig. 8.7 is the simplest form that allows an easy understanding of the concept.

Practical considerations in the design of the SC coil and the high-voltage AC coils have led to other designs. Early designs used a single superconducting coil to saturate the iron in all three phases. Several devices of this design have been installed and tested at low voltage as a prototype on utility systems. This geometry is often referred to as the 'spider' design.

The number of ampere-turns in the superconducting coil must be sufficient to maintain iron saturation so long as the AC current is at or below the nominal operating values. At any instant this is only an issue in the coil whose field opposes the field induced by the superconducting coil. When a high-current fault occurs and the current in the AC coil exceeds a critical level during a part of a cycle, the iron in the coil that is in opposition falls out of saturation. While the current is above that critical level the device appears as a reactor and has a significant voltage drop, or back electromotive force (emf), which causes the fault current to be decreased. Because the value of the inductance is quite variable on a millisecond basis, the waveform of the voltage across the device is not a simple sinusoid. The results of a test of a model saturated core FCL during a fault are shown in Fig. 8.8. The abscissa on the plot is time, with each large division representing 10 ms. The two curves show the measured voltage and the voltage calculated in a model of the device. The differences are not significant in terms of device performance.

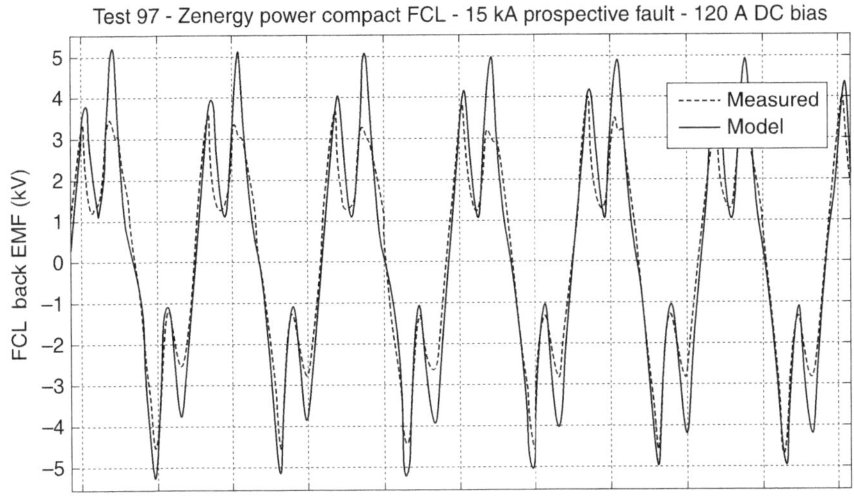

8.8 Results of a test of one phase of a saturated core SFCL during a fault. The abscissa is time, with the large divisions representing 10 ms. Both measured and calculated voltages are shown. The back emf of an FCL is a measure of its current limiting capability.

The important advantages of the saturated core SFCL are:

- The superconducting coil is fully isolated from the high voltage of the power grid,
- The high-voltage coils and their containment are based on conventional transformer design.

8.2.6 Commercial activities

Although there have been many pre-commercial SFCL activities, there are two companies that have several ongoing installations. In many regards their technologies are representative of the general field of FCL development, while at the same time they are have active patents that safeguard their particular developments. The first of these companies is Nexans, which has built and installed several resistive SFCLs on the power grid, including the first commercial resistive SFCL (Dommerque *et al.* 2010). The device described in this paper received no government support for design, construction, or installation. The second company is Zenergy Power, which has built and installed several saturated core SFCLs on utility power grids.

Nexans is a major manufacturer of power and communication cables for various applications. Over the past decade they have designed and produced several resistive SFCLs for grid deployment. Their particular design consists of an arrangement of superconducting coil elements that transition to the normal state when their critical current is exceeded. These elements consist of ceramic BSCCO-2212 bulk tubes and a conventional metallic conductor in parallel, as shown in Fig. 8.9. These tubes are manufactured using a sophisticated melting, casting, and processing (MCP) technology. The number, size, and arrangement of these elements in a given FCL are determined by the operating voltage and current at the installed site and by the expected and allowable fault currents.

The bulk superconducting material is soldered on the inside of a normal conducting metal tube and equipped with suitable contacts at each end. The assembly is then machined to form a spiral coil and the resulting element is insulated. The spiral allows a 1 m length element to have an effective superconducting length of several meters. The normal conducting bypass is necessary for the protection against hot regions, which is especially important in the case of low current faults where an entire element might not transition to the normal state. A critical part of the element

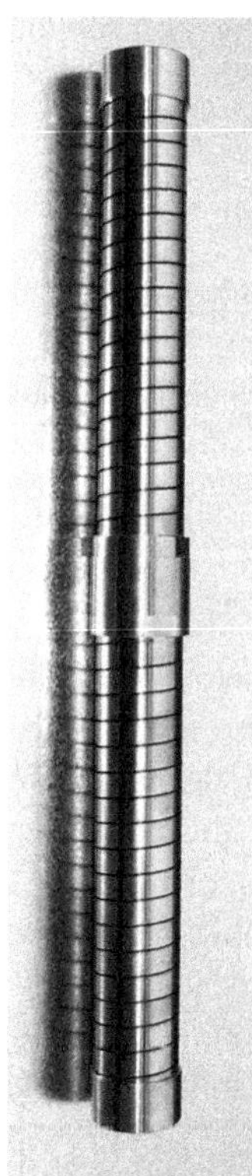

8.9 A Nexans HTS SFCL element produced by a melt casting process using BSCCO-2212. The typical element is somewhat less than 1 m long and has a diameter of about 6 cm. (Courtesy of Nexans.)

fabrication is to assure complete electrical contact between superconductor and the normal metal bypass over the whole conducting length. A critical parameter in the overall SFCL design is the voltage drop per centimeter of length along the superconductor. The choice of this value is a trade-off among several factors including reliability and overall cost, and can vary from one installation to another.

A large number (a few tens) of such elements are connected in series. This structure has a given current and voltage capability. If the current requirements for the FCL are greater than the capability of individual elements, then two or more may be connected in series as needed. A concept of the assembly of one phase of the resistive SFCL is shown in Fig. 8.10. This figure shows the fabrication of a module consisting of many superconducting elements, and its installation into a cryostat. Nexans has recently initiated a program to use YBCO tapes in their FCLs in an effort to reduce cost. As this book is going to press Nexans is installing the first such device on a utility in Germany.

Zenergy Power (Zenergy) is a multinational company with subsidiaries in Germany, the US, and Australia. Saturated core SFCL technology is one of Zenergy's focus areas. Zenergy will install two FCLs on utility systems within the next two years. One will be at the AEP Tidd substation in Ohio, which was shown in Fig. 8.3; and one at a CE Electric UK substation near Scunthorpe, England.

The Zenergy SFCL design has gone through several stages, progressing from some initial concepts for modest devices to more robust systems that can operate at transmission-level voltages of 130 kV or greater. A model SFCL employing four copper DC coils is shown under test at the Lane Cove facility in Sydney, Australia in Fig. 8.11. The similarity with the concept shown in Fig. 8.7 is clear.

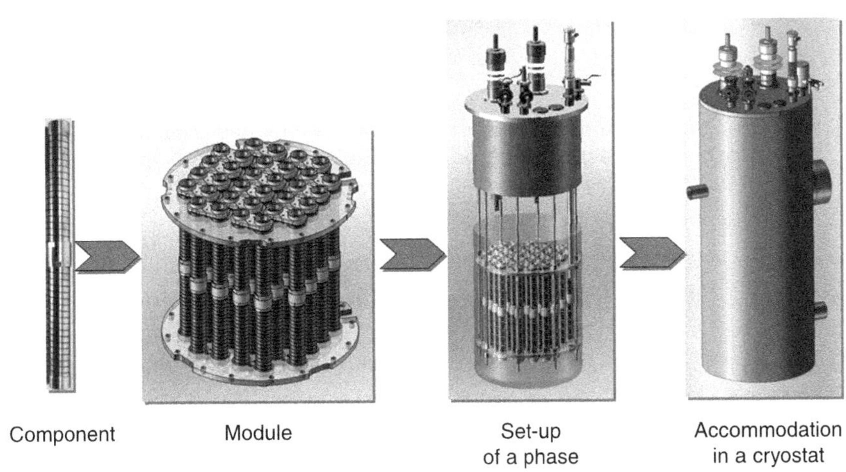

8.10 Assembly of multiple elements in series and parallel as needed to match system voltage and current requirements. Also shown are the set up and installation of a single phase into a cryostat. (Courtesy of Nexans.)

8.11 A test set up for a model Zenergy SFCL. The coils that saturate the iron are conventional and the heat that is generated is such that each test can only last a few seconds. Full height of the model is about 2 m. (Courtesy of Zenergy Power.)

Zenergy Power has plans to install two SFCLs over the next year. The first will be in the CE Electric UK network in the UK. The oil tank for this device is shown in Fig. 8.12, and one of the two superconducting coils is shown in Fig. 8.13. Installation of this device should be completed before publication of this book. The second SFCL will operate at transmission voltages and is scheduled to be installed on the AEP power system at the Tidd substation, which was shown in Fig. 8.3. This device is under construction. Because of the high voltage, it will use three single-phase SFCLs, each in a separate oil tank. Figure 8.14 is an artist's concept of the installed device. For a size comparison, each of the three units will be approximately the same size as the circuit breakers shown in Fig. 8.4.

8.12 Tank for the Zenergy SFCL to be installed at CE Electric. This will be a three-phase SFCL so that three different high-voltage coil sets and iron cores will be installed in the tank. They are made using typical transformer technology. When installed, the axis of the tank will be vertical and it will be filled with oil of the type that is used in transformers. (Courtesy of Zenergy Power.)

8.13 One of two Zenergy superconducting magnets for the SFCL to be installed at CE Electric in the UK. The coil is made of BSCCO-2223 from American Superconductor. It can produce a magnetic field over 2 T and fully saturates the iron cores in the tank shown in the previous slide. This picture shows power leads for the coil and the flanges where two refrigerators will be installed. (Courtesy of Zenergy Power.)

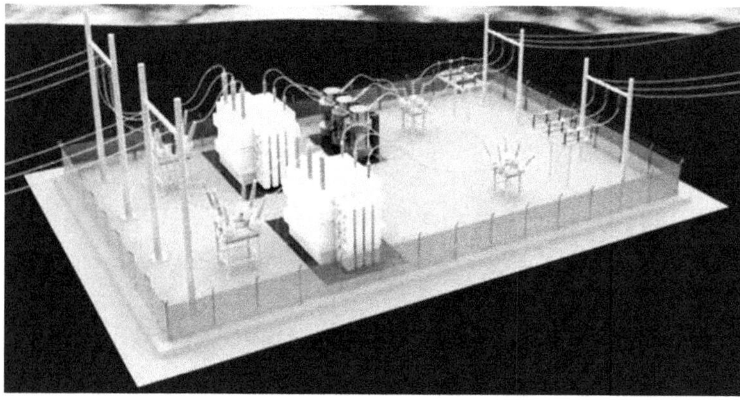

8.14 An artist's concept of the three single-phase SFCLs installed at AEP's Tidd substation, which was shown in Fig. 8.2. Though approximately 3 m tall, the FCLs are dwarfed by the other components of the substation, only part of which is shown. (Courtesy of Zenergy Power.)

8.3 Superconducting power cables

8.3.1 Background on conventional cables

Conventional underground cables are used for power transmission and distribution in a variety of situations. Urban underground and undersea power cables are the most highly developed and widely distributed of all cable applications. The most important drivers for these and other cable applications are rights of way, environmental concerns, and effectiveness in niche markets. Undersea power cables were first used in locations where islands are near a mainland and where cables coupled with the use of large and efficient power plants were more effective and reliable than local, island-based power plants, which needed fuel on a continuous basis. The urban use of power cables has a longer history and is typical in all cities having a high population density, particularly in industrialized countries. Cable technology for three-phase power systems is highly developed. In general, the length of single cables is limited by total shipping weight and shipping spool diameter. Figure 8.15 shows a cable in place on a spool and Fig. 8.16 shows the transport of three such spools on a highway between the manufacturing facility and the site of installation. The rationale for the historical development of various conventional cable designs, their installation, and the critical issue of splicing two cables together are described in detail in the EPRI publication referred to as the *Green Book* (EPRI 2006).

The key to the effectiveness of cables compared to elevated power lines is that the voltage breakdown capability (V/m) of the insulators used for cables is 100–1000 times greater than that of air. Thus, the power flow along underground cables in a restricted corridor can be considerably greater than is possible with overhead

8.15 A typical high-voltage power cable made for underground installation. Drum diameter and length are approximately 3 m and 4 m respectively. Shipping weight, in this case about 60 tonnes, and dimensions determine the maximum length of cables of a given diameter. Usually cable lengths are 700–2000 m. (Courtesy of CCI Ltd.)

8.16 Three drums similar to that shown in Fig. 8.15 during transit to an installation site. Three generally identical cables are needed to form a three-phase power system. (Courtesy of CCI Ltd.)

lines along the same right of way. A further advantage is that the stray electric field from a three-phase underground cable system is negligible compared to that of overhead power lines. Both overhead power lines and underground cables transmitting AC electric power are affected by the inductances among their conductors. In both cases the inductance causes the relative phase of the current and the voltage to shift as the distance from the power source increases. The real power, P, that can be used is equal to the product of the current, the voltage and the cosine of the phase between them; $P = I_{rms} V_{rms} \cos(\alpha)$.

The phase shift α depends on the geometry of the lines or cables and also depends on the magnitude of the current. The electric utilities install capacitors along AC lines and cables to oppose the effect of the inductance. One disadvantage of AC cables compared to overhead AC lines is their higher inductance, which requires phase correction every 30 km or so, instead of 500 km for overhead power lines.

Superconducting AC cables having the same cable diameter as conventional cables will be able to carry much greater currents and, depending on the design, will have a reduced inductance so that the phase variation with length will be reduced. Thus, they will have a capability of carrying more power in a given corridor than conventional cables. Most conventional AC cables are buried and have significant losses. Thus, since cooling in the underground environment is limited, they tend to operate at elevated temperatures when the loads are high. Superconducting cables may be more efficient than conventional cables even considering the refrigeration load, they may have longer lives, and the life cycle cost may be less. In the not-too-distant future, superconducting cables will, in some niche applications, replace both overhead lines and conventional cables.

8.3.2 History of superconducting power cables

Before addressing various programs that are in effect today, it is of interest to review some of the more significant developments in the history of superconducting power cables. It is of note that before the early work on superconducting cables, there were several programs on the development of cryoresistive cables. In particular, in the 1960s, teams in the US, France, Germany, and Russia considered the possibility of using the decreased resistivity of high purity aluminum at cryogenic temperatures to produce a more efficient power cable. The rationale for this approach was that the resistivity of both aluminum and copper decrease more rapidly than the efficiency of refrigeration so that there would be a net increase in overall efficiency of cables using conventional conductors and operating at a lower temperature. Depending on the magnetic field and the purity of the material, the resistivity of copper at 4 K can be less than 0.5% of its value at room temperature, and the resistivity for aluminium can be as low as 0.1% of its value at room temperature. This was a strong driver for the development of cryogenic power cables. Though there is not space to discuss cryogenics here, its availability in the 1960s was the result of developments within the industrial gas industry, including their work on cryogenic refrigeration systems in the 1940s and 1950s. It is that effort that has provided an enabling technology for the work on superconducting power cables.

Although it is generally thought that superconducting power cables should be very efficient, it was not until the late 1960s that (Long and Nottaro 1971) and others carried out the experiments needed to show that the losses (including the refrigeration power requirement) in superconductors could in fact be lower than those of cooled normal conductors. Making these early loss measurements was a challenge, and the researchers in this area had to develop both cryogenic and electronic equipment as well as establishing measurement procedures in order to definitively determine the actual losses. Some of those measurements and many other early observations are described in Brookhaven National Laboratory (1968), noting that the emphasis of that study was particle accelerators for physics experiments. Several reports of the time (Hammel and Rogers 1970, Hammel *et al.* 1970) summarized work in that area. Particle accelerator technology was instrumental in the improvement of the characteristics of superconducting materials more than any other technology until the invention and commercial development of magnetic resonance imaging.

Many attendees at the 1968 study were already considering power applications of various types. Aggressive development of superconducting power cables had begun a few years earlier when Nb-Ti and Nb_3Sn became available on a near-commercial scale. It was also in the 1967–9 timeframe that multifilamentary Nb-Ti superconductors became available. Prior to that time, however, Klaudy (1967) in Austria had determined many of the criteria needed for the development of superconducting AC cables. Early work on model and prototype superconducting cables began in Europe (Klaudy *et al.* 1981, Moisson and Leroux 1971) and later in the US (Laquer *et al.* 1977, Forsyth and Thomas 1986) and in Russia (Dolgosheyev *et al.* 1979). Significant development programs existed for AC and DC superconducting cables. These programs addressed the significant issues of superconducting cable technology; and the research in the 1970s and early 1980s is generally applicable to many of the technical issues associated with HTS power cables today. The most critical issue that led to their disfavor was the need for an extensive refrigeration system to maintain the operating temperature. The utilities were concerned with operating and maintenance costs and other issues of a superconducting cable with an extensive refrigeration system that would not be completely under utility control.

The US had the largest power cable efforts, which included Brookhaven National Laboratory's AC cable program (Forsyth and Thomas 1986) and the Los Alamos Scientific Laboratory's DC cable program (Laquer *et al.* 1977). At the time, electric power use was increasing by up to 10% per year, and it was generally thought that energy costs would go down as more and more nuclear reactors were built and grouped in farms with 10 or more GW capacity. History has shown us differently, but the need for long-distance high-power transmission was already affecting several corridors. In particular, a 1.4 GW DC line termed the Pacific DC Intertie had been completed in 1970. It moved power from the hydro resources in the Columbia basin (specifically the Celilo converter station near The Dalles, Oregon) to Southern California. Recognizing the effectiveness of DC power transmission and the extended hydro resources in places like Siberia and the James Bay region of Canada, Garwin and Matisoo (1967) proposed superconducting DC cables with the capability of carrying 10 GW to transmit power from these areas to load centers in more populated areas.

1986 saw renewed interest in superconducting power cables. Since the refrigeration issue had turned out to be the show-stopper for LTS cables, the fact that HTS-based power cables would require a factor of 10 or so less refrigeration encouraged many cable manufacturers and utilities to embrace the new materials. Though capital and operating costs are less, operating and maintenance costs are still an issue and were the subject of an electric utility workshop (EPRI 2004). Though the refrigeration requirements for HTS cables are less than for LTS cables, there are several barriers to the wide scale implementation of HTS cable technology. The two most important are:

- They are more expensive per unit of current carried.
- The perovskite materials resemble mica and are brittle rather than robust as is Nb-Ti. The latter material is malleable and has mechanical characteristics that in some regards are better than steel.

In many locations AC power cables are pulled into ducts that are installed in multiples of three. In some cases the utilities plan for the future expansion by having additional holes that can be filled with additional cables when power demand increases. Because superconductors can carry orders of magnitude more current than conventional conductors of the same cross-section, a likely use for HTS AC cables will be their installation as additional or retrofit cables in some existing paths where the existing power cables are operating at or near their safe limits.

8.3.3 Types of superconducting power cables

Superconducting power cable designs are, for the most part, modifications of the designs that have been used for over a century for conventional cables. There are, however, some adaptations to take advantage of the characteristics of superconductors. Many of those adaptations were proposed during the development of LTS cables (Laquer *et al.* 1977, Forsyth and Thomas 1986). The most important differences take advantage of some of the characteristics of superconductivity.

The simplest, and the design first used for superconducting cables, used a superconductor in a cryostat with a conventional electrical insulation on the outside. The HTS wires are wrapped around a copper core and are in a tube filled

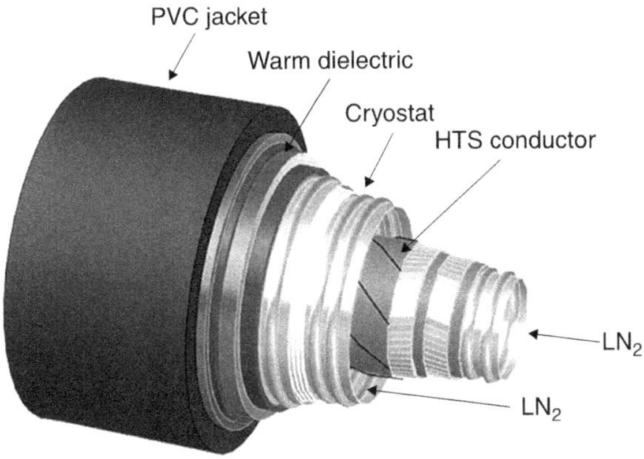

8.17 One configuration of a single-phase warm dielectric cable. The coolant flows in one direction on the inside of the flexible core tube and in the other direction outside of the superconductor. This arrangement may have a significant counterflow of heat between the two fluid paths, which limits the length of cables made of this design. A conventional extruded dielectric is used on the outside of the cable. Not shown is a copper layer that is a protective element when the critical capabilities of the superconductor are exceeded.

with flowing coolant, typically liquid nitrogen. This design is shown in Fig. 8.17 and is referred to as a warm dielectric design. The copper at the center is a protection element that carries the current for short periods when the critical current of the superconductor is exceeded. It uses the least amount of HTS conductor for a given level of power transfer. However, it has a high inductance and an external magnetic field. If the three phases are in close proximity, the magnetic field produced by one phase causes hysteresis losses in the other phases.

Because of the high losses associated with the warm dielectric cable, a cable with two layers of superconductor, with the outer layer acting as a magnetic shield for adjacent cables, was proposed. The two concentric HTS layers are insulated from each other by a cold dielectric as shown in concept in Fig. 8.18. This cold dielectric design has been used for most superconducting cable installations. Liquid nitrogen coolant may contact both layers, providing both cooling and dielectric insulation between the center conductor layer and the outer shield layer. Compared to the warm dielectric design, it has a lower inductance, a higher current carrying capacity, reduced AC losses, and very low stray magnetic fields. It also uses more superconductor than either of the other designs.

Perhaps the most effective AC cable design is the three-conductor concentric triplex design shown in Fig. 8.19. Because each of the three layers of insulation must be fairly thin for the cable to be flexible, it seems to be most effective for low- to medium-voltage levels. The design has many advantages because it has no stray magnetic fields, a very low inductance, and it uses less superconducting

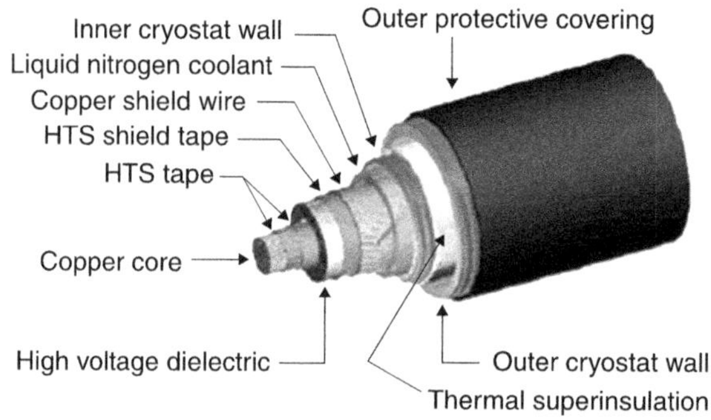

8.18 One phase of a cold dielectric cable system. Because the dielectric is cold it must be a flexible structure, such as a wound tape that is permeated with the liquid nitrogen.

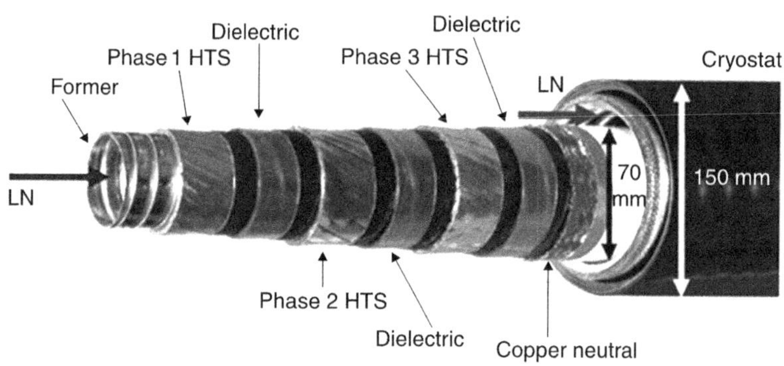

8.19 A triaxial three-phase cable showing the three layers of HTS tape and the three layers of electrical insulation. This particular cable is a section of the cable installed by Southwire at AEP's Bixby substation. (Courtesy of Southwire.)

material than the other designs. However, it does require special wound insulation that is permeated with liquid nitrogen.

There has been little interest in HTS DC cables because the power levels and lengths necessary to make a DC power system affordable are well above the capabilities of the manufacturing capabilities of HTS materials at the present time. To explain this, the cost of the converter stations at the terminals of a DC line cost as much as a hundred or more kilometers of the line itself. Similarly, the cooling requirements of the terminations of a high-power, high-current DC cable are greater than the cooling requirements of several kilometers of cable. Thus the topic was discussed, but not addressed in detail until EPRI began a detailed study (EPRI 2006).

The developing interest in renewables such as wind and solar have increased the possibility of the need for massive power transfers, e.g. greater than 10 GW over distances of 1000 km or more. This requirement must be compared to the AC cables that carry at most 250 MW and are generally less than 1 km in length. Most of those HTS AC cables use many short cryostats that are permanently evacuated. An installation of 1000s of km must use a different approach because of the reliability issues associated with 10 000 cryostats that must work in concert. Avoiding this difficulty requires a different concept for the overall cable system design. Figure 8.20 shows the EPRI concept for a superconducting DC cable that is designed to transmit 10 GW over distances greater than 1000 km (Hassenzahl *et al.* 2009). Refrigeration stations are required every 10–20 km Liquid nitrogen below 70 K and at a pressure of 20–30 bar seems to be the ideal coolant. Because of the long length, the elevation changes several hundred meters along the path between refrigerators. The high pressure was chosen to accommodate these effects.

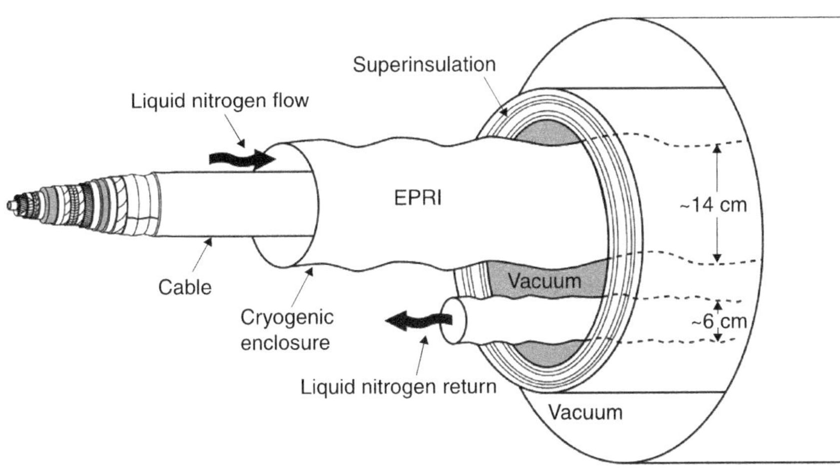

8.20 A superconducting DC cable designed to carry 10 GW over distances greater than 1000 km. The cable is vacuum insulated and uses multilayer insulation to keep the total heat input to the coolant to less than 1 W/m length. Two flow paths for the cryogen are used to allow long-distance operation without a counterflow heat exchange effect. The outer wall can be made of a conventional steel pipe of the type that is being used for high-pressure gas pipelines. The cable shown inside the cryogenic enclosure contains two layers of superconductor and two protective layers of copper. Insulation between the two layers of superconductor is rated at the peak value of the system with safety margins. A thinner layer of insulation is placed between the outer layer of superconductor, which is nominally at ground potential and an outer corrugated steel shield. (Reproduced with kind permission from EPRI.)

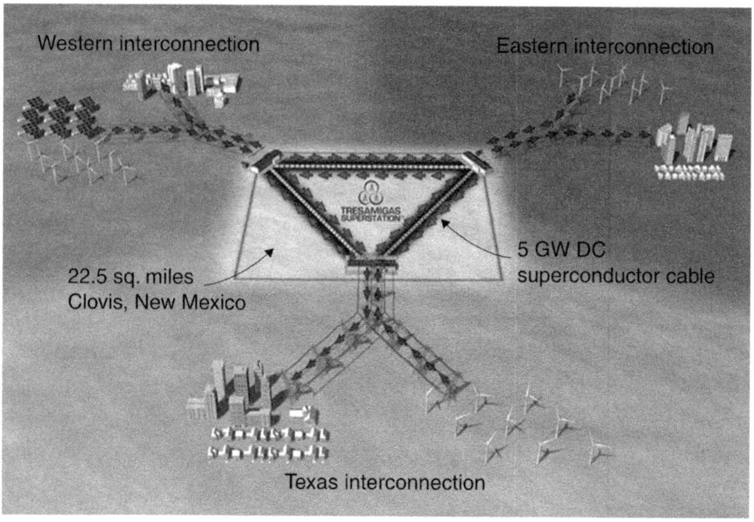

8.21 An artist's concept of the Tres Amigas power interconnection between the three power areas of the United States. Tres Amigas will be located in southwestern New Mexico.

Not all power systems that are in close proximity operate at the same frequency, and even if they are at the same frequency they need not be fully synchronous. Japan uses both 50 Hz and 60 Hz power in different parts of the country. The Itaipu dam produces power at both 50 Hz and 60 Hz to satisfy users in Brazil and Paraguay. The United States power grid is separated into three distinct control areas: western, eastern, and Texas. These areas all operate at 60 Hz, but are not synchronized. As a result, where it is necessary to transfer power from one area to another, it must be converted to DC and then back again into AC. The three control areas abut in the southeastern part of New Mexico. The Tres Amigas facility has been proposed to interconnect the three areas, as shown in Fig. 8.21. Because the land requirement for the 5 GW power converters is so large, the proposal is to separate them by about 10 km. A DC line or cable is needed for the interconnection and the consortium developing Tres Amigas is considering superconducting DC cables.

8.3.4 Installed superconducting power cables

Several superconducting power cables have been installed around the world over the past decade or so. Some of them have operated for a considerable time. A few of those cables are shown in Figs 8.22, 8.23 and 8.24. The captions for those figures indicate location and some characteristics of the cable installation.

8.22 Terminations for the three phases of the LIPA cable produced by Nexans and AMSC. The installation is on Long Island in New York. The insert shows a cross-section of the cold dielectric cable. (Courtesy of AMSC and Nexans.)

8.23 The first long-term superconducting cable installation at the Southwire facility. Southwire is a producer of copper conductors for transformers, cables, etc. This three-phase cable installation delivered power to the plant for over 26 000 hours. (Courtesy of Southwire.)

8.24 Test installation of the three phases of the Changtong warm dielectric HTS cable at the Changtong cable plant in Baiyin, China. (Courtesy of Changtong Cable.)

8.4 Conclusions

Superconductivity power applications have been under development since the 1960s. Early analyses and progress with power applications using LTSs provided a basis for work during the last two decades on power applications based on HTSs. There has been sufficient progress in several of the technologies that they are capable of satisfying many of the electric power utilities' technical requirements. Unfortunately, the cost of superconducting material has not decreased to a sufficient level where they can be economically competitive with conventional technologies. However, there are a significant number of niche applications where cost is not as important as performance. For example, superconducting cables will be used in areas where increased power must be delivered in critical corridors to urban areas where energy use is growing rapidly and environmental and other concerns do not allow the installation of additional power cables or transmission lines. Another niche application is the use of superconducting FCLs to avoid upgrading existing equipment. Several FCLs are under development and they show promise in demonstration projects around the world. Not mentioned in the text above is the recent suggestion that MgB_2 may be an ideal material for near-term development of FCLs that will operate at distribution level voltages (15–50 kV) that exist on the power grid in locations close the ultimate power user.

8.5 References

Bartlit J, Edeskuty F, and Hammel E (1972) 'Multiple use of cryogenic fluid transfer lines', *Proc. Int. Cryogenic Eng. Conf.*, Eindhoven, pp. 176–179.

Boenig H, and Paice D (1983), 'Fault current limiter using a superconducting coil', *IEEE Trans. On Magnetics*, **Mag-19**, 3 pp. 1054–1058.

Brookhaven National Laboratory (1968) *Summer Study on Superconducting Devices and Accelerators*, Brookhaven National Laboratory Report BNL-50155 C-55 1, 2, and 3.

Dam Q (2009) *Preserving Circuit Breaker Adequacy in Electric Power Systems*, Saarbrücken, Lambert Academic Publishing.

Dolgosheyev P, Peshkov I, Svalov G, Bortnik I and Karapazyuk V (1979) 'Design and first stage test of 50-meter flexible superconducting cable', *IEEE Transactions on Magnetics*, **Mag-15**(1), pp. 150–154.

Dommerque R, Skramer S, Hobl A, Bohm R, Bludau M, Bock J, *et al.*. (2010), 'First commercial medium voltage superconducting fault current limiters – production, test and installation', *Supercond. Sci. and Tech.*, **23** 034020.

EPRI (1977) 'New concepts in fault current limiters and power circuit breakers', *EPRI Symposium Proceedings*, EL-276-SR.

EPRI (2004) *EPRI Cryogenic O&M Workshop—Proceedings*, ed. W. Hassenzahl, EPRI Report 1008699 2004.

EPRI (2006), *Underground Transmission Systems Reference Book*, EPRI product number 1012334.

Eto J H, Undrill J, Macklin P, Daschmans R, Willaims, B, *et al.*, (2010) 'Use of frequency response metrics to assess the planning and operating requirements for reliable integration of variable renewable generation', LBNL Report LBNL-4142E.

Forsyth E and Thomas R (1986) 'Performance summary of the Brookhaven superconducting power transmission system', *Cryogenics*, **26**(11), pp. 599–614.

Garwin L and Matisoo J (1967), 'Superconducting lines for the transmission of large amounts of electrical power over great distances', *Proceedings of the IEEE.* **55**(4), pp. 538–548.

Gouge M, Demko J, Mcconnell B and Pfotenhauer J (2002) 'Cryogenics Assessment Report' ORNL Report.

Haldar P and Abetti P (2011) 'Superconductivity's First Century', *IEEE Spectrum,* March 2011.

Hammel E and Rogers J (1970) 'Cryogenics and nuclear physics – Part I,' *Cryogenics*, June, pp. 5–13.

Hammel E, Rogers J and Hassenzahl W (1970) 'Cryogenics and nuclear physics – Part II,' *Cryogenics*, June, pp. 186–195.

Hassenzahl W (2000) 'The all superconducting substation: a comparison with a conventional substation', EPRI Tech. Rep. TR-1000915.

Hassenzahl W (2005) 'Electricity storage: believe in it', *IEEE Power and Energy*, March/April, pp. 20–21.

Hassenzahl W (2006) 'Cryogenics: A Utility Primer', EPRI report 1010897.

Hassenzahl W, Hazelton D, Johnson B, Komarek P, Noe M *et al.*, (2004) 'Power applications of superconductivity', *Proceedings of the IEEE*, **92**(10), pp. 1655–1674.

Hassenzahl W, Gregory B, Eckroad S, Nilsson S, Daneshpooy A, *et al.*, (2009) 'A superconducting dc cable', EPRI report 1020458.

Hassenzahl W and Young M (2010) 'Superconducting power equipment: technology watch 2010', EPRI report 1019995.

Hingorani N and Gyugyi L (1999) *Understanding FACTS: Concepts and Technology of Flexible AC Transmission Systems*, Wiley-IEEE Press.

Hongesombut K, Mitani Y, and Tsuji K (2003) 'Optimal location assignment and design of superconducting fault current limiters applied to loop power systems', *IEEE Trans On Appl. Supercond*, **13**(2) pp. 1828–1831.

Hoshino T, Salim K M, Nishikawa M, Muta I, and Nakamura T (2001) 'DC reactor effect on bridge type superconducting fault current limiter during load increasing', *IEEE Trans. Appl. Supercond.*, **11**(1), pp. 1944–1947.

Kalsi S, Weeber K, Takesue H, Lewis C, Neumueller H *et al.*, (2004) 'Development status of rotating machines employing superconducting field windings', *Proceedings of the IEEE*, **92**(10), pp. 1688–1704.

Kirtley J and Edeskuty F (1989) 'Application of superconductors to motors, generators, and flywheels', *Proceedings of the IEEE,* **77** 8, pp.1143–1155.

Klaudy P (1967) 'Some remarks on cryogenic cables', *Advances in Cryogenic Engineering*, **11**, pp. 684–693.

Klaudy P, Gerhold I, Beck A, Rohner P, Scheffler E *et al.*, (1981) 'First field trials of a superconducting power cable within the power grid of a public utility', *IEEE Transaction on Magnetics*, **17**, pp. 153–156.

Laquer L (1962a) 'Superconductive electric switch', US patent 3145284, filed Oct. 2, 1962.

Laquer L (1962b) 'Incremental electrical method and apparatus for energizing high current superconducting magnets', US patent 3150291, filed Oct. 2, 1962.

Laquer L (1968) 'Automatic superconducting switches', *Journal of Applied Physics*, **39**(6) pp. 2639–2640.

Laquer, Carroll J and Hammel E (1966) 'Automatic superconducting pump', US patent 3414777, filed June 1, 1966.

Laquer H, Dean J and Chowdhuri P, (1977) 'Electrical, cryogenic, and systems design of a dc superconducting transmission line', *IEEE Transactions on Magnetics*, **Mag-17**, pp. 182–187.

Leung E, Rodriguez I, Albert G, Burley B, Dew M *et al.* (1997) 'High temperature superconducting fault current limiter development', *IEEE Trans. Appl. Supercond.*, **7** 2, pp. 985–988.

Long H and Nottaro J (1971) 'Design features of ac superconducting cables', *Journal of Applied Physics*, **42**(1), pp. 155–162.

Moisson F and Leroux J (1971) 'Development of a superconducting cable for transmission of high electric power', *Journal of Applied Physics*, **42**(1), p 154.

Moran M and Shapiro H (2003) *Fundamentals of Engineering Thermodynamics*, 5th ed., Chichester, John Wiley & Sons.

Nagata M, Tanaka K, and Tanaguchi H (2001) 'FCL location in large scale power system', *IEEE Trans On Appl. Supercond*, **11**(1) pp. 2489–2494.

Noe M and Oswald B (1999), 'Technical and economical benefits of superconducting fault current limiters in power systems', *IEEE Trans. Appl. Supercond.*, **9**(2), pp. 1347–1350.

Noe M and Steurer M (2007), 'High-temperature superconductor fault current limiters: concepts, applications, and development status', *Supercond. Sci. Technol.* **20** pp. R15–R29.

Rogers J, Boenig H, Chowdhuri O, Schermer R, Wollan J *et al.* (1983), 'Superconducting Fault Current Limiter and Inductor Design', *IEEE Trans. On Magnetics*, **Mag-19**(3) pp. 1051–1053.

Waynert J, Boenig H, Mielke C, Willis J, and *et al.*, (2003) 'Restoration and testing of an hts fault current controller', *IEEE Trans. Appl. Supercond.*, **13**(2), pp. 1984–1987.

Wilson M (1983) *Superconducting Magnets*, Oxford, Oxford University Press.

Xiao L (2009) 'A 35 kV, 1.5 kA spider type of FCL is installed at the Puji substation in China', Chinese academy of Sciences private communication.

Young M and Hassenzahl W (2009) 'Superconducting fault current limiters: technology watch 2009', EPRI report 1017793.

Zhang G, Dai S, Song N, Guo W, Zhang J *et al.*, (2011) 'The construction progress of a high temperature superconducting power substation in China', Proc. 2010 ASC, to be published.

9 Superconducting magnetic energy storage (SMES) systems

P. TIXADOR, Grenoble INP/Institut Néel – G2Elab, France

Abstract: Superconducting magnetic energy storage (SMES) is one of the few direct electric energy storage systems. Its specific energy is limited by mechanical considerations to a moderate value (10 kJ/kg), but its specific power density can be high with excellent energy transfer efficiency. This makes SMES promising for high-power and short-time applications. So far, SMES uses NbTi due to its low cost. Excepting the cost, YBCO appears very attractive (current capacity, mechanical performances and 'high' operating temperatures). Since the first studies in the 1970s, successful demonstrations have been carried out but their high cost prevents commercialization.

Key words: SMES, storage devices, large-scale superconductivity, magnet.

9.1 Introduction

A short-circuited superconducting magnet stores energy in magnetic form due to the flow of a persistent direct current. The current remains constant due to the zero DC resistance of the superconductor (except in the joints). The current decay time is the ratio of the coil's inductance to the total resistance in the circuit. File and Mills performed measurements of the persistent current decay and determined decay time constants of 10^5 years[1]. The equivalent resistivity is around 10^{-25} Ω m. The resistivity of copper at room temperature is 1.7 10^{-8} Ω m. Thus, the decay time for a copper coil at room temperature of the same dimensions and inductance would be less than 0.1 ms. Superconductors are thus indispensable for magnetic energy storage systems, except that some might store energy for very short storage durations (lower than 1 s). This storage system is known as Superconducting Magnetic Energy Storage (SMES)[2,3]. This rather simple concept was proposed by Férrier in 1969[4].

The magnetic stored energy (W_{mag}) is determined by a coil's self inductance (L) and its current (I) or, equivalently, by the magnetic flux density and field integrated over all space (Fig. 9.1):

$$W_{mag} = \frac{1}{2} L I^2 = \frac{1}{2\mu_0} \int\int\int_{Space} B\, H\, dxdydz \qquad [9.1]$$

Ferromagnetic materials cannot store a great deal of energy since their magnetic field is close to zero. However, they increase the magnetic flux density in the air that is adjacent to them. This tends to increase the total magnetic energy. Unfortunately, the weight of the ferromagnetic component increases the total

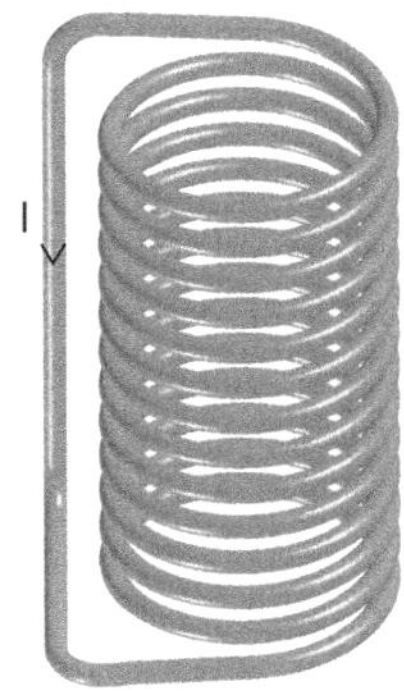

9.1 SMES concept (short-circuited SC winding).

weight so that the specific energy (energy/mass) increases with their use. As a result, they are nearly never used.

When the short circuit is opened, the stored energy is transferred in part or totally to a load. The current of the coil decreases via a negative voltage whereas a positive voltage charges the magnet. SMES is thus a current source. It is the 'dual' of a capacitor, which is a voltage source (Table 9.1). The loads in which the capacitor or a SMES are discharged are not the same. For a SMES the load should be a resistance or a capacitor but not only an inductance. For a capacitor the load should be a resistance or an inductance, but not just a capacitor. The source and the load cannot be of the same type. The load fixes the current for the capacitor and the voltage for the SMES. The resistance of the loads should be different. A low resistance load leads to a long discharge time (L/R) with a SMES and to a

Table 9.1 Comparison of capacity and coil for energy storage

	Storage mode		Discharge mode	Time constant*	Source
SMES	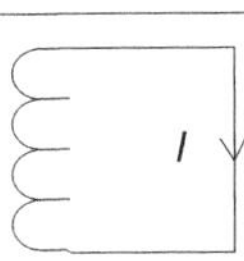 Short-circuit $V = 0$	$w = \frac{1}{2} L I^2$	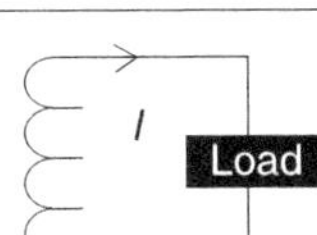	$\frac{L}{R}$	Current
Capacitor	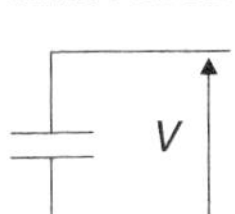 Open-circuit $I = 0$	$w = \frac{1}{2} C V^2$	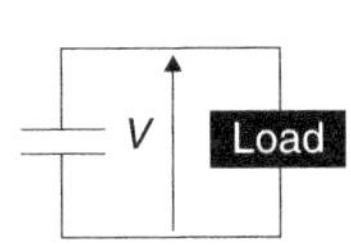	RC	Voltage

* For resistive load.

short one (RC) with a capacitor. In storage mode SMES does not offer any danger in terms of voltages since the voltage is zero, unlike capacitors, which are often high voltage in storage mode.

Energy is lost when a capacitor is discharged into another capacitor. In the same way, if a SMES (inductance L_1) is discharged into another inductance (L_2), energy is lost in the switch, which should simultaneously withstand voltage and current during switching. The current cannot vary instantaneously in the external inductance L_2. With an inductance L_2 with initially no load, the lost energy is expressed by[5]:

$$\Delta W = \frac{L_2(L_2 + 2L_1)}{(L_1 + L_2)^2}\frac{1}{2}L_1 I_1^2 \qquad [9.2]$$

The load inductance should be much lower compared to the SMES inductance to reduce energy losses in the switch.

Figure 9.2 shows the schematic drawing of two main applications of a SMES, connected to an AC grid and directly to a load. A SMES system consists of four main components or subsystems, shown schematically in Fig. 9.2:

- Superconducting (SC) magnet and its mechanical supporting structure and current leads (electric connections between the SC magnet and the room temperature circuit).
- Cryogenic system (cryostat, vacuum pumps, cryocooler, etc.).
- Power conditioning system (interface between the superconducting magnet and the load or electric grid).
- Control system (electronics, cryogenics, magnet protection, etc.).

A rectifier/inverter, a power electronic circuit, is typically part of the power conditioning system, as required to convert the direct current (DC) of the

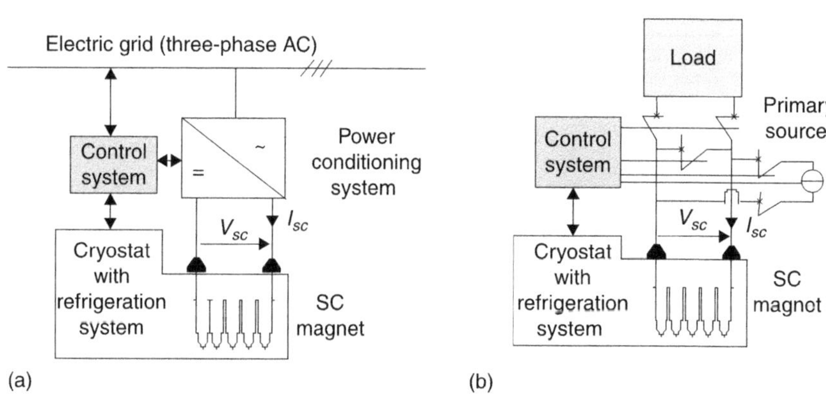

9.2 Main applications of SMES: (a) connected to AC grid (b) connected to a load.

superconducting coil to alternating current (AC) and vice versa since the majority of grids operate in AC (Fig. 9.2).

The power converter has three possible topologies (Fig. 9.3)[6]:

- A thyristor bridge. The reactive power is then linked to the exchanged active power and cannot be independently controlled:

$$P = \frac{3\sqrt{2}}{\pi} U I_{sc} \cos\alpha \, ; \quad Q = \frac{3\sqrt{2}}{\pi} U I_{sc} \sin\alpha = P tg\alpha \qquad [9.3]$$

  where U is the voltage amplitude (RMS value) between the three phases and I_{sc} is the current in the SC magnet, α is the firing angle of the thyristors. This converter induces a lot of harmonics and some AC losses in the SC magnet.
- A voltage source converter in series with a chopper using insulated gate bipolar transistor (IGBT) components; a capacitor is required for the DC bus between the two converters. This topology offers independent control of active and reactive power and low AC distortion. Voltage ripples across the SC magnet lead to some AC losses; of course only the chopper is required if a DC bus is already available.
- A current source converter. Active and reactive power can also be independently controlled. The SMES is directly connected to the DC side (no chopper); the harmonic content may be as low on the AC side as the AC losses in the SC magnet.

In these three cases a transformer, not shown in Fig. 9.3, is generally placed between the grid and the converter.

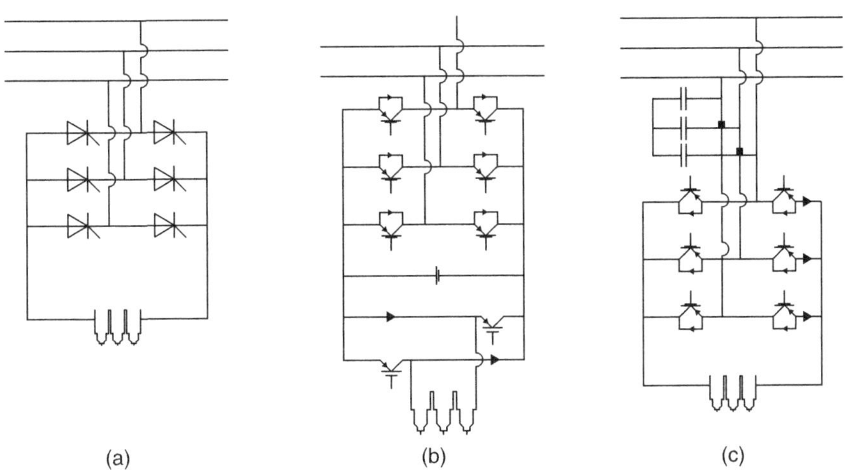

9.3 The three main power converters for a SMES connected to a three-phase AC grid[6]. (a) Thyristor bridge, (b) voltage converter and chopper, (c) current converter.

The cost of the power converter is a significant part of the SMES cost. The power conditioning system is very simple for a direct-load supply; basically comprising switches (Fig. 9.2).

9.2 Superconducting magnetic energy storage (SMES) systems limitations

SMES is an emerging energy storage technology, which has to be compared with other alternatives. The main characteristics of an energy storage device are:

- Specific energy (volume, mass),
- Specific power (volume, mass)/discharge duration,
- Investment and running costs (per kWh, per kW),
- Operation security/fail characteristics,
- Lifetime/number of cycles,
- State of the technology (demonstration/emerging/mature),
- Environmental considerations.

The importance of the different items depends of the application. Nevertheless the energy and the power densities are fundamental. Energy storage systems are classically compared using the Ragone chart, which plots the specific power versus the specific energy (Fig. 9.4). SMES is in terms of energy density between conventional capacitors and batteries, comparable to supercapacitors/ultracapacitors but with higher power density capabilities.

For SMES, the grey zone indicates the presently attained values. The black zone covers theoretically possible ranges, but requires much more research and development.

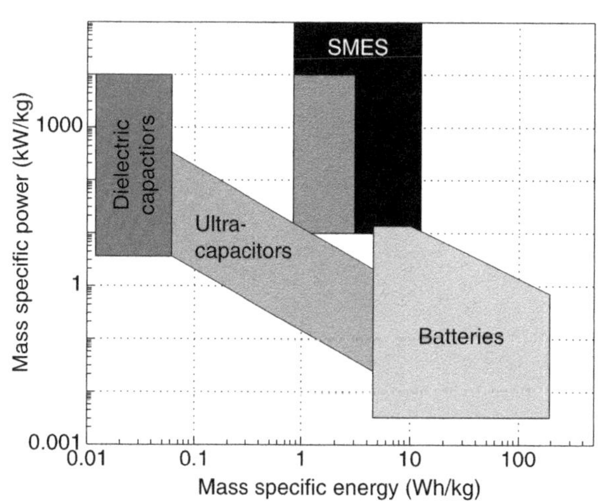

9.4 Ragone plot for SMES, batteries and capacitors.

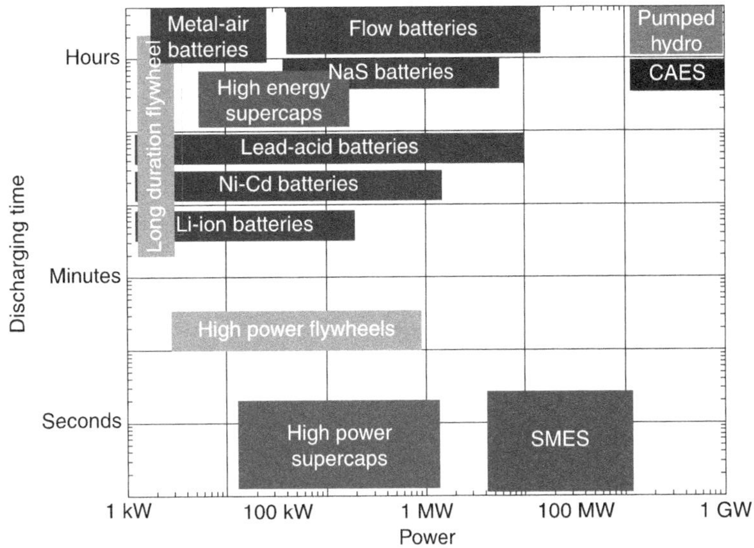

9.5 Discharging time versus power for various energy-storing devices[7,8].

Power characterizes how quickly the energy can be released. The mean power is the ratio of the energy to the discharging time. Figure 9.5 plots the ranges of power and discharge time for different storage technologies. Short discharging times correspond to power quality applications and long discharging times to energy management applications.

Compressed air energy storage (CAES) uses underground reservoirs (salt caverns, old hard rock mines, etc.), to pressurize large volumes of air and then release the stored volume to recover the energy. Pumped hydro storage (two water reservoirs at different elevations) and CAES are the only available technologies for very large energy storage systems.

A Ragone plot (Fig. 9.4) shows that SMES is more a power source than an energy source such as batteries. This is why SMES is suitable for high power, up to 100 MW, and short durations (Fig. 9.5) under a few seconds.

It is important to compare the specific energy of storage systems to classical sources of energy (Table 9.2) to see that their energy performances are very low. One kWh is a lot of energy: it is the kinetic energy in a 10-ton truck travelling at a speed of 100 km/h.

9.2.1 Specific energy and power

Although the attainable magnetic flux density limits the energy per unit volume (Eq. 9.1) ($B^2/2\mu_o$), the real limit of the energy stored in a SMES is mechanical. The virial theorem[9] clearly shows that the magnetic energy is related to the

Table 9.2 Comparison of the specific energy of various systems

Source of energy	Energy density (kWh/kg)
Coal	8
Wood	4
Oil	10–12
Natural gas	10–14
Enriched uranium	600 000
Waterfall (1000 m)	0.003
Pb/Li-ion battery	0.03–0.2
SMES	0.003

structural mass. For one-dimension stresses (tension and compression stresses), the relation between the masses of the mechanical structure in tension and in compression ($M_{tension}$, $M_{compression}$) and the stored energy (W_{mag}) is[10]

$$W_{mag} = \frac{\sigma}{d}(M_{tension} - M_{compression}) \qquad [9.4]$$

where σ is the working stress (tension and compression) and d the structural material density. The total structure mass becomes:

$$M_{Structure} = M_{tension} + M_{compression} = 2M_{compression} + W_{mag}\frac{d}{\sigma} \qquad [9.5]$$

The ultimate limit is when the full mechanical structure only works in tension:

$$M_{Min} = W_{mag}\frac{d}{\sigma} \qquad [9.6]$$

This ultimate limit is theoretically approached for force-balanced or force-reduced coils[11,12] but the winding design and construction remain very complex. 'Force-free' configurations, in which the current flows are always parallel to the magnetic field, have been proposed. However they are physically impossible. The intent of the proposers has been to devise a geometry that would cancel the Lorentz force[13]. In principle, cancellation might be possible in some infinite structures, which are inherently impossible as well. Thus, Eq. 9.6 gives the absolute minimum for real structures. In the fundamental formation of energy and force, the arguments for magnetic systems apply equally for any mechanical system, including kinetic energy storage (flywheels) and compressed air.

Since the virial theorem[9,14] is very important for SMES, let us consider an example to find again the expression. A simple case is the thin infinite solenoid with a thickness e, small compared to the inner radius R. The maximum hoop stress in such a solenoid, considering that turns act dependently, is:

$$\sigma = \frac{J\,B_{\max}\,R}{2} \quad (B_{\max} = \mu_0\,J\,e) \tag{9.7}$$

Energy is only stored inside a thin infinite solenoid:

$$W_{mag} = \frac{1}{2\mu_0} B_{\max}^2 \quad \pi R^2\,h \tag{9.8}$$

where h is the unit height of the solenoid. The magnetic energy per solenoid volume is simply expressed by the hoop stress:

$$W_{mag} = \frac{\sigma}{2} Vol_{solenoid} \tag{9.9}$$

This expression does not take into account the compressive force that reduces the stored energy[15] and the exact relation for an infinitely thin solenoid becomes with the compressive contribution:

$$W_{mag} = \frac{\sigma}{3} Vol_{solenoid} \tag{9.10}$$

This is one third of the ultimate virial limit. The limit of the volume specific energy for a solenoid[11] is about $\sigma/2$. Eyssa[10] gives the specific energy for single-layer solenoids with different aspect ratios. The expression for the stress in a solenoid (Eq. 9.7) shows that at a given stress two main options are possible:

- high magnetic flux density and small radius/volume,
- low magnetic flux density and large radius/volume.

Both magnets will have similar weights for a similar magnetic energy. With low magnetic flux densities, the operating temperature may be much higher. Assuming a reasonable working stress of 100 MPa, Eq. 9.6 gives for a magnet with steel structure working in full tension a maximum value of 12.5 kJ/kg (3.5 Wh/kg) for the stored energy per unit mass. The working stress of 100 MPa may be increased somewhat, but the mass specific energy will still be limited to some tens of kJ/kg.

The Compact Muon Solenoid (CMS)[16] magnet of the LHC collider almost reaches this value for its cold mass (2.6 GJ/225 tons, 11 kJ/kg). The record specific energy reaches 13.4 kJ/kg[17] for a thin NbTi solenoid for particle astrophysics.

The mechanical design of a SMES is a challenging issue; the magnet conductor must be designed to contain tremendous stresses and deformations without degradation of superconducting properties.

One method used to contain the large Lorentz forces of a SMES is the self-supported/cold concept[3]. The cold structure of the magnet resists the magnetic forces. The functions have to be combined to optimize the specific energy: the conductor should provide adequate strength for both current transport and mechanical support. Nevertheless as the stored energy increases additional structures are often required to contain the forces, which increase with size.

Another approach is to transfer the forces outside the magnet to an ambient temperature support. For example, the magnet could be installed in an underground cavern or in reinforced surface trenches. Such a solution might be possible for huge energies, higher than GWh[2,18].

In summary, the specific energy is limited by mechanical stresses to values of tens of kJ/kg (0.003 kWh/kg), much lower compared to batteries or other energy sources (Fig. 9.4, Table 9.2).

If a SC magnet does not store, relative to its size, a huge energy, it can release it in a very short time. The power per unit mass does not have a theoretical limit but often has little sense from a physical point of view. It can, in any case, be extremely high (100 MW/kg). The power is the product of the magnetic current (I_{sc}) by the terminal voltage (V_{sc}) across it. High powers thus require large currents and excellent electric isolation for high voltages.

The maximum power is also determined by the power conditioning system. As energy is removed from the magnet, the current decreases since the energy is proportional to the square of the current (Eq. 9.1). The rated power cannot then be fixed for the maximum operating current I_{sc}^{max}. It refers typically to about half of I_{sc}^{max} and the magnet operates then between $I_{sc}^{max}/2$ and I_{sc}^{max}. Energy always remains in the magnet ($W_o/4$ for $I_{sc}^{max}/2$). This is similar to a flywheel, which never stops. For a discharge at constant power P, the fraction f of the energy used is given by:

$$f = 1 - \left(\frac{P}{V_{sc}^{max} \; I_{SC}^{max}} \right)^2 \qquad [9.11]$$

where V_{sc}^{max} is the maximum voltage across the magnet terminal.

Figure 9.6 gives the voltage and energy versus current for a discharge at constant power from $I_{sc}^{max}/2$ to I_{sc}^{max}. The voltage has a maximum value at the end of the discharge to compensate for the current reduction. However, an SMES can be totally discharged but not at constant power.

For ultrafast discharging times, under 100 µs, the series and derivative capacities of the coil should be taken into account.

The relationship between the stored energy and the SC volume[19] is

$$W_{mag} \propto Vol_{SC}^{3/2} \qquad [9.12]$$

This expression assumes the same magnetic flux density and current density in the superconductor, but not the same mechanical stress, which increases with size. It shows that contrary to many other energy storage systems, the SMES active volume does not increase linearly with stored energy. It implies cost advantages for large energy SMES. However, when the energy increases, the engineering

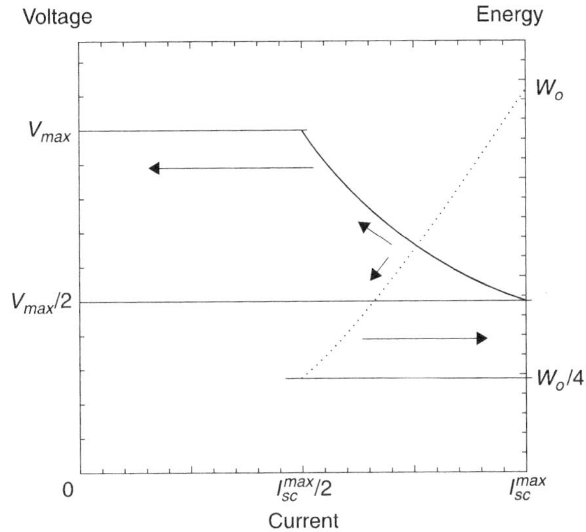

9.6 Voltage and energy evolutions versus current during a discharge at constant power.

current density must in general decrease for stability and protection reasons. Equation 9.12 should then be taken with some care.

9.2.2 Energy transfer efficiency

During current changes (charge and discharge) some energy is lost due to the AC losses in the superconducting coil and to eddy current losses in the cryostat. These two contributions can be kept to a very low level (some percentage of the stored energy) thanks to a suitable design of a low-AC-loss superconducting conductor and of the cryostat. Therefore, SMES shows excellent energy conversion efficiencies, greater than 95%. This value is very high compared to other storage systems (batteries 70–90%, pumped hydro up to 70%). This inherently high energy efficiency is due to the absence of energy conversion to and from another form, mechanical or chemical. For the same reason, the capacitors also show a high energy conversion factor of 90–95%.

However, this efficiency is important to adjust the SMES to the load energy, but it is not the real energy efficiency of the system, which may be irrelevant depending upon the application. The energy required by the cryogenic system should then be taken into account and degrades the high transfer efficiency.

Charging the magnet must be less rapid than its discharge. The operating margins do indeed decrease during charging, while they increase during discharge. The temperature margin is the difference between the operating (T_o) and current sharing (T_{cs}) temperatures. At T_{cs} the superconductor starts to be dissipative and

T_{cs} decreases with the transport current. It reaches T_o for the critical current at T_o and T_c for no current. During charges and discharges the operating temperature increases due to the AC losses but T_{cs} increases or decreases during a discharge and a charge, respectively.

The number of charge–discharge cycles can be very high since it is mainly limited by mechanical fatigue of the support structure. A SMES may then be used for repetitive and rapid bidirectional exchanges of power with the load.

9.2.3 Summary and main SMES applications

The SMES is a short time/pulse current source. The energy can be directly usable without any conversion. The stored energy density is limited by mechanical considerations but can be released in a very short time with little energy losses. It is basically a transient power source, not an energy source. The charge can be fast. Without moving parts except in the refrigeration system, SMES requires only very reduced maintenance. It is an environment-friendly device.

The characteristics of SMES make them a promising candidate for pulse power sources in different fields, such as flexible AC transmission systems (FACTS), rail electromagnetic launchers[20,21] or catapults (aircraft launch), magnetic forming (use of electromagnetic forces to form metal)[22], and possible others.

The rail electromagnetic launcher uses a pair of conductive rails with a moving part with sliding/rolling contacts between. Supplied by huge currents, the velocities may reach several km/s. Surpassing the velocities of conventional guns, a possible application of the electromagnetic launcher is the rail gun. It could also be used to launch small payloads into suborbital or orbital altitudes[23].

A flexible AC transmission system (FACTS) is a device that enhances the operation (security, capacity and flexibility) of power grids. FACTS cover various grid applications and SMES suit those characterized by a high power and a moderate energy such for transient stability, voltage sag smoothing and flicker mitigation.

Table 9.3 characterizes three different SMESs intended for different power and energy ranges. Only the smallest was constructed and operated.

Table 9.3 Some SMES characteristics

	SMES plant[2]	SMES/ETM[24]	5 MVA SMES[25]
Power (MW)	1 000	400	5
Magnet diameter	1 000	129	0.648†
Magnet height (m)	19	7.5	0.7†
Current (kA)	200	200	2.657
Superconductor	NbTi	NbTi	NbTi
Operating temperature (K)	1.8	1.8	4
Status	Design only	Abandoned	Used for voltage dips

† (4-pole configuration)

9.3 Superconducting magnet

The heart of the SMES, the SC magnet must be designed to:

- minimize the SC volume for a given magnetic energy taking into account the mechanical stresses,
- ensure proper cooling and mechanical support of the electromagnetic forces.

The magnet must fulfil the specified electromagnetic signature, how far should the 0.5 mT line be, for example. It must be protected in case of a quench, which should be avoided by proper design. Nevertheless, quench is always a possible event, which must not degrade the magnet.

9.3.1 Magnet topologies

There are two main magnet topologies: solenoidal and toroidal (Fig. 9.7). The solenoid has a simple structure and the electromagnetic forces are easier to handle than in a toroid. A toroid is, indeed, submitted to a large net radial force towards the central axis, in addition to the transverse and longitudinal forces. The quench of a toroidal coil is problematic since it creates an imbalance in the force distribution. The main advantage of a toroid is its natural low stray field since the field is contained only within the magnet bore. Its disadvantage is that it stores only about half of the energy per unit conductor stored by the non-shielded solenoid. Nevertheless, the conductor quantity per unit stored energy becomes nearly the same when compared to an actively shielded solenoid. Active shielding uses compensated coils around the main magnet (Fig. 9.7) in order to cancel the magnetic field outside.

A hexagonal arrangement of solenoids with opposite polarity for adjacent coils (Fig. 9.8) can be used to reduce the stray field from solenoids[26]. Moreover this topology offers a modular design with elementary 'small' solenoids. The production of several identical coils leads to cost savings.

The energy per unit conductor volume is optimized in a non-shielded solenoid with a diameter to height ratio of about five but without stress constraint.

9.3.2 Cryogenics

Figure 9.9 shows that the minimum theoretical number of watts required to remove 1 W at low temperature (Carnot efficiency) increases by a factor of 5.3 at 20 K and 14.8 at 50 K when compared to 4 K operation. A higher operating temperature only slightly reduces the cost of the cryostat, but it reduces very significantly the cost of the cryocooler, in a way even more pronounced than the Carnot efficiency. To illustrate this, Fig. 9.9 also shows the cold power capacity

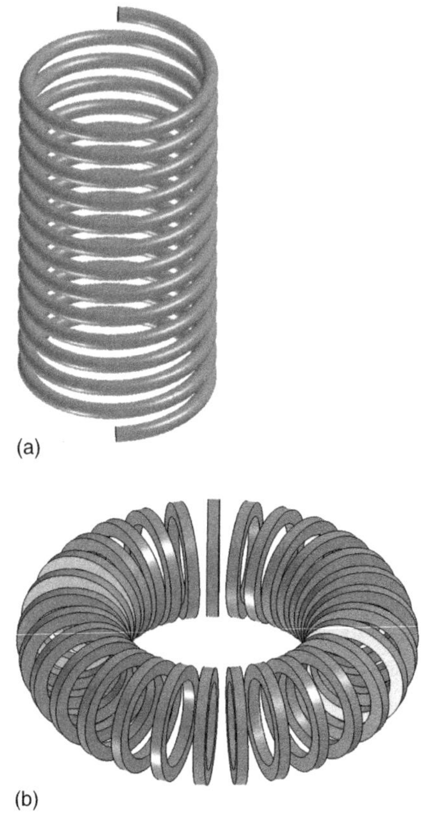

9.7 The two basic magnet topologies for an SMES system: (a) solenoid, (b) toroidal.

from a 7 kW compressor, which approximately fixes the cryocooler cost. The inverse of this cold capacity gives then an approximate cost per watt (Fig. 9.9). It drastically decreases when the operating temperature increases (Fig. 9.9).

However, in large systems, the cryogenic cost is only a limited (even if not very small) fraction of the total cost. Therefore, higher operating temperature cannot bring a very strong reduction of the system total cost and then a tremendous modification of the SMES competitiveness. Nevertheless, operation at higher temperatures brings a lot of advantages for the magnet itself (Section 9.3.5).

9.3.3 Magnet conductor

The superconducting conductor for the magnet winding must meet several requirements:

- high engineering current densities (overall, i.e., the superconductor and normal matrix/shunt/substrate) in large magnetic fields,

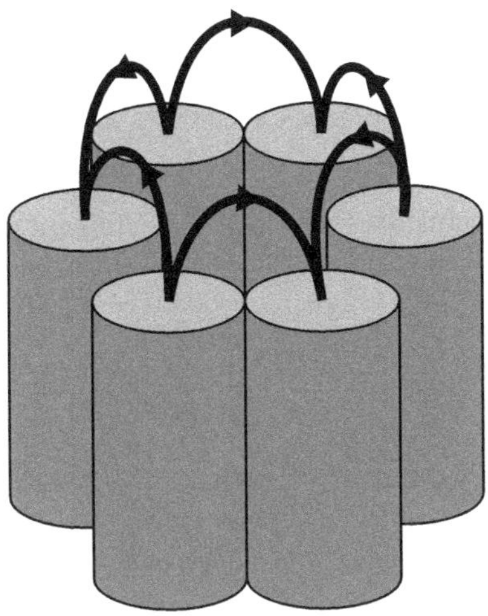

9.8 Schematic drawing of hexagonal arrangement of solenoids for stray field reduction[26].

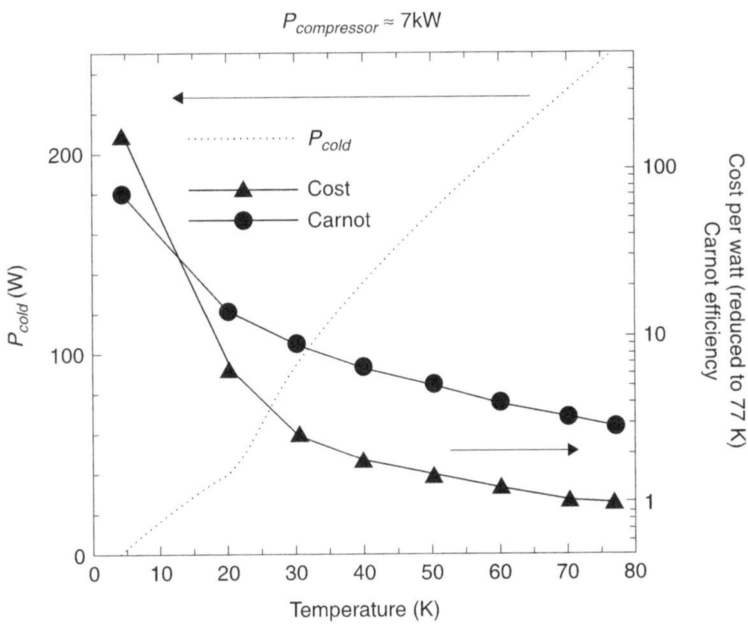

9.9 Cold capacity from a 7 kW compressor, cost per cold power reduced to the 77 K value and Carnot efficiency.

- support of mechanical stresses/deformations to integrate the mechanical function,
- low cost since the real present SMES bottleneck is its cost,
- operating temperature as high as possible.

The cable-in-conduit (CIC)[27,28] conductor, for example, combines several functions: the external sheath contains the Lorentz forces and is used as a cooling fluid vessel.

9.3.4 LTS, NbTi

At present, only NbTi conductors meet the first three requirements, but its operating temperature is unfortunately low, at or slightly above 4.2 K, the liquid helium temperature. Four K cryogenics have been mastered, but this remains very expensive in terms of capital investment cost and, to a much lesser extent, operational costs.

NbTi magnets benefit from steady advances in cryogenic cooling. The gradual improvements of large cryocoolers extend the maintenance cycle and reduce the refrigeration electrical load. The introduction of high-critical-temperature (HTS) current leads was another important improvement. These significantly reduce the related losses (1/10 at 4 K, 1/3 total at 300 K), a large contribution to the total loss[29].

With low temperature superconductors (LTSs), due to stability, protection and stress containment reasons, the engineering current density decreases with energy[30]. For energies above the MJ level (0.3 kWh), the overall current density should be lower than 100 MA/m^2. Protection considerations limit the current density to values much lower compared to the possible current densities in superconductors. This is why the current density is limited by protection issues, not by the superconductor critical characteristic.

9.3.5 HTS materials

HTS materials offer very interesting opportunities for SMES. Firstly, they offer the possibility of operating at higher temperatures, which reduces the cryogenic operating cost and above all the cryocooler cost (Fig. 9.9). Operation at higher temperatures brings a very significant enhancement of the material specific heat (Fig. 9.10) (factors of 80 and 1000 between 4 K and 20 K and 50 K respectively), then a much more stable operation, so less sensitivity against external perturbations. The minimum quench energy (MQE) characterizes the energy density deposited in the magnet required to quench it, then its stability. In adiabatic conditions, the expression of the MQE per unit volume may be estimated by[31]

$$MQE = H(T_{cs}) - H(T_o) \approx c_p(T_o)[T_{cs} - T_o]\left[1 - \frac{I_{sc}}{I_c}\right] \qquad [9.13]$$

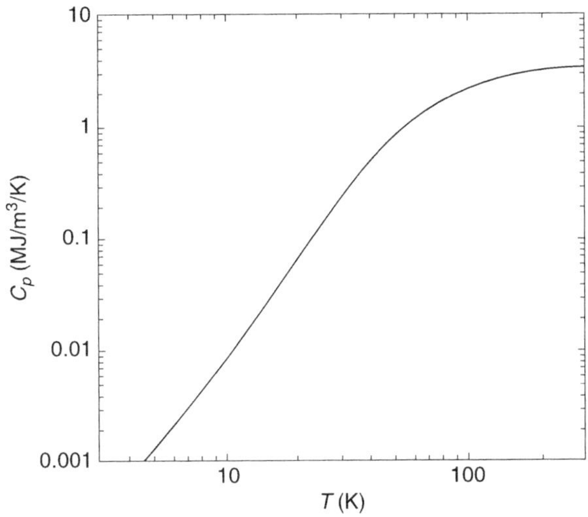

9.10 Typical specific heat versus temperature (copper example).

The large increase of the specific heat (Fig. 9.10), in addition to the larger temperature margin ($T_{cs} - T_o$) enhances the MQE a lot.

The constraints relative to stability are much lower compared to LTSs and higher overall current densities may be foreseen. Since SMES are in general characterized by high powers, constraints linked to protection are not a limitation.

The increased specific heat, however, is not favorable for magnet protection[32]. The zone, which has lost its SC state, propagates very slowly and may experience high-temperature excursions and therefore damage. Quench detection becomes indeed more difficult: the voltage to be detected is proportional to the quenched zone. Improved detection methods have been proposed[33].

The YBCO-coated conductor or second generation (2G) HTS wire is particularly promising for SMES. Their current capacity under high magnetic flux densities keeps high values even when temperature increases (Fig. 9.11), overstepping 1000 MA/m^2 at 20 K under 13 T. The 2G wires continuously experience remarkable improvements and advances; these values will still increase in the future. Their mechanical performances are very high, opening interesting perspectives for SMES, whose design is mainly based upon mechanical constraints (virial theorem). IBAD 2G tapes withstand more than 600 MPa without any degradation.

Nevertheless, 2G wires are anisotropic thin tapes. Magnet designers should reduce the transverse magnetic flux density, which affects the current capacity a lot (Fig. 9.11). High-current cables from 2G wires remain an issue due to their geometry but some ways are under investigation such as Roebel bars[34].

If 2G HTS wires offer exciting prospectives, their present cost is the real bottleneck, but they are still in development stage with reduced production. They

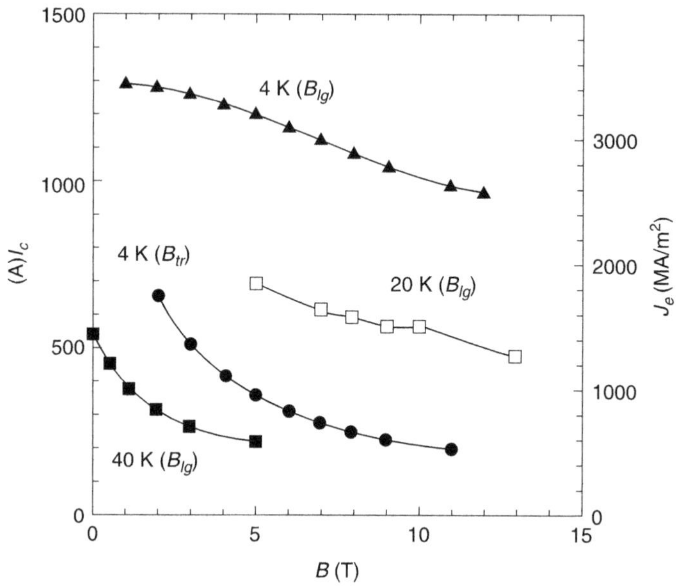

9.11 Overall tape current densities at various temperatures versus longitudinal magnetic field exept at 4.2 K (transverse characteristic as well). SuperPower YBCO tape, 4 mm × 0.1 mm. (Source: data from CEA and CERN.)

are today at least 20 times more expensive compared to NbTi, whereas NbTi SMES are already too expensive. The HTS cost will decrease a lot in the future. So HTS SMES offer great possibilities.

The MgB_2 wires do not have the potential for a very significant breakthrough even at low conductor cost. The operating temperature increase is not significant enough compared to NbTi (10–15 K instead of 4 K). The current capacities at higher temperatures decrease rapidly with magnetic flux density.

9.3.6 Magnet protection

To protect a magnet is to design it so that the hot spot temperature after a quench does not reach values that may damage the magnet. The hot spot temperature is overestimated from the thermal adiabatic equation:

$$\rho(T)\, j(t)^2 = c_p(T)\frac{dT}{dt} \Rightarrow j(t)^2\, dt = \frac{c_p(T)}{\rho(T)}dT \qquad [9.14]$$

The integration of this equation gives the hot spot temperature (T_{max}):

$$\int_0^\infty j(t)^2\, dt = \int_{T_0}^{T_{\max}} \frac{c_P(T)}{\rho(T)}dT = F(T_{\max}) \qquad [9.15]$$

The classical technique is to discharge the magnet into an external discharge resistance (R_d) when the quench is detected (t_{det} after the real quench). The magnet resistance is supposedly negligible compared to the discharge resistance. The time integration of the square of the current is then very simple and the equation can be rewritten introducing the relevant quantities:

$$F(T_{\max}) = J_o^2\left[t_{\det} + \frac{W_o}{I_o\ V_{\max}}\right] = J_o^2\left[t_{\det} + \frac{W_o}{P_{\max}}\right] \qquad [9.16]$$

W_o is the energy for the SC current I_o or current density J_o. V_{max} is the maximum voltage across the SC magnet. This equation shows that a magnet becomes more difficult to protect when the current density or/and the energy increase. The sensitivity of the quench detection, which defines the detection delay (t_{det}), plays an important part. But detection is difficult with HTS magnets. Nevertheless, since SMES are above all attractive as power sources, their low ratio W_o/P_{max} is favorable for their protection.

Let us consider some orders of magnitude. Superconductors experience no damage for temperatures of 400 K, with some margins. Some YBCO tapes have indeed reached 720 K without any significant degradation[35] in fault current limit (FCL). For magnets, thermal differential contraction stresses have to be studied. Taking 400 K for T_{max} and 100 ms for t_{det} as well as for W_o/P_{max} the overall current density must be lower than 450 MA/m^2 for SCS4050 SuperPower tape[36]. Reducing the tape resistivity (thicker copper stabilizer) may increase this already rather high value.

9.4 SMES uses and history

Table 9.4 summarizes the main applications of SMES.

9.4.1 SMES in power grid: flexible AC transmission system (FACTS)

The idea of SMES first appeared in 1970; the motivation was to level the diurnal load in the French electricity network[4]. The energies required (thousands of

Table 9.4 Main applications of SMES

Application	UPS	FACTS	Pulse power source
Use	Voltage–power quality and security for sensitive/critical loads.	Active (and reactive) power exchanges with the grid to improve its operation (stability, capacity enhancement, etc.)	Devices with high power/short time requirements: • Electromagnetic launcher, • Magnetic forming

MWh) led to huge magnets (1 km in diameter, see Table 9.1) with a lot of realization difficulties. Furthermore, SMES is not the best solution in this case due to its moderate energy density. Pumped hydroelectric offers an alternative, which is generally chosen.

With energies of some to tens of MJ, SMES can be used in the electricity networks as a flexible AC transmission system (FACTS) device. A SMES system operating in part as a FACTS device was the first superconducting application installed on a real power grid. In 1976, the Los Alamos Scientific Laboratory began the development of a 30 MJ SMES for use on the Bonneville Power Authority's Tacoma substation[37] to damp low-frequency power oscillations. This SMES operated in real grid conditions for a very short time with only a few hundred energy transfers during a test, evaluation and commissioning period. The main operational issue was the refrigerator, which had been damaged during shipping from the manufacturer. No problems occurred with the NbTi magnet and the cryostat. An alternative solution for power flow variations was found while the 30 MJ coil was under construction. That approach was to replace some of the rectifier diodes on the northern terminus of the existing DC link between the Pacific Northwest and Southern California with SCRs. That solution successfully damped the oscillations. As a result, the LASL-BPA SMES operation was stopped. However, this experience was the first successful and conclusive demonstration and field test of a large SMES operation in a real power grid.

More recently, in 2000, the American Superconductor company installed six SMES units at key points in the grid in northern Wisconsin, USA, to enhance its stability[38]. This grid experienced voltage instability problems with large momentary voltage depressions, which could lead to the grid's collapse. The six SMESs at different key locations in the grid injected real and reactive powers into the grid to boost the voltage and they increased the power transmission capabilities by 15%. Each SMES could provide continuously 2.8 MVAR (5.6 MVAR during 1 s) and 2 MW during a short duration. These SMESs were packaged in standard trailers for an easy and rapid deployment. The commissioning of a 345 kV line some years later has solved the voltage instability problems so that these SMES units were disconnected.

But electric grids require much more reactive than active powers to enhance their operation. An active power requirement is rather rare. A SMES is only required for active power in a FACTS. For reactive power a FACTS using capacitors with a relatively low value is sufficient: dynamic VAR (D-VAR) for example[39]. Such a FACTS meets most of the grid requirements.

The development of renewable energies for electricity grids poses the problem of storage since they often are intermittent. If SMES cannot solve the issue regarding the management of energy or load levelling it may bring an attractive solution for short durations, up to seconds for power quality or bridging power. SMES may assure the continuity of service before the starting of other storage technologies better suitable for long durations. So, associated with other storage devices, SMES covers the full range of duration. Badel[5] and Xue *et al.*[40] review SMES in power systems.

9.4.2 SMES for local power conditioning: uninterruptible power supplies (UPS)

Several SMESs have been operated in uninterruptible power supplies (UPS) with power rating on the order of MW. These are used locally for critical loads requiring ultra-clean power for sensitive processing (such as semiconductor chip fabrication facilities, for example) or military and research laboratory applications. The superconducting magnet replaces the batteries classically used. These SMES are mainly provided by American Superconductor[41]. They have led to a considerable amount of test experience. From 2000, this company had accumulated more than 35 unit-years of operation.

One of the first systems was installed in 1993 for an ammonia production furnace. Another 1.4 MVA 2.4 MJ SMES was installed at the Brookhaven National Laboratory (USA) to offer high quality power for a synchrotron source[42]. It provides power during voltage sags or momentary interruptions to avoid beam loss. Owens Corning's extrusion and production lines in North Carolina have been protected by SMES from voltage sags[43]. Likewise, in South Africa, a SMES has protected a paper machine against 72 dips in 11 months[44].

In Japan several SMESs have been built. One objective is the protection of sensitive load against voltage dips. In 2003 a 5 MW 7 MJ SMES was fabricated usinga NbTi solenoid in a 4-pole configuration[25]. It has compensated for voltage dips in a liquid crystal manufacturing factory. Another SMES system with a rating of 10 MVA was built in 2005. A national Japanese program is carried out on SMES for load fluctuation compensations. They have initiated the development of high-T_c SMES and tested a 1 MVA SMES using a Bi-2212 PIT wire, but operating at 4 K[45].

9.4.3 SMES for pulse power sources

Another SMES class is pulse-power sources for dedicated applications having a very high power demand over a short time.

In the 1980s, a large SMES development program was carried out in the US under the Strategic Defence Initiative (SDI)[46,47]. The main objective was a power source for the Free Electron Laser, but utility applications were studied as well. As an intermediate stage, SMES Engineering Test Model (ETM) aimed to design and build a 20–30 MWh (72–108 GJ) 400/1000 MW magnet (see Table 9.1, middle column). Two teams (Bechtel and Ebasco) were in competition in the design phase and they proposed rather different solutions. One used a 60 kA Al stabilized NbTi conductor with helium bath cooling. The second was based on a 200 kA NbTi CICC. It was developed and tested up to 303 kA under 5 T at 1.8 K, which still is a world record in terms of current capacity. Abandoning the SDI resulted in the termination of this SMES project.

Pulse power SMES sources using toroidal coil magnets have been also studied in Russia[48,49].

The Délégation Générale pour l'Armement (DGA), the French institution in charge of research for the army has launched in the first century of this decade a program to develop an HTS SMES for rail electromagnetic launchers/guns[50,51] with Institute Néel and G2Elab among others. Since this application requires a pulse of current, SMES is the ideal source with a very simple implementation (Fig. 9.2). Today, power capacitors are used, but a dissipative conversion circuit is required to transform the voltage source (capacitor) in the current source necessary for the rail gun. The huge currents (100–1000 kA) required by a rail gun remain a challenging issue for a SMES. Using an SMES, the current can be lowered compared to a capacitor-based source since the supplied current can be more constant during the shot[5]. The XRAM concept (MARX spelt backwards; the MARX generator was invented by E. O. Marx in 1923 and is used to generate a high voltage. This is why the high current generator is known as XRAM) may be used to increase the discharge current: the magnet is made with several coils charged in series and discharged in parallel. Figure 9.12 shows the successful tests carried out by ISL (Institut Saint Louis) with the XRAM generator[46].

This DGA program has led *inter alia* to an HTS SMES (Fig. 9.13) designed for 800 kJ (425 kJ achieved) based on a BiSrCaCuO conductor from Nexans. It

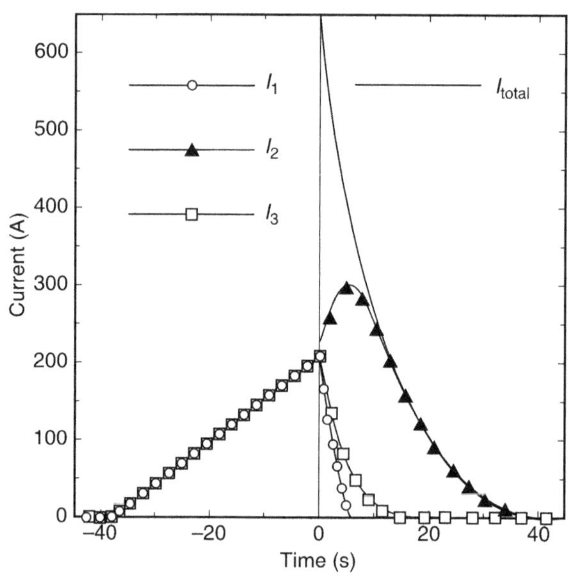

9.12 XRAM generator based on a SMES with three coils, charge in series, discharge in parallel.

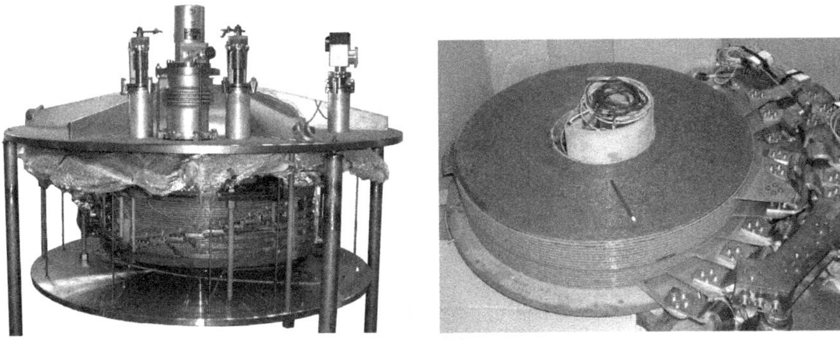

(a) (b)

9.13 800 kJ 20 K BSCCO conducting cooling SMES built in Grenoble.

operates at 20 K using only conduction cooling from cryocoolers for user-friendly cryogenics. Taking into account the severe military environment, HTS suits much better than LTS. The purpose was to qualify several technological solutions on a representative level and to acquire essential operational experience feedback. It has been followed by a reflection phase with ISL to optimize the complete system. The use of YBCO is analysed.

9.4.4 Summary

In summary, several SMES have proved their operational capabilities for short-term (seconds) power at MW scale. They are commercially available and the field test experience is very large in the US and Japan. Nevertheless, the number of sold SMES remains very low and does not increase, in fact it is even decreasing. The major reason is the high initial cost while in competition with more mature technologies. Moreover, SMES addresses niche applications having high active power and short time demands. Research and development currently mainly concerns HTS SMES (Table 9.5).

9.5 Conclusions

SMES is particularly suitable for power sources of short duration, because the power density is much higher than the stored energy density. SMES is thus an excellent solution for applications such as pulse power sources, especially current sources, UPS and FACTS for power grids. A number of SMES units have been installed and operated successfully over many years, demonstrating their very satisfying performance. The bottleneck for widespread commercialization of SMES remains the high capital cost.

The deregulation of the electricity market and the requirement to enhance the power capacities of the present grids bring the opportunity for FACTS using SMES.

Table 9.5 Main HTS SMES systems throughout the world

Organization	Country	Year	Specification	SC	Application
Chubu[45,52]	Japan	2004	1 MVA, 1 MJ	Bi-2212	Voltage stability
CAS[53]	China	2007	0.5 MVA, 1 MJ	Bi-2223	Power quality
KERI[54]	Korea	2007	0.6 MJ	Bi-2223	Power, voltage quality
DGA/CNRS[50]	France	2007	0.8 MJ	Bi-2212	Pulse application/ rail gun
KERI[54]	Korea	2011	2.5 MJ	YBCO	Power quality
SuperPower, ABB, BNL, TcSUH	USA	2011–2013	20 kW, 2.5 MJ	YBCO	MV distribution networks

The need for pulse power sources for emerging applications such as electromagnetic launchers also offers very good chances for SMES since they offer a high-performances, well-adapted solution.

Several characteristics of HTS are very attractive for SMES, especially YBCO-coated conductors: important current capacity under large magnetic flux densities at temperatures much higher than 4 K, and mechanical strength. But these very favorable and promising properties will be used only if these conductors are produced at low cost.

9.6 Acknowledgments

This work is performed under a DGA program and an ANR project. The author warmly thanks A. Badel for fruitful discussions.

9.7 References

1 J. File, R.G. Mills, 'Observation of persistent current in a superconducting solenoid', *Physical Review Letters*, **10**, pp. 93–96 (1963).

2 W. Hassenzahl, 'Superconducting magnetic energy storage', *IEEE Transactions on Magnetics*, **25**, 750–758 (1989).

3 C. A. Luongo, 'Superconducting storage systems: an overview', *IEEE Transactions on Magnetics*, **32**, 2214–2223 (1996).

4 M. Ferrier, 'Stockage d'énergie dans un enroulement supraconducteur', *Low Temperature and Electric Power*, Pergamon Press, 425–432 (1970).

5 A. Badel, 'High critical temperature SMES as pulse source', PhD, Grenoble INP, 2010.

6 M. H. Ali, B. Wu, R. A. Dougal, 'An overview of SMES applications in power and energy systems', *IEEE Transactions on Sustainable Energy*, **1**, pp. 38–47 (2010).

7 Electricity Storage Association, http://electricitystorage.org/.

8 J. Kondoh, I. Ishii, H. Yamaguchi, A. Murata, K. Otani *et al.* 'Electrical energy storage systems for energy networks', *Energy Conversion & Management*, **41**, pp. 1863–1874 (2000).
9 F. C. Moon, 'The virial theorem and scaling laws for superconducting magnet systems', *Journal Applied Physics*, **53**, pp. 9112–9121 (1982).
10 Y.M. Eyssa, 'Design of single layer superconductive energy storage magnets', *Journal Physics D: applied Physics*, **13**, pp. 1719–1726 (1980).
11 S. Nomura, N. Watanabe, C. Suzuki, H. Ajikawa, M. Uyama *et al.*, 'Advanced configuration of Superconducting Magnetic Energy Storage', *Energy*, **30**, pp. 2115–2127 (2005).
12 J. L. Smith, 'Field analysis for a SMES magnet with radial force balance', *IEEE Transactions on Applied Superconductivity*, **5**, 357–360 (2005).
13 G.E. Marsh, *Force-Free Magnetic Field: Solutions, Topology and Application*, World Scientific, ISBN: 978-981-02-2497-4.
14 A. A. Kuznetsov, *Soviet Physics Technical Physics*, No 6, December 1961.
15 I. N. Sviatoslavsky, W.C. Young, 'Structural design features for commercial fusion power reactor', *Nuclear Engineering and Design*, **58**, pp. 207–218 (1980).
16 D. Campi, B. Curé, A. Gaddi, H. Gerwing, A. Hervé *et al.* 'Commissioning of the CMS magnet', *IEEE Transactions on Applied Superconductivity*, **17**, pp. 1185–1190 (2007).
17 A. Yamamoto, Y. Makida, H. Yamaoka, H. Ohmiya, K. Tanaka *et al.*, 'A thin superconducting solenoid magnet for particle astrophysics', *IEEE Transaction on Applied Superconductivity*, **12**, pp. 438–441 (2002).
18 C. Rix, C. Luongo, W. Bingham, A. Bulc, K. Cooke *et al.*, 'A self-supporting superconducting magnetic energy system (SMES) concept', *Cryogenics*, **34**, Supplement 1, pp. 737–740 (1994).
19 W. V. Hassenzahl, D. W. Hazelton, B. K. Johnson, P. Komarek, M. Noe *et al.*, 'Electric power applications of superconductivity', *Proceedings of the IEEE*, **92**, pp. 1655–1674 (2004).
20 H. D. Fair, 'Electric launch science and technology in the United States', *IEEE Transactions on Magnetics*, **39**, pp. 11–17 (2003).
21 P. Lehmann, 'Overview of the electric launch activities at the French–German Research Institute of Saint-Louis (ISL)', *IEEE Transactions on Magnetics*, **39**, pp. 24–28 (2003).
22 T.E. Motoasca, H. Blok, M.D. Verweij, P.M. Van Den Berg, 'Electromagnetic forming by distributed forces in magnetic and non magnetic materials', *IEEE Transactions on Magnetics*, **40**, pp. 3319–3330 (2004).
23 P. Lehmann, O. Bozic, H. Grobusch, J. Behrens, 'Electromagnetic railgun technology for the deployment of small sub-/orbital payloads', *IEEE Transactions on Magnetics*, **43**, pp. 480–485 (2007)
24 R.J. Loyd, T.E. Walsh, E.R. Kimmy, 'Key design selections for the 20.4 MWh SMES/ETM', *IEEE Transactions on Magnetics*, **27**, pp. 1712–1715 (1991).
25 S. Nagaya, N. Hirano, M. Kondo, T. Tanaka, H. Nakabayashi *et al.*, 'Development and performance results of 5 MVA SMES for bridging instantaneous voltage dips', *IEEE Transactions on Applied Superconductivity*, **14**, pp. 699–704 (2004).
26 W. Weck, P. Ehrhart, A. Müller, G. Reiner, 'Superconducting inductive pulsed power supply for electromagnetic launchers: design aspects and experimental investigation of laboratory set-up', *IEEE Transactions on Magnetics*, **33**, pp. 524–527 (1997).

27 M.O. Hoenig, Y. Iwasa, D.B. Montgomery, 'Supercritical-helium cooled 'bundle conductors' and their application to large superconducting magnets', *Proceedings of the 5th International Conference on Magnet Technology*, pp. 519–524 (1975).

28 C.A. Luongo, K.D. Partain, J.R. Miller, G.E. Miller, M. Heiberger *et al.*, 'Quench initiation and propagation study (QUIPS) for the SMES-CIC', *Cryogenics*, **31**, supplement 1, pp. 611–614 (1994).

29 J. R. Hull, 'High-temperature superconducting current leads', *IEEE Transactions on Applied Superconductivity*, **3**, pp. 869–875 (1993).

30 M.N. Wilson, 'Stabilization, protection and current density: some general observations and speculations', *Cryogenics*, **31**, pp. 449–503 (1991).

31 L. Dresner, *Stability of Superconductors*, Plenum Press, ISBN 0-306-45030-5.

32 T. Effio, U. P. Trociewitz, X. Wang, J. Schwartz, 'Quench induced degradation in $Bi_2Sr_2CaCu_2O_{8+x}$ tape conductors at 4.2 K', *Superconductor Science and Technology*, **21**, 045010 (10pp) (2008).

33 A. Badel, P. Tixador, G. Simiand, O. Exchaw, 'Quench detection system for twin coils HTS SMES', *Cryogenics*, **50**, pp. 674–681 (2010).

34 W. Goldacker, A. Frank, A. Kudymow, R. Heller, A. Kling *et al.*, 'Status of high transport current ROEBEL assembled coated conductor cables', *Superconductor Science and Technology*, **22**, 034003 (10pp) (2009).

35 M. Schwarz, C. Schacherer, K-P. Weiss, A. Jung, 'Thermodynamic behaviour of a coated conductor for currents above T_c', *Superconductor Science and Technology*, **21**, 054008 (4pp) (2008).

36 Superpower Inc., http://www.superpower-inc.com/

37 H.J. Boenig, J.F. Hauer, 'Commissioning tests of the Bonneville power administration 30 MJ superconducting magnetic energy storage unit', *IEEE Transactions on Power Apparatus and Systems*, **104**, pp. 302–312 (1985); and W. V. Hassenzahl private communication [WVH4].

38 T.R. Abel, 'D-SMES for Wisconsin', Modern power systems, **19**, 1999, n°10, pp. 28–29.

39 H.G. Sarmiento, G. Pampin, J. Diaz De Leon, 'Feasibility studies for dynamic VAR and STATCOM applications to prevent a fast voltage collapse', *Power System Conference and Exposition*, IEEE PES, pp. 275–282 (2004).

40 X.D. Xue, K.W.E. Cheng, D. Sutanto, 'A study of the status and future of superconducting magnetic energy storage in power systems', *Superconductor Science and Technology*, **19**, R31–R39 (2006).

41 American Superconductor, http://www.amsuper.com/.

42 J. Cerulli, 'Operational experience with a SMES device at Brookhaven National Laboratory, New York', *IEEE Power Engineering Society 1999 Winter Meeting*, pp. 1247–1252 (1999).

43 J. Cerulli, G. Melotte, S. Peele, 'Operational experience with a superconducting magnetic energy storage device at Owens Corning Vinyl Operations, Fair Bluff, North Carolina', *IEEE Power Engineering Society Summer Meeting*, pp. 524–528 (1999).

44 R. Schottler, R. G. Coney, 'Commercial application experiences with SMES?', *The IEE Power Engineering Journal*, **13**, pp. 149–152 (1999).

45 K. Shikimachi, H. Moriguchi, N. Hirano, S. Nagaya, T. Ito *et al.*, 'Development of MJ-class HTS SMES system for bridging instantaneous voltage dips', *IEEE Transactions on Applied Superconductivity*, **15**, pp. 1931–1934 (2005).

46 R.L. Verga, 'SMES and other large-scale SDI cryogenic application programs', *Advances in Cryogenic Engineering*, **35**, Plenum Press, pp. 555–564 (1990).

47 G. W. Ullrich, 'Summary of the DNA SMES development program', *IEEE Transactions on Applied Superconductivity*, **5**, pp. 416–421 (1995).
48 E.P. Polulyakh, L.A. Plotnikova, V.A. Afanas'ev, M.I. Kharinov, A.K. Kondratenko *et al.*, 'Development of toroidal superconducting magnetic energy storages (SMES) for high-current pulsed power supplies', *Advances in Cryogenic Engineering*, **48A**, Plenum Publ., pp. 721–727 (2000).
49 E.Yu. Klimenko, E.P. Polulyakh, 'Closed flux winding for SMES', *Advances in Cryogenic Engineering*, **47A**, Plenum Publ., pp. 323–328 (2002).
50 P. Tixador, B. Bellin, M. Deleglise, J.C. Vallier, C.E. Bruzek *et al.*, 'Design of a 800 kJ HTS SMES', *IEEE Transactions on Applied Superconductivity*, **17**, pp. 1707–1710 (2007).
51 A. Badel, P. Tixador, K. Berger, M. Deleglise, 'Design and preliminary tests of a twin coil HTS SMES for pulse power operation', *Superconductor Science and Technology*, **24**, 055010 (2011).
52 K. Shikimachi, N. Hirano, S. Nagaya, H. Kawashima, K. Higashikawa, T. Nakamura, 'System coordination of 2 GJ Class YBCO SMES for power system control', *IEEE Transactions on Applied Superconductivity*, **19**, pp. 2012–2018 (2009).
53 X. Liye, W. Zikai, D. Shaotao, Z. Jinye, Z. Dong *et al.*, 'Fabrication and tests of a 1 MJ HTS magnet for SMES', *IEEE Transactions on Applied Superconductivity*, **18**, pp. 770–773 (2008).
54 K. C. Seong, 'An introduction of HTS-SMES project in Korea', *IEEE/CSC & ESAS European Superconductivity News Forum (ESNF)*, no. 7, January 2009.

10

Application of high temperature superconductor (HTS) technology in rotating machinery

R. FAIR, Converteam UK Ltd, UK

Abstract: This chapter discusses the application of high temperature superconductors (HTS) to generators and motors. It first reviews the benefits and rationale for applying HTS technology to generators and motors. It then describes compromises affecting the electromagnetic, mechanical and thermal design for such machines. The different types of machine topology possible, their respective advantages and disadvantages and their key performance characteristics compared to equivalent conventional machines are then outlined. Issues with developing such machines within an industrial production environment and their installation and operation within a demanding commercial environment are discussed. Challenges facing the design, build, testing and commissioning of such machines are reviewed, with a discussion of future machine and equipment requirements and trends.

Key words: high temperature superconductivity, BSCCO, YBCO, cryogenics, vacuum, superconducting rotating machinery.

10.1 Introduction

High temperature superconductor (HTS) wire is presently very expensive but forecasts from the major wire manufacturers indicate that in time these novel materials could cost less than copper! For example, the cost to carry 1000 A over 1 m using copper (at the current densities used in conventional high power density motors) is about US$15/kA m (with copper prices at US$6.5/kg), whereas present HTS (1G) wire has only just fallen to below US$100/kA m with 2G wire (a key contender for energy applications in the 45–75 K operating temperature range) still at over US$350/kA m at the end of 2010.

HTS machines can be 30% of the size of conventional machines, leading to potential savings in weight, volume and material costs. The most attractive markets are those where high power density and particularly high torque is required by utilising higher magnetic flux densities and reduced iron in the magnetic circuit – both achievable with HTS materials. Ship propulsion motors and wind turbine generators are potential candidates for such a technology.

High power density is the result of the high current density that can be obtained in HTS coils, reducing the space required for the rotor field coils. The copper coils in a conventional machine typically operate with a current density between 3–5 A/mm^2, while the current density in the wire in an HTS coil can operate at 200 A/mm^2 or more[1]. In HTS wire this is known as the engineering current density, which is the current density in the full cross-section of the wire, but as the HTS material only forms a small proportion of the cross-sectional area of the wire, the current density in the HTS material itself is much higher. Additionally, the ability to place many ampere turns of field winding in a small volume, well beyond that which could be achieved using copper without unacceptable losses, can be used to increase the air gap flux density.

Until fairly recently, all commercially available HTS wire was based on $Bi_2Sr_2Ca_2Cu_3O_x$ (more commonly referred to as BSCCO-2223), also known as first generation (1G) wire. The manufacturing process for this type of wire was inherently expensive and efforts have been ongoing to find a cheaper processing method. The new second generation (2G) wire currently being developed is based on $YBa_2Cu_3O_x$ normally referred to as YBCO-123 or simply YBCO, and has the potential for volume production at lower cost.

Electrical efficiency improvements of several percent are possible, leading to substantial operational cost savings over the lifetime of the machine. A rotor with an HTS field winding has almost zero ohmic losses – the power consumed by the cryogenics is a small fraction of the normal losses in a conventional rotor – about 25 kW compared to 300 kW in a 25 MW machine. These higher efficiencies are available even with the machine operating under partial load conditions.

Higher steady-state system stability is also attainable due to the inherently low reactances associated with certain designs of HTS machine which employ large magnetic air gaps. A lower synchronous reactance permits operation of these HTS machines at smaller load angles compared to conventional machines. Another consequence of a low synchronous reactance is a high short-circuit ratio, which might be beneficial if the machine operates within 'weak' electrical power grids with low short-circuit power. An additional benefit of an HTS machine with decreased armature reaction is its somewhat 'stiff' behaviour within an electrical power network, producing only small voltage changes with changing electrical loads. Machines having a 'low synchronous reactance' can be used very effectively as dynamic synchronous condensers providing or absorbing reactive power (VARs) to or from the grid. Conventional synchronous condensers have been used extensively but they tend to suffer from reliability problems, mainly due to the heating from the large rotor currents that must be applied to reach rated VAR levels. HTS synchronous condensers can alleviate these problems.

Certain designs of HTS machines can supply both leading and lagging reactive power up to their full rating as shown in Fig. 10.1. This figure shows the entire safe operating region (shaded area) of a typical HTS hydroelectric synchronous generator that covers a much larger area when compared to conventional (non-HTS) generators of a similar rating (Fig. 10.2). Conventional machines have a limit on the right-hand side of the figure related to field coil heating and a limit on the left-hand side of the figure related to machine stability – these limits are not present for an appropriately designed HTS synchronous generator. These results have also been confirmed by a series of studies carried out by the Korea Electrotechnology Research Institute[2].

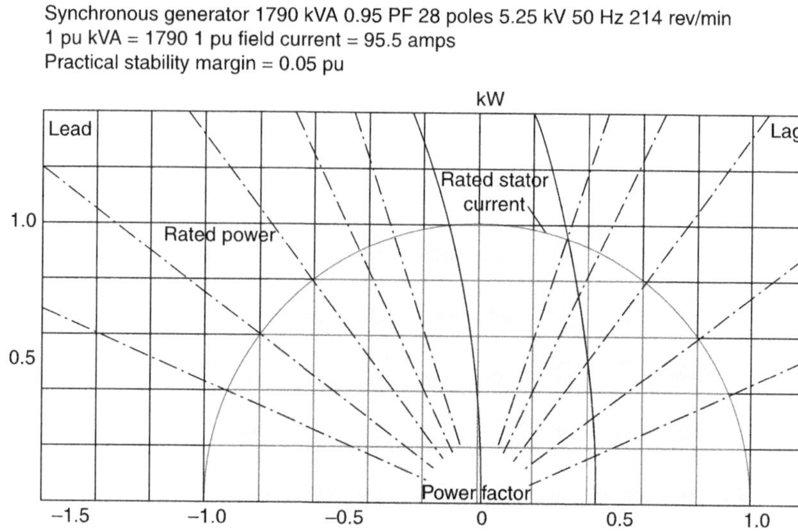

10.1 Hydrogenie HTS generator operating chart (safe operating region shown shaded). (Reproduced with permission from Converteam UK.)

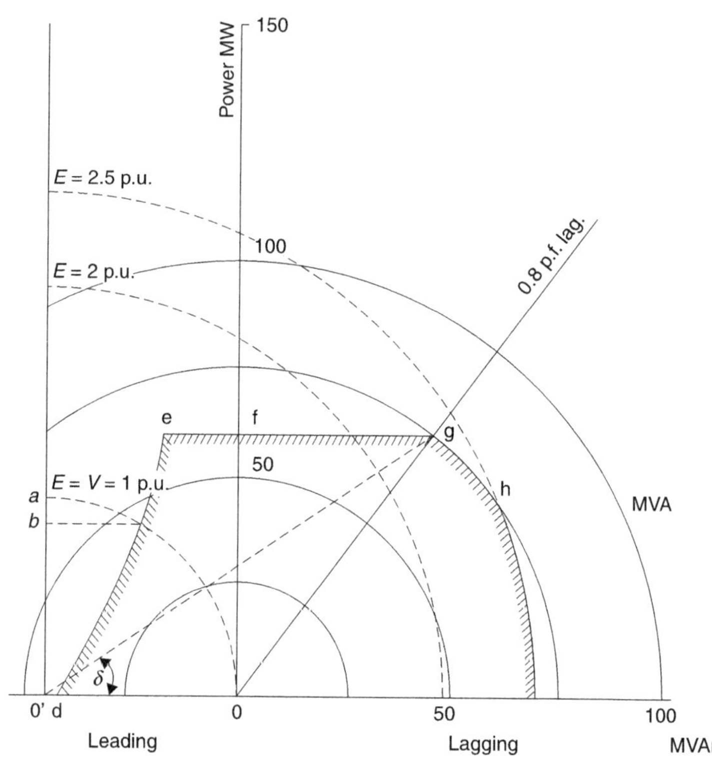

10.2 Conventional generator operating chart (safe operating region shown within hashed boundary).

Thermal cycling of conventional copper rotor field windings can cause winding failures in large power generators as the ground and turn insulation deteriorates due to mechanical wear and thermal ageing. A rotor using HTS field coils will undergo fewer thermal cycling events as the winding will always be at the same cryogenic temperature. All thermal contraction and expansion of the winding and its suspension system will only occur during cool-down and warm-up of the rotor. However, it must be borne in mind that although the core technologies (i.e. HTS and composite materials) have improved over the years, the same may not always be true for the ancillary equipment (e.g. vacuum pumps, cold heads and cryocoolers, etc.) required to make the application of HTS technology a success. This equipment necessarily needs to be reliable and robust, especially if it is to be installed within demanding environments, e.g. on ships or offshore wind turbine towers.

10.2 Types, properties and performance of high temperature superconductor (HTS) generators and motors

There have been many variations of HTS machine design over the years. A few of the more recent developments are reviewed and reported here. Figure 10.3 illustrates the variety of HTS machines worked on worldwide since about 1999.

Between 2001 and 2006, the General Electric Company worked with the US Department of Energy and the National Renewable Energy Laboratory on the design and development of a 100 MVA HTS generator for a commercial application[3]. The generator was based on a warm iron rotor concept in which a cold HTS coil is wound round a warm magnetic iron pole. A detailed design of the 100 MVA generator and the design and test of a 1.5 MW demonstrator HTS generator was completed. However, the program was shut down in 2006 due to the continuing uncertainty of the economic viability of such a machine at that time.

Just prior to 2005, a Japanese research and study committee was established composed of members from industrial, governmental and academic fields[4]. The purpose of this group was to evaluate the feasibility of HTS synchronous machines by considering promising application fields for such machines. Designs were carried out for these machines and their merits and demerits compared, especially in relation to HTS field coils and refrigeration systems.

A 3 kVA superconducting motor employing HTS coils and iron plates cooled by solid nitrogen was reported by Young-Sik Jo *et al.* from the Republic of Korea in 2002[5]. Ag-sheathed Bi-2223 HTS tape was used to wind the field coils, and a liquid helium-liquid nitrogen heat exchanger system was used to produce the solid nitrogen which served as the primary cooling medium for this machine. This cooling system was chosen primarily to avoid vibrations due to thermal shrinkage, which are inherent in systems employing coaxial cryogen transfer tubes and ferrofluid seals.

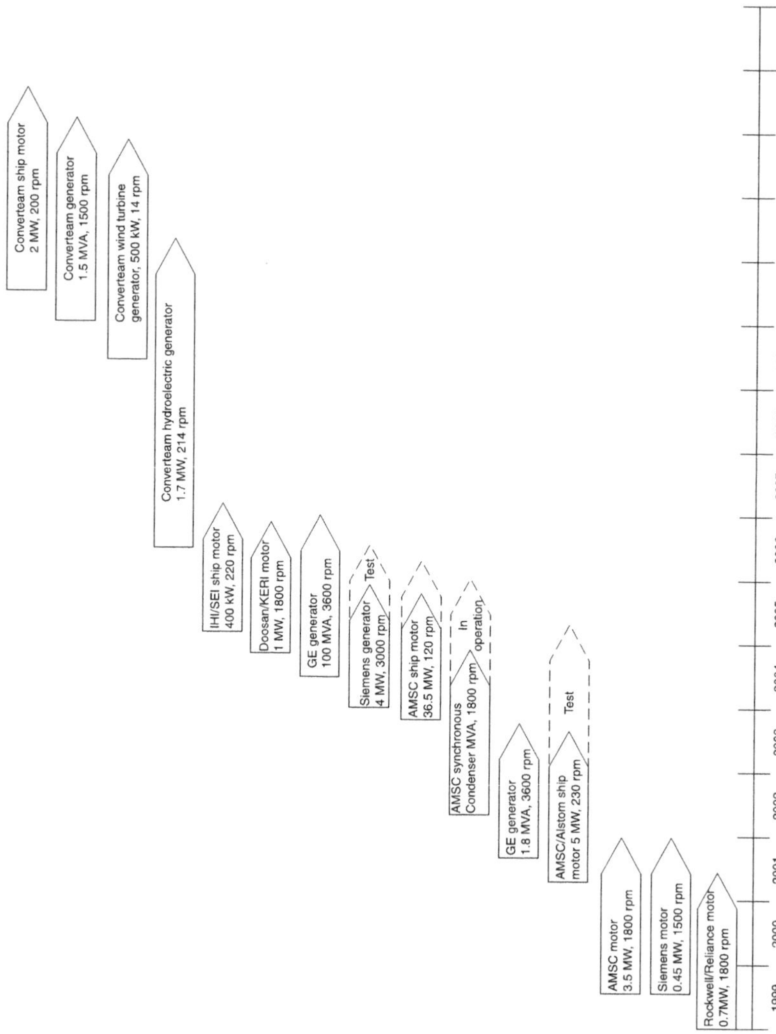

10.3 Timeline for HTS rotating machinery projects.

In 2002, Siemens AG reported some long-term operational experience with a 400 kW HTS machine[6]. This machine consisted of an HTS rotor and air-cored stator; rotor cooling was provided by a Gifford-McMahon cryocooler. Tests on this machine demonstrated the improved electrical stability of such a design when connected to an electrical network and the enhanced efficiency even after allowing for the rotor HTS cooling requirements.

Siemens AG once again reported in 2006 on the development work on a 4 MVA HTS synchronous machine[7]. This 2-pole 60 Hz 3600 rpm machine utilised a high temperature superconducting field winding for a ship-borne application. The HTS coils were cooled using several cold heads. The stator winding was of the air-gap design and the stator core teeth were manufactured from a composite material.

More recently, Sivasubramaniam *et al.* of the General Electric Global Research Centre reported work on the development of a high-speed HTS generator for airborne applications[8]. The generator comprises a stationary HTS field excitation coil, a solid rotor forging and an advanced but conventional stator. The generator was load-tested up to 1.3 MW and provides power densities greater than 8.8 kW/kg.

The above body of work, together with developments reported by other researchers in the field, can be used to steer the design decisions involving the use of HTS materials. The high current density of HTS materials allows the use of designs without iron or with a partial iron magnetic circuit. The four options that are usually considered for HTS machine designs are as follows:

1. Conventional stator with iron teeth plus rotor with magnetic pole bodies (warm or cold).
2. Conventional stator plus rotor with non magnetic pole bodies. More HTS wire needed, but lower cold mass, otherwise similar to A.
3. Air-gap stator winding plus rotor with magnetic pole bodies. Rotor iron can be operated highly saturated. Allows significant reduction in size and mass.
4. Air-gap stator winding plus rotor with non-magnetic pole bodies. Uses more HTS wire than (3), depending on flux density. Significant reduction in size and mass.

The machine-equivalent circuit parameters, i.e. reactances and resistances, can vary widely depending on which HTS topology is selected. As an illustration, Table 10.1 demonstrates the differences for key machine parameters for a conventional salient pole synchronous generator and an HTS generator designed around the concept of option (1) (warm iron) described above. Both machines have similar power ratings. These differences can lead to a vastly different operating area for each type of machine. As described earlier Figure 10.1 illustrates the wide operating region of a typical HTS synchronous generator as indicated by the shaded area. By comparison, Fig. 10.2 shows that a conventional machine

Table 10.1 Comparison of typical parameters (similar rated conventional salient pole synchronous generator vs HTS generator)

	Conventional generator	HTS generator		
Parameter	Value	Value	Unit	Description
X_d	1.587	0.213	Per unit	Synchronous reactance *d*-axis
X_q	0.937	0.178	Per unit	Synchronous reactance *q*-axis
X'_d	0.349	0.107	Per unit	Transient reactance *d*-axis
X''_d	0.262	0.074	Per unit	Subtransient reactance d-axis
X''_q	0.388	0.073	Per unit	Subtransient reactance *q*-axis
T'_d	2.61	40	Seconds	Transient time constant d-axis
T''_d	0.023	0.040	Seconds	Subtransient time constant d-axis
T''_q	0.057	0.068	Seconds	Subtransient time constant *q*-axis
x_2	0.279	0.074	Per unit	Negative sequence reactance
r_2	0.195	0	Per unit	Negative sequence resistance
x_0	0.207	0.005	Per unit	Zero sequence reactance
r_0	0.497	0	Per unit	Zero sequence resistance

(Reproduced with permission from Converteam UK.)

would have a reduced operating region limited by the stator and rotor field heating on the right boundary and the stability of the machine on the left boundary.

Due to the inherently smaller synchronous reactances of some HTS machines, this can lead to much higher currents and torques under certain fault conditions, when compared to conventional machines. Figures 10.4 and 10.5 illustrate the

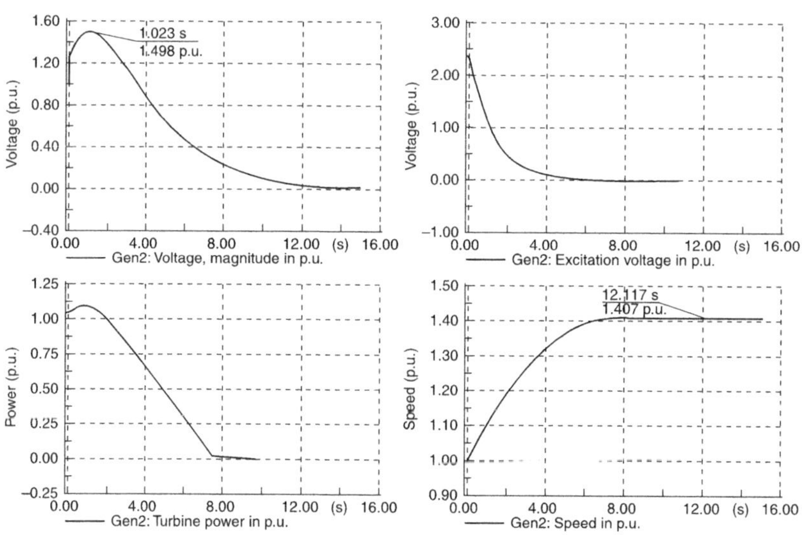

10.4 Load rejection (conventional generator). (Reproduced with permission from KEMA, Netherlands and Converteam.)

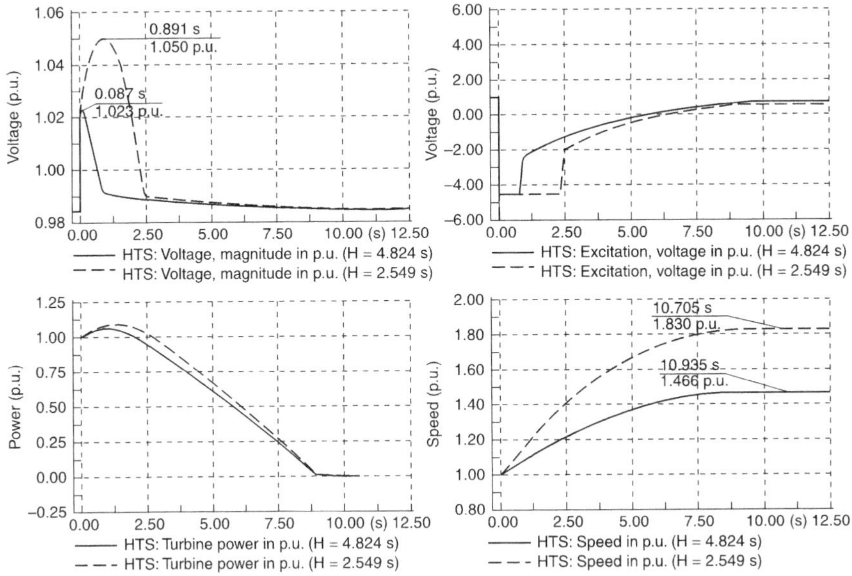

10.5 Load rejection (HTS generator.) (Reproduced with permission from KEMA, Netherlands and Converteam.)

differences in behaviour between a typical conventional generator and an HTS generator under a load rejection condition. Studying the curves for the generator output voltage, it can be seen that for the conventional generator, the voltage peaks at a higher value compared to that for the HTS generator ($H = 2.549$ s) and dies away to zero after about 14 s whereas the HTS machine's voltage remains at about 0.985 per unit (p.u.) Differences can also be noted in the way that the excitation voltage behaves for the two machines.

10.3 Fundamental design and development issues for HTS generators and motors

The Hydrogenie project (part funded by the European Union) is used as an example for the following discussion of the application of HTS to a commercial machine. The Hydrogenie machine is a horizontal-shaft hydroelectric generator, with its specifications tabulated in Table 10.2. This machine would be classified as a 'small' machine in terms of its power rating and is believed to be the first of its kind to be put into commercial service to generate electricity for a national grid system. Hydroelectric generators require high inertia; hence reduction of mass was not a key driver for this particular project. Although a hydroelectric generator may not be the ideal vehicle to publicise the application of HTS technology, it has nonetheless proven to be an excellent technology demonstrator and has allowed Converteam, the lead partner on this project, to develop and demonstrate the

Table 10.2 Key specifications for the Hydrogenie HTS hydrogenerator

Parameter	Value
Output (kVA)	1789.5
Output (kW)	1700
Line voltage (V)	5250
Power factor (overexcited)	0.95
Poles	28
Speed	214
Overspeed (rpm)	320
Turbine runaway speed (rpm)	450
Efficiency (excluding cryocoolers) (%)	98
Efficiency (including cryocoolers) (%)	97
Shaft torque (kN m)	77.3
Rated field current (A)	109
Field current (at open circuit rated volts) (A)	98
Generator inertia (kg m^2)	18 200

(Reproduced with permission from Converteam UK.)

necessary expertise and relevant technology building blocks to further develop HTS machines.

Design type A was selected primarily to satisfy the requirement to be able to synchronise the machine directly to the electrical grid (an air-gap winding at 28 poles, 50 Hz would have resulted in most of the HTS ampere turns being wasted as leakage). The selected machine construction was therefore fairly conventional, i.e. iron stator and copper winding with 1G HTS (Bi-2223) field coils on a more-or-less conventional rotor using magnetic iron at room temperature. Figure 10.6 indicates the main warm and cold components of the Hydrogenie generator.

Keeping the rotor iron at room temperature reduces the overall cold mass of the superconducting rotor. This reduces the required cryogenic cooling power during cool-down and allows faster cool-down and warm-up times. Special steel suitable for operation at cryogenic temperatures, although available, was thus not necessary for this generator. However, if a cold iron design had been selected, this would mean less warm surface area and thus less thermal radiation losses but possibly higher conduction losses as the thermal links would have to be more massive.

The majority of the load torque is transferred from the rotor shaft and magnetic iron poles directly to the stator teeth and then to the stator core frame and finally to the power station floor with minimal torque acting on the superconducting coils, resulting in a more simplified suspension system.

The complete rotor is housed within a vacuum chamber that is sealed to, and rotates with, the shaft. The HTS coils are cooled using gaseous helium at about 30–40 K at 20 bar, from a Stirling-type cryocooler, which is pumped up the shaft and radially outwards to the coils. As the rotor iron is at room temperature, the

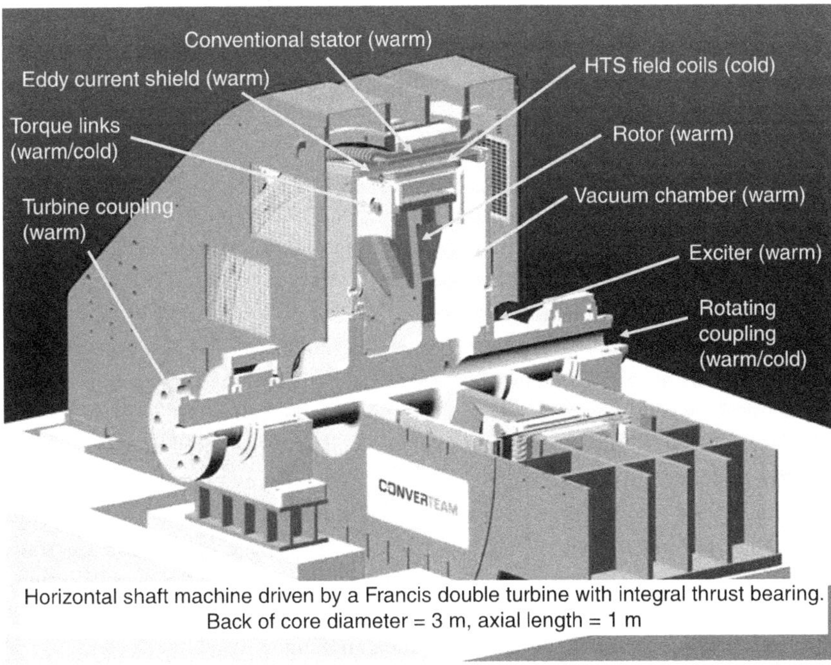

10.6 The Hydrogenie superconducting generator (warm = room temperature, cold = cryogenic temperature). (Reproduced with permission from Converteam UK.)

individual HTS field coils 'float' over the iron poles and are wrapped in multi-layer insulation (MLI or superinsulation) as are all the other cryogenically cold components.

Field excitation for the HTS coils is provided via electronic circuitry located on the outside wall of the vacuum chamber. The temperature and voltage of each HTS coil is monitored continuously and is fed into the control and protection circuitry also located on the outside of the vacuum chamber. Communication between the HTS machine and the automatic voltage regulator (AVR) is provided via a bespoke wireless system.

10.3.1 HTS winding design and construction

The maximum current carrying capacity (I_c, critical current) of HTS wire is normally quoted with its self-field at a temperature of 77 K, i.e. with no externally applied magnetic field. If an external field is applied the critical current reduces. HTS flat conductor is more sensitive to fields in the direction perpendicular to the broad face compared to the narrow face[9]. Minimising exposure to this particular magnetic field vector is therefore a key part of the HTS coil design.

The critical current density decreases as the stress or strain in the wire increases – this can be very difficult to quantify in practice. Recommendations from HTS wire manufacturers with respect to minimum bend radii and wire strains must be adhered to.

The coil and former design must be able to cope with the relatively high rotational inertial forces that also contribute to the overall stress in the HTS conductors. For this particular design most of the torque is experienced by the field pole iron rather than the coils; however, the coil design still needs to allow for worst-case short-circuit conditions when the field current, and therefore coil forces, can increase significantly.

Coil winding and impregnation processes can influence HTS coil performance significantly, as can soldering temperatures while making joints. Differential thermal contraction between the coil winding pack and the coil former can also contribute to the stress within a coil and can lead to delamination of the winding pack.

True superconductivity is only achieved today with direct current. Alternating current and time-varying fields produce losses in the HTS wire necessitating the use of eddy current shields. These shields are only truly effective for a certain range of frequencies – therefore the performance of the coil and the overall system heat budget must make allowance for additional internal coil heating produced during the various phases of operation of the generator.

The AC losses within the HTS coils also limit the rate at which the field current is allowed to change. The AC losses for the Hydrogenie generator were quantified via a collaborative project between Converteam and The University of Cambridge.

10.3.2 Cooling and coolant transfer

Low temperature superconductors (LTS) require cooling to sub-30 K temperatures, normally using liquid helium at 4 K. With helium costs ranging from £3–6/litre, this can be prohibitively expensive. High temperature superconductors can be cooled by liquid nitrogen at 77 K which is much cheaper at about £0.70/litre. However, to get any useful work out of the current range of HTS materials, they still need to be cooled to lower temperatures of about 30–40 K.

Commercially available cryocoolers can provide about 100 W or more of cooling power at 25–40 K and the relative cooling costs are lower when compared to the cost of liquid helium. Figure 10.7 shows the energy consumption in normalised units versus required cooling temperature.

Although it is widely accepted that the ideal Carnot efficiency can never be attained in practice, present cryocooler designs are still a very long way off this ideal – a typical work input/output ratio for the temperature range 300–30 K is approximately 100:1 – i.e. to obtain 100 W of cooling power at 30 K, an electrical input power of about 10 kW is required.

For rotating machinery, one of the significant challenges lies in transferring the cryogenic coolant from a stationary cryocooler to a rotating component. A rotating

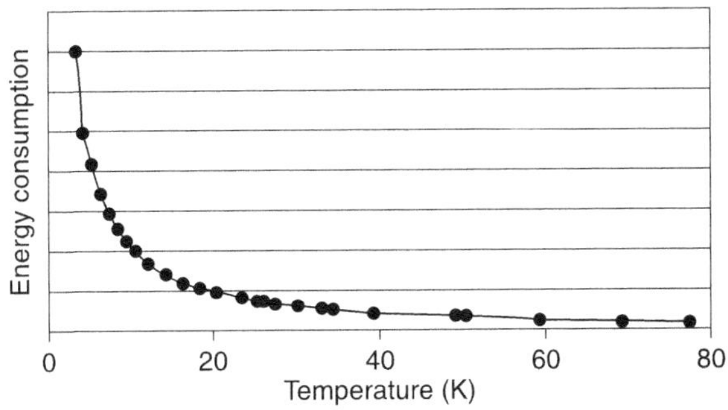

10.7 Refrigeration energy consumption, Carnot thermodynamics. (Reproduced with permission from Converteam UK.)

coupling is usually the solution and in its simplest form uses two concentric thin-walled tubes, one rotating inside the other. Seals used can be either polytetrafluroethylene (PTFE) or ferrofluid or a combination of the two. At high speeds and torques, ferrofluid seals can generate a considerable amount of heat, which must be removed. Alignment, robustness and reliability of these transfer couplings remain key concerns.

Transferring the 'cooling' to where it is needed on the HTS coils poses its own challenges and is dependent on the heat transfer mechanism employed – be it via convection or conduction. For conduction cooling methods, choice of high thermal conductivity materials, e.g. annealed high purity copper as well as material surface conditions can be critical for adequate thermal transfer – oxidation of thermal transfer surfaces should be kept to a minimum[10,11,12].

Soldered thermal connections are best but need to be assessed for integrity. Correct choice of solder for operation in vacuum and at cryogenic temperatures is essential to ensure even solder flow between jointed parts (without void formation) and the prevention of cracking when subjected to repeated thermo-mechanical cycling. Bolted connections are also a possible solution but, again, thermal differential contraction of materials must be allowed for, usually by employing the use of disc springs with bolts torqued to appropriate values. On a rotating machine, some vibration is always present and can loosen bolts and fixings if not secured appropriately. A suitable high-vacuum thermal grease such as Apiezon N® is recommended for bolted connections but should be used sparingly to avoid build-up of thick layers that could crack and flake at cryogenic temperatures, producing a bad thermal connection.

Depending on the cooling mechanism employed, different parts of the HTS machine can cool down at different rates, as shown in Figure 10.8. This can lead to differential stresses between welded and bolted components, necessitating the use of flexible connections such as braids, bellows and disc springs.

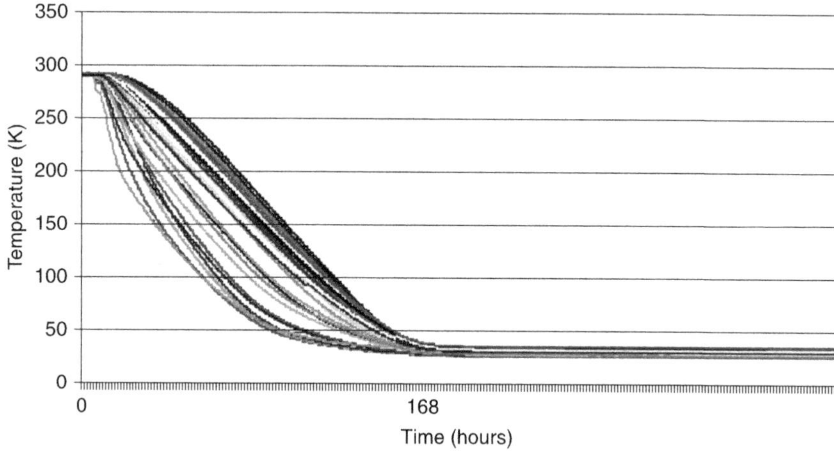

10.8 Typical cool down profile of different cryogenic parts of a machine. (Reproduced with permission from Converteam UK.)

10.3.3 Heat loads and vacuum level

The Hydrogenie machine heat budget is split fairly evenly between conduction and radiation heat loads, with a small amount linked to internal coil heat generation during operation (Fig. 10.9). Conduction loads were reduced by the appropriate use of low thermal conductivity materials, e.g. glass-fibre reinforced plastic (GFRP) or carbon-fibre reinforced plastic (CFRP) for the linkages between cryogenic components and components at room temperature.

A vacuum space was used to minimise conduction and convection heat loads, and it was critical that this space was kept scrupulously clean. A vacuum level of about 1×10^{-5} mbar was found to be adequate for the generator. Multi-layer

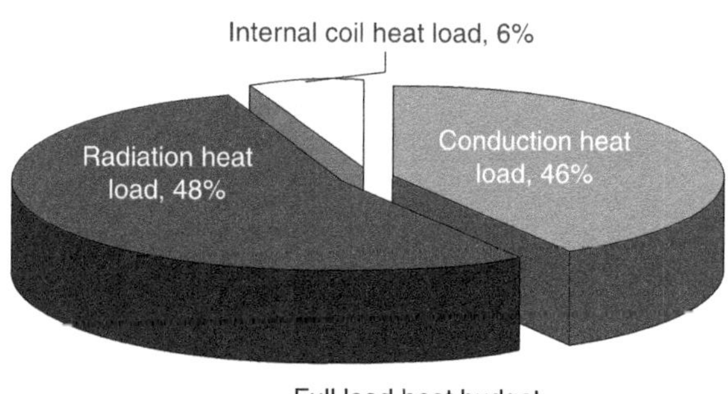

10.9 Hydrogenie heat budget. (Reproduced with permission from Converteam UK.)

insulation performs best at vacuum levels below 1×10^{-4} mbar. Rotating machinery construction involves the use of many non-vacuum friendly materials such as mild steel, plastic and paper insulation materials, impregnation resins, lubricating oils and greases, etc. It was therefore essential that, wherever possible, materials were cleaned to at least high vacuum standards and clean handling methods and manufacturing areas were employed during the machine build.

Rust is extremely hygroscopic and corrosion was therefore kept to a minimum by the plating of large components. Certain key components were also vacuum-baked to remove moisture. A controlled pump-down and cool-down process was also adopted to produce a clean moisture-free system. Differentially pumped vacuum seals greatly reduced the 'in-leak' into the vacuum system.

'Trapped volumes' were eliminated by the careful design of all vacuum sub-assemblies – otherwise these volumes would have manifested themselves as 'leaks' thereby degrading the vacuum level and limiting the operational lifetime of the machine. Appropriate fasteners, for example vented bolts and washers, and vacuum-compatible weld preparations were used for all sub-assemblies within the vacuum chamber.

Outgassing and residual gas analyses were carried out for each individual material used within the vacuum chamber to enable the proper design of vacuum seals, cryopumping surfaces and getters. Vacuum degassing of the rotor shaft steel was carried out while still in its molten state, primarily to remove hydrogen. Figure 10.10 illustrates a typical set of spectral peaks obtained from a sample of nickel-plated mild steel. It is clear that water is the main constituent and steps must be taken during the machine build and the pump-down process to eliminate its presence.

Cryogenically cold surfaces can behave as 'cryopumps' and act to 'remove' certain contaminants within the vacuum system, thereby improving its overall vacuum level – typically by an order of magnitude. The screenshot in Figure 10.11 shows the vacuum level in mbar on the *y*-axis versus time on the *x*-axis. It can be seen that at $t = 0$, when the vacuum pumps are turned off, the vacuum pressure rises from about 1.34×10^{-5} mbar to about 1.75×10^{-5} mbar, where it reaches the pumping speed provided by the cold cryopumping surfaces. From this point onwards the vacuum level actually starts to improve – indicated by the falling pressure with time.

In order to assess the efficacy of any cryopumping surface, the temperature of the said surface and its surrounding vacuum pressure level must be established[13,14,15]. For example, if there is a leak from atmosphere into a vacuum chamber operating at 1×10^{-5} mbar, in order to successfully cryopump N_2, the major constituent of air, the cryopumping surface temperature needs to be lower than 30 K. Figure 10.12 illustrates this particular example.

The emissivities of the various materials employed for the machine build were assessed in order to define the radiation heat load[16]. Multi-layer insulation (MLI or superinsulation) was utilised to reduce radiation heat loads onto cryogenically cold components. A typical HTS coil being wrapped in its pre-cut blanket is shown in Fig. 10.13. These pre-cut MLI blankets were appropriately supported

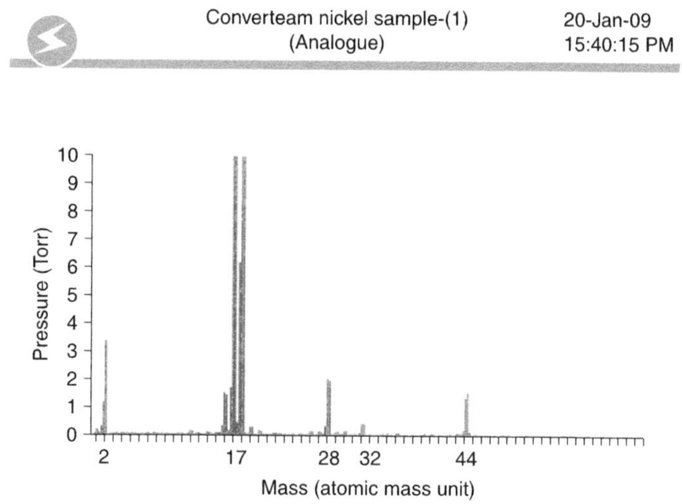

10.10 Spectral peaks from outgassing tests (2 amu = H_2, 17 or 18 amu = H_2O, 28 amu = CO, 32 amu =O_2, 44 amu = CO_2). (Reproduced with permission from Converteam UK.)

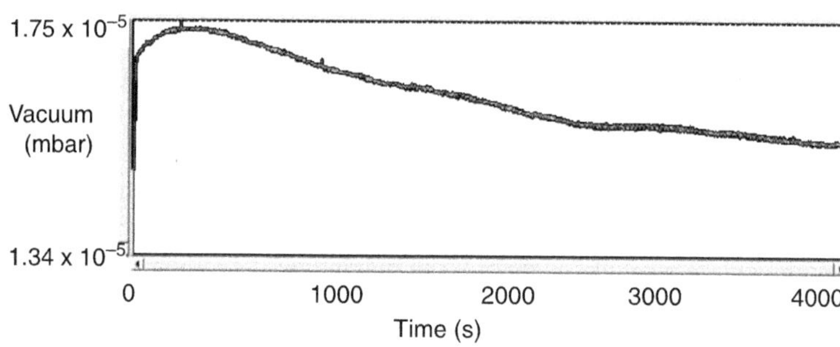

10.11 Improvement of vacuum level due to cryopumping (vacuum pumps turned off). (Reproduced with permission from Converteam UK.)

against rotational forces. During the application of these blankets care was taken not to thermally short the individual layers of the blanket. Vulnerable areas are the joints between individual blankets, and sufficient overlap area was employed.

An estimation of the radiation heat load transmitted through a typical (uncompressed) multi-layered superinsulation blanket can be obtained using[12]:

$$\frac{Q}{A}=\frac{\sigma.e.\left(T_2^4-T_1^4\right)}{n+1} \qquad [10.1]$$

where Q/A = watts transmitted per unit area (W/m^2), σ = Stefan–Boltzmann constant = 5.67×10^{-8} W/(m^{2}K^4), e = emissivity, T_2 = 'hot' surface temperature,

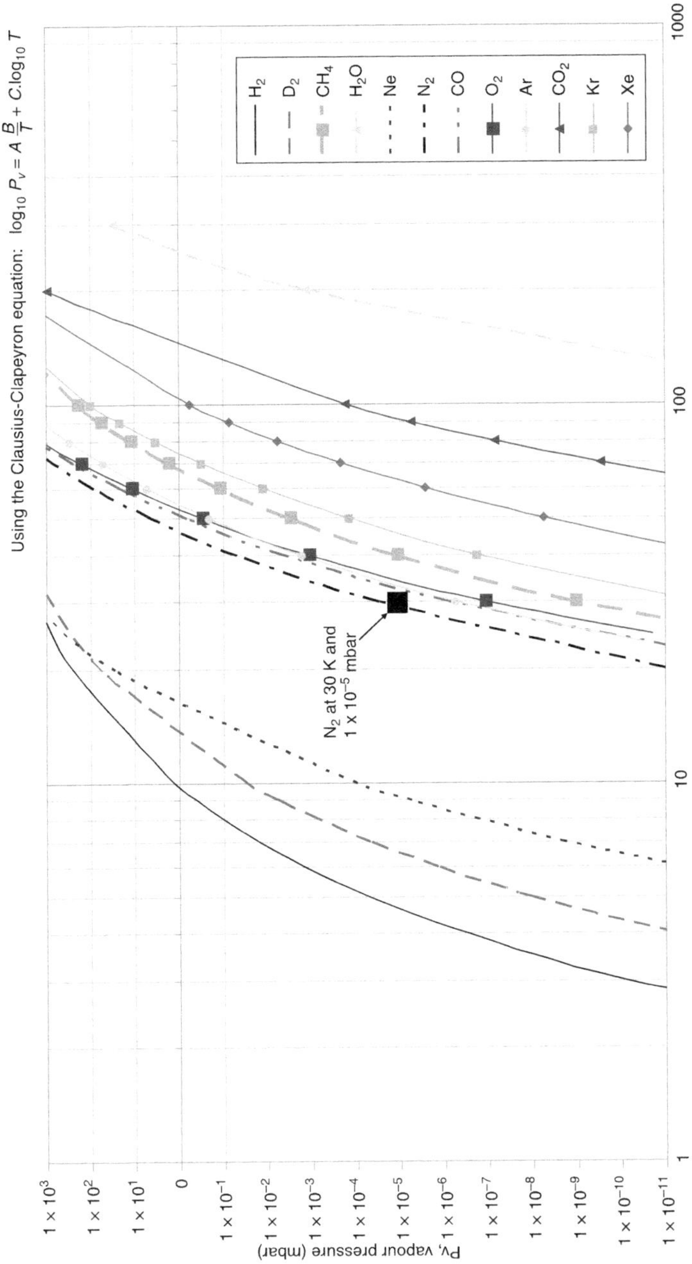

10.12 Vapour pressure of some common gases. (Reproduced with permission from Converteam UK.)

10.13 MLI blanket fitted to an HTS coil.

T_1 = 'cold' surface temperature, n = no. of layers of superinsulation. It is essential that these blankets are not over-compressed or their excellent radiative properties will be compromised by conduction heat loads through the compressed blanket walls. Figure 10.14 was used to select the number of layers for the coil blankets taking into account the heat budget and the space available for superinsulation.

Special consideration was also given to the design of the HTS coil current leads. The leads were profiled in order to minimise the conduction heat load by varying the cross-section of the leads along their length. The leads together with all the diagnostic wiring were also thermally dumped to further reduce conduction heat loads.

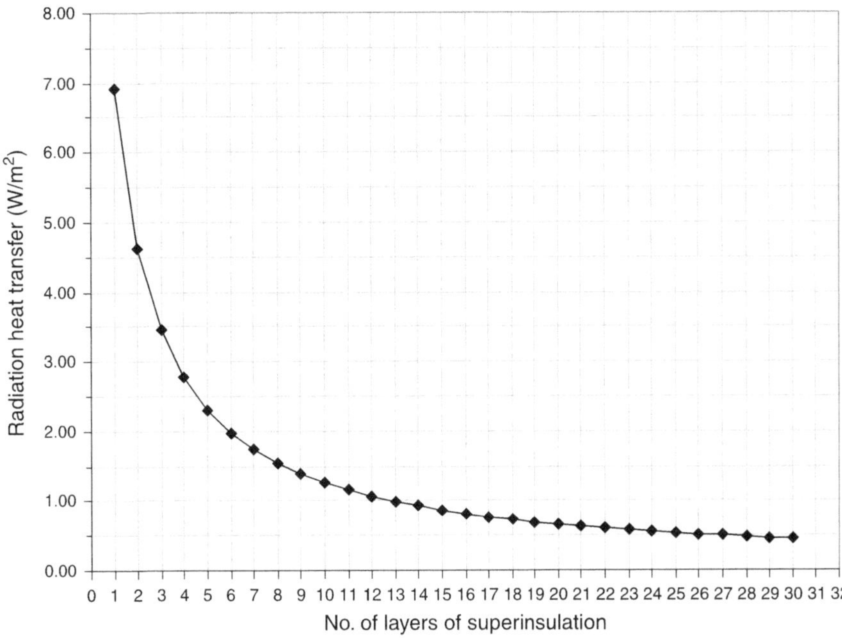

10.14 Radiation heat transfer vs no. of layers of superinsulation. (T_2 = 300 K, T_1 = 30 K, e = 0.03 for aluminised Mylar.) (Reproduced with permission from Converteam UK.)

10.4 Key requirements for successfully applying HTS technology

It is important for a business to develop key 'technology building blocks' in this particular field. Some obvious and perhaps not so obvious building blocks are:

- machine design utilising HTS materials,
- HTS materials knowledge,
- cryogenic cooling design,
- high vacuum design,
- leak detection methods,
- multi-layer insulation blanket design and application,
- clean room best practice,
- vacuum best practice,
- cryogenic best practice,
- liquid cryogen transfer (nitrogen),
- high-pressure cylinder gas handling,
- health and safety including COSHH,

- residual gas analysis – equipment use and results interpretation,
- outgassing/gas load calculations/getter sizing,
- pump-down curves,
- pressure Equipment Directive (PED)/BS5500/ASME VII,
- cryogenic and vacuum instrumentation.

Experimental test facilities to support the application of HTS materials should comprise as a minimum:

- clean areas to prepare and conduct experiments,
- vacuum chamber or chambers,
- cold head/s or cryocooler/s with appropriate fan/s (dependent on cooling mechanism employed),
- high-vacuum pumps,
- high-pressure gas (dependent on cooling mechanism employed),
- leak detection equipment,
- pressure gauges, vacuum fittings, feedthroughs, etc.

A typical vacuum and cryogenic test facility set up by Converteam UK Ltd in Rugby, England is shown in Fig. 10.15.

When a company starts down the road of developing HTS machines, ready access to information databases and relevant expertise is crucial. In terms of skilled personnel, they can be sourced from industries long associated with the field of low temperature superconductivity, e.g. superconducting magnets for bioscience and research, MRI scanners and so on. The magnet, cryogenic and vacuum skills utilised for low-temperature systems can be applied very

10.15 Test vacuum chamber and associated equipment. (Reproduced with permission from Converteam UK.)

successfully to the world of HTS. Suppliers to these low-temperature industries have vast reserves of useful information and skilled in-house experts – most of whom will be quite happy to provide advice and services. Another set of key resources are the universities and national laboratories scattered around the country. Most universities would be keen to collaborate with industry, and funding is usually available for such liaisons to fund postgraduate students and related research activities. The national laboratories usually charge for their services but their information databases are vast and second to none. Their experts are usually unrivalled in their respective fields of expertise and can prove to be invaluable in making the right decisions – thus enabling companies to 'short-circuit' the normal trial and error processes associated with these complicated fields of research.

The provision of training for the end-customer on the operation of the machine is essential for a 'demonstrator' machine. There may be various manual checks (more than usual) that need to be performed before the machine can be used. There may be additional or new health and safety issues that need to be de-risked and managed. For example, the use of cryogens in a power station is unusual and measures must be put into place for their safe handling and use.

10.5 Challenges and future trends

HTS material is now commercially available in relatively long lengths with the required performance characteristics, albeit still at a very high price. Figure 10.16

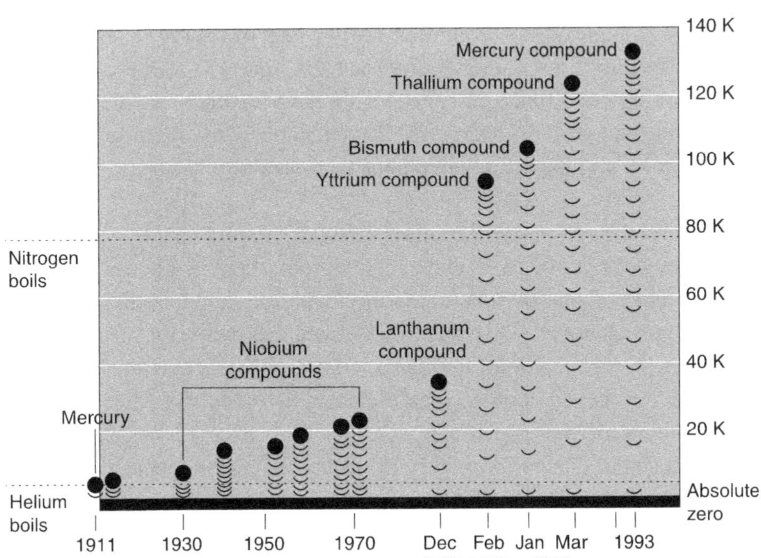

10.16 The trend for critical temperatures. (Reproduced with permission from the National Renewable Energy Laboratory[18].)

charts the progress in the discovery of these superconducting materials. As critical temperatures rise, it will become even more imperative for the growth of HTS applications that these novel materials are turned into commercial products at affordable prices.

However, in order to ensure that HTS applications have a secure future in the industrial world, there are still several key obstacles to overcome. The commercial world's natural conservatism and scepticism of the reliability and robustness of HTS technology is understandable. HTS material and conductor performance have come a long way since their discovery in the mid-1980s. The ancillary equipment that is so necessary to enable HTS technology still requires significant development work. There are five key areas for further development that are crucial to the successful application of HTS materials to rotating machinery.

10.5.1 Industrialisation of cryocoolers

A cryocooler system incorporates one or more cold heads, a reciprocating mechanism and a means for transferring the cold cryogen to its point of use. These systems are normally used in clean static environments such as laboratories. They are not really suited to an offshore or marine environment where they may be exposed to corrosive conditions and large displacement movements and vibrations. Most cryocoolers have a mean time between maintenance (MTBM) of approximately 10 000 hours (about 14 months) when seals have to be replaced. However, if the cryocooler is operating at varying angles, for example on a ship, seals will tend to wear faster. Cooling power can also be reduced if the cold heads are operated at large angles away from their normal axis of operation. Some cold head systems make use of a separate compressor unit that will also be subjected to the same operating conditions. There is insufficient operational data in harsh environments over long periods of time for these machines so one has to assume that they will require significant development to meet these stringent operating conditions.

For low-speed HTS machines, for example direct drive wind turbine generators, where rotational speeds are of the order of 10–12 rpm, there is a possibility that the cryocoolers, and perhaps their compressors, could be mounted directly onto the rotating vacuum chamber. This would eliminate the necessity of having to transfer cold cryogen across a rotating interface. However, this may pose an even harder challenge for the cryocooler manufacturer. To keep the forces due to rotation to a minimum, the cryocooler systems may need to be completely redesigned to reduce the size and weight of the individual components, new types of seals may need to be developed, gas bearing systems may need to be incorporated to reduce wear and tear, and so on.

The efficiencies of present-day cryocoolers are still very far from the ideal calculated Carnot efficiency and it is accepted that this cannot truly be achieved in practice. However, even an improvement of 10–20% on current figures would still be an acceptable target for the foreseeable future. Some of this improvement could possibly be achieved by developing an improved regenerator material although this is easier said than done!

10.5.2 Industrialisation of rotating cryogen transfer couplings

These couplings link the stationary cryocooler to the HTS machine's rotor enabling transfer of cryogenic coolant under rotating conditions. These couplings are usually constructed from concentric long, thin-walled stainless steel tubes rotating at several hundred rpm (and perhaps even higher speeds) incorporating a combination of PTFE and ferrofluid seals. Radial clearances are usually kept to a minimum on these devices to reduce surface area and the inevitable radiation heat load. Long, thin-walled tubes are used primarily to minimise conduction heat loads. Once again, these assemblies are normally designed for use on stationary platforms, i.e. not continually swaying as on a ship. Bearing in mind that the rotational speed of HTS machines will vary during normal operation, producing high acceleration and deceleration forces and possible axial movement of several millimetres, not to mention thermal growth of the machine's shaft system, rotating couplings need to allow for all these conditions, including possible misalignment, without compromising the cooling. These devices must also not introduce unduly high heat loads into the cooling system and the HTS machine. Due to the high rotational speeds of some machines, couplings can experience high temperatures and may need to be cooled themselves, perhaps via water cooling. This can complicate the design of these devices and needs to be considered together with the HTS machine's operating scenarios when developing new reliable and robust systems.

10.5.3 Industrialisation of instrumentation and vacuum components

All components (e.g. pressure gauges, getter modules) fitted to the rotating vacuum chamber must necessarily be able to withstand the forces due to acceleration and continue functioning faultlessly. Once again, there are not many commercial off-the-shelf (COTS) devices available on the market. Converteam have had to carry out tests on pressure gauges[17], getter and electronics modules in order to prove their reliability under rotational conditions. Accelerations of up to 150 g were expected on the Converteam HTS hydrogenerator and all components had to be tested accordingly. Electronics modules in particular had to be designed almost from first principles to ensure that individual components on circuit boards were mounted so that the components were pushed back into the board under rotational conditions. Soldered joints and cable strain relief mechanisms had to be developed to allow for these rotational forces.

10.5.4 Industrialisation of vacuum pumps

As for the cryocoolers discussed earlier, vacuum pumps need to be low-maintenance and able to operate faultlessly in similarly harsh environments. Although rotating vacuum chambers would ideally require no re-pumping for periods of up to six months or longer – assuming that any employed cryopumps and getters are effective – it would be better to have mechanical pumps physically attached to the chamber. This could perhaps only be possible for low-speed machines, for example direct-drive wind turbine generators operating at 10–14 rpm. Once again, if these pumping systems are to be utilised in such a manner, they need to be low-mass, low-profile and ultimately robust and reliable. This would of course apply to both the backing pump as well as the turbomolecular pump.

10.5.5 Supplier chain and test facilities

Working in such a specialised field as high temperature superconductivity requires access to appropriate supplier chains, in particular suppliers who have a good understanding and appreciation of what is required in terms of vacuum cleanliness and good cryogenic practice. It is essential that such relationships are forged at the start of an HTS project so that suppliers are involved from the very beginning. This is especially important if the supplier is new to the field and needs to be informed of the requirements of this technology.

Having ready access to vacuum and cryogenic test facilities is essential for such projects to be successful. It is always advisable for the research and development organisation to have its own test facilities and expertise to carry out such test programmes. It is of course possible to utilise test facilities at suppliers' premises, at universities or even national laboratories. However, not only will this be a very expensive way to carry out research, it will not provide a very flexible means of carrying out test programmes. Converteam UK Ltd decided to specify, design, build and equip its own cryogenics laboratory in Rugby, England, and this has proved to be absolutely the right thing to do. It provided the business with tailormade testing facilities to efficiently carry out the necessary range of vacuum and cryogenic tests within a clean and controlled environment.

10.6 Conclusions

It is clear that the electromagnetic design of an HTS machine cannot be carried out in isolation from the cryogenics and vacuum design. These areas need to be considered as early on as possible in the design process in order to home in on an optimised design.

In order to be able to apply HTS technology successfully, a company's infrastructure must be able to support it. For example, appropriate testing facilities are crucial as are suitable industrial and academic partners. An understanding of good vacuum and cryogenic practice is essential for the successful production of these machines as is the availability of 'clean' manufacturing facilities.

Databases of experience and knowledge are essential in order to qualify the materials and processes applied within the dual fields of cryogenics and vacuum technology. National laboratories such as the Rutherford Appleton Laboratory and Daresbury Laboratory have vast amounts of expertise and knowledge databases that can be tapped. Similarly, certain specialist suppliers to various industries, notably aerospace and motor sport can also be of enormous assistance.

If the application of superconductivity to industrial machinery is handled selectively and appropriately, it will undoubtedly offer great potential for the future. Cryogenics and vacuum technologies (such as superconductivity itself) are sometimes treated like the 'black arts'. However, this stems from the fact that for every HTS-related project, the number of variables is high and they are all interlinked – any one of them can compromise a cryogenics or vacuum system if not treated appropriately. It is for this reason that most present HTS machines are over-designed. Until we can begin to understand better the limits of this technology, only then will we be able to make more educated design compromises.

The application of superconductivity to industrial machinery is an extremely exciting but demanding field that will no doubt come into its own in the near future. For its application to be truly successful in the industrial world, the development and long-term operation of HTS research machines is essential in order to build up confidence in this technology and to allow the various manufacturers of ancillary equipment to develop and ruggedise their equipment.

10.7 References

1 Lewis C. (2010) 'Direct drive superconducting wind generators', in *Wind Power Generation and Wind Turbine Design*, ed. W. Tong, Southampton, Boston, WIT Press, ISBN 978-1-84564-205-1.

2 Baik S.K., Sohn M.H., Lee E.Y., Kwon Y.K., Moon T.S., *et al.*, (2006) 'Effect of synchronous reactance and power factor on HTS synchronous machine design and performance' *IEEE Transactions on Applied Superconductivity*, **16**(2), 1489.

3 Ackermann R.S., Alexander J.P., Bray J.W., Gadre A.D., Huang X. *et al.*, (2006) *Design and Development of a 100 MVA HTS Generator for Commercial Entry* (DOE Co-operative Agreement No. DE-FC36-02GO11100).

4 Maki N., Takao T., Fuchino S., Hiwasa H., Hirakawa M., *et al.*, (2005) 'Study of practical applications of HTS synchronous machines' *IEEE Transactions on Applied Superconductivity*, **15**(2) 2166.

5 Jo Y-S., Kwon Y-K., Sohn M-H., Kim Y-K., Hong J-P., (2002) 'High temperature superconducting synchronous motor' *IEEE Transactions on Applied Superconductivity*, **12**(1) 833.

6 Klaus G., Nick W., Neumuller H.W., Nerowski G. *et al.*, (2006) 'Advances in the development of synchronous machines with high-temperature superconducting field winding at Siemens AG Meeting 2006'. IEEE, Digital Object Identifier: 10.1109/PES.2006.170964B. ISBN: 1-4244-0493-2.

7 Frank M., Frauenhofer J., Van Hasselt P., Nick W., Neumuller H.W. *et al.*, (2003) 'Long term operational experience with first Siemens 400 kW HTS machine in diverse configurations' *IEEE Transactions on Applied Superconductivity* **13**(2) 2120.
8 Sivasubramaniam, K. Zhang, T. Lokhandwalla, M. Laskaris, E.T. Bray, J.W. B. *et al.* (2009) 'Development of a high speed HTS generator for airborne applications' *IEEE Transactions on Applied Superconductivity*, **19**(3) 1656.
9 Young M.A., Demko J.A., Gouge M.J., Pace M.O., Lue J.W. *et al.*, (2003) 'Measurements of the performance of BSCCO HTS tape under magnetic fields with a cryocooled test rig' *IEEE Transactions on Applied Superconductivity*, **13**(2), 2964.
10 British Cryoengineering Society, (2002) *Cryogenic Fluids Databook*, first ed., British Cryogenics Council ISBN 0-9543030-0-8.
11 Reed R.P. and Clark A.F. (1983) *Materials at Low Temperatures* (National Bureau of Standards, Boulder, Colorado).
12 Ekin J.W. (2006) *Experimental Techniques for Low Temperature Measurements* National Institute of Standards and Technology, Boulder, Oxford University Press.
13 Barnes C. (1968) 'Cryopumping', *Proceedings of 1968 Summer Study on Superconducting Devices and Accelerators*, Brookhaven, New York, CVI Corporation, Columbus, Ohio, 230–248.
14 Grogner, O. *Vacuum and Cryopumping* (Schmiedgasse 5, Innsbruck, A–6020, Austria).
15 Haefer R.A. (1989) *Cryopumping – Theory and Practice*, Oxford Science Publications.
16 Musilova V., Hanzelka P., Kralik T., Srnka A., (2005) 'Low temperature radiative properties of materials used in cryogenics', *Cryogenics* **45** 529–536.
17 Grzesik B., Stepien M., Bodzek K., Fair R., Ingles M., (2010) 'Tests of a vacuum gauge for an HTS hydrogenerator', in *14th International Power Electronics and Motion Control Conference*, EPE-PEMC.
18 National Renewable Energy Laboratory, (1998) 'At the frontiers of Science – Superconductivity and its Electric Power Applications', Produced for the U.S. Department of Energy (DOE) by the National Renewable Energy Laboratory, a DOE National Laboratory Report Reference: DOE/GO-10098-434.

11

High temperature superconductors (HTS) in fusion technologies

J.-L. DUCHATEAU, CEA, France

Abstract: In the International Thermonuclear Experimental Reactor (ITER), HTS current leads enable cryogenic refrigeration power savings of about 20% at 5 K. They have the potential to play a significant role in future fusion reactors. Existing HTS materials (BSCCO and YBCO) potentially allow operation between 20 and 30 K and high magnetic field; operation in liquid nitrogen (65–77 K) is probably still very far off. HTS materials may allow simplification of the cryogenics of very large systems. Indirect cooling could replace forced-flow cooling, as used in superconducting tokamaks. Indirect cooling, the traditional method of cooling superconducting magnets, could be selected for the refrigeration of these large systems thanks to the thermal gradients that are now achievable as a result of using HTS materials.

Key words: HTS magnets, fusion reactor, cryogenic refrigeration, cable-in-conduit conductor.

11.1 Introduction

Superconducting magnets are mandatory for future fusion reactors. If resistive magnets were to be used for a fusion reactor, the power consumption linked to the magnets would be greater than the fusion power of the reactor itself! The International Thermonuclear Experimental Reactor (ITER) is still a pulsed machine, but the electrical power necessary to energize the magnet system in the case of a solution with resistive magnets ($P = 2$ GW for 500 s) cannot be reasonably obtained from the electrical grid.

The next step after ITER is the DEMOnstration power plant (DEMO). The DEMO fusion reactor should provide 1000 MW direct current (DC) electrical power. Construction is planned to start 20 years from the time of writing (see Fig. 11.1, 11.2). One of the key questions facing researchers is whether the DEMO magnet system[1] should be an extrapolation of the ITER magnet system (which has been in preparation for the last 20 years) or whether a technological revolution is needed; would this technological revolution be achievable within the next 20 years?

The same question is being discussed for the heliotron-type reactors, the force free helical reactors (FFHRs)[2], based on existing experience with the superconducting Large Helical Device (LHD) machine, which has been in operation for the last 12 years. Researchers are considering whether the conventional NbTi and Nb_3Sn low-critical-temperature (T_c) low temperature superconducting (LTS) materials used

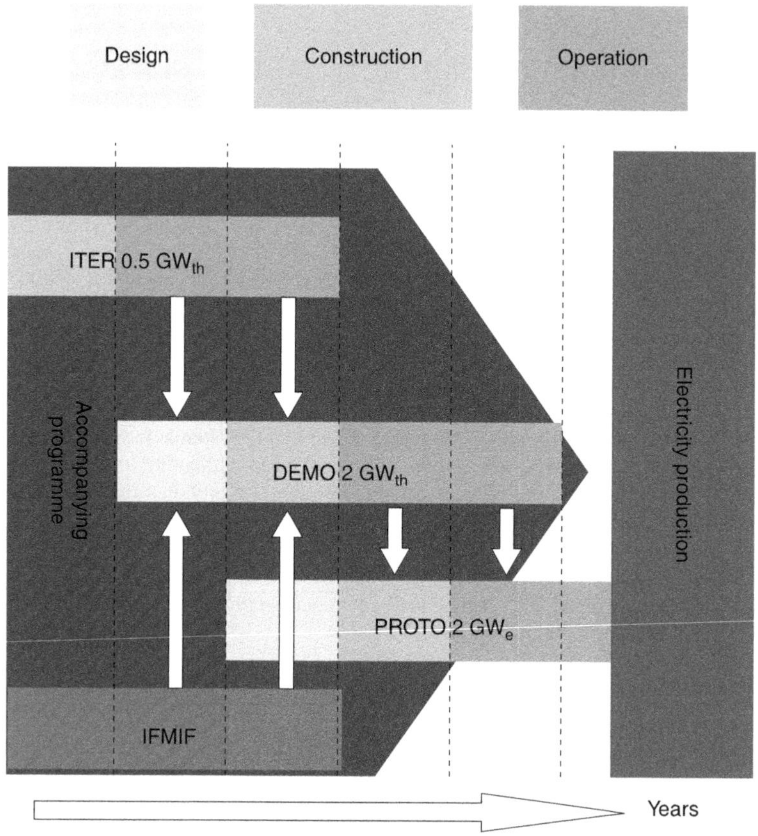

11.1 Schematic roadmap from ITER to DEMO and PROTO.

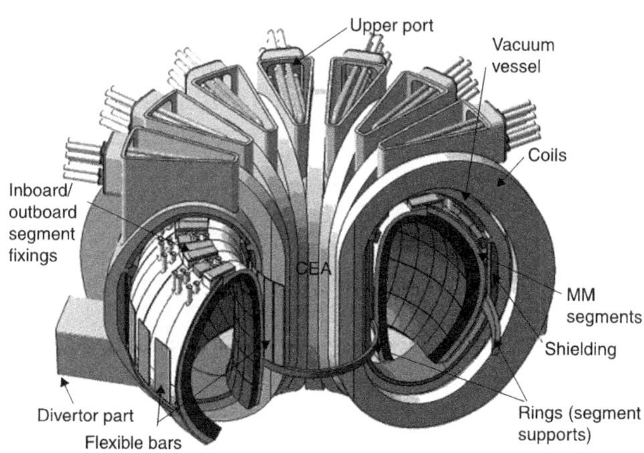

11.2 3D view of a version of DEMO reactor.

(a) (b)

11.3 Conductors of the two ITER model coils. (a) CS Model Coil (51 × 51 mm, 40 kA) (b) TF Model Coil (exploded view) (diameter 40.7 mm, 80 kA). (Courtesy of ENEA.)

in the ITER cable-in-conduit conductor (CICC), operating at 5 K (see Fig. 11.3), are able to fulfil the technological specifications of DEMO magnet systems or whether it is necessary to use the emerging high-T_c superconducting materials (HTS materials) operating at 20–50 K.

Looking even further into the future, it cannot be taken for granted that CICC, such as that developed for ITER, is the best-adapted conductor for future fusion reactors. This question has recently been raised by teams at the National Institute for Fusion Science in the framework of their studies regarding FFHRs[3].

The above are typical questions that have to be answered through a development program that has to be defined.

Due to the high cost of the magnet system in a fusion reactor, representing typically 30% of the cost investment in the case of ITER, it is quite clear that the selection of the best option is crucial for the success of the reactor. Central plasma parameters, such as major radius (R) and plasma magnetic field (B_t), are major dimensioning parameters for the reactor and also for the magnet system. It is therefore becoming evident that the optimization of the reactor and of the magnet system cannot be carried out independently.

The design of future fusion reactors has to be initiated within a consistent new approach[1] combining the scaling laws of physics and the structural dimensioning of the magnet system – the basis of which has yet to be clarified. Within this approach, the potential benefit of HTS materials for reactors has to be examined, taking into account investment cost, operational cost, efficiency and reliability.

For tokamaks, the discussion will mainly focus on the toroidal field (TF) system, which is the most important magnet system for tokamaks (41 GJ stored energy for the ITER TF system versus 6.4 GJ for the ITER central solenoid (CS) and 4 GJ for the ITER poloidal field (PF) system).

11.2 Magnetic field and magnet current density considerations for a fusion reactor

Figure 11.4 clearly demonstrates the benefits that HTS materials could bring in comparison with classical niobium materials such as NbTi and Nb_3Sn. The figure shows the critical (for LTS materials) or irreversible (for HTS materials) magnetic field as a function of temperature. This means that for a given LTS or HTS material, the plotted line represents the limit at which the critical current density starts to differ from zero. In order to be used in a given application, for a given superconducting material, the practical limit should be parallel to this plotted line and, unfortunately, lower, corresponding to the critical current density that is needed for the application. This parallel curve is available for niobium materials (LTS), but it is less clear for HTS, these materials being currently still under development. Figure 11.4 can therefore be taken only as showing the potential for HTS materials as compared with LTS materials.

As is well known, HTS materials open two new potential windows for the application of superconductivity:

- For magnetic fields larger than 20 T where the use of Nb_3Sn is not possible;
- For operation at higher temperatures, as has to be considered for fusion magnets with the aim of decreasing the circulating power of the fusion reactor (see Section 11.3.1).

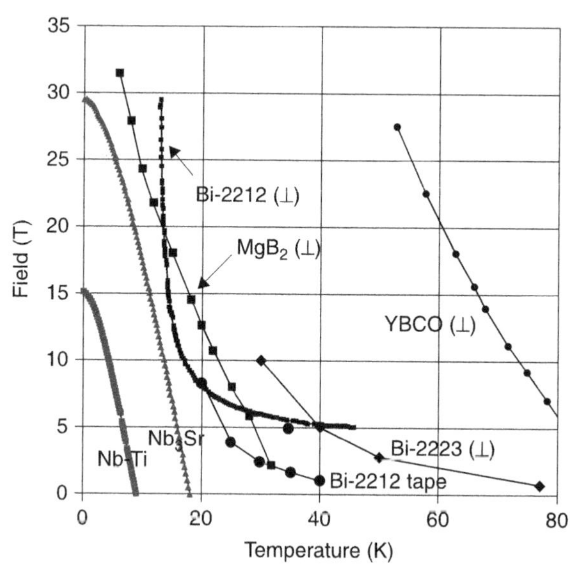

11.4 Critical magnetic field (LTS) or irreversible magnetic field (HTS) as a function of temperature. (Private communication from David C Larbalestier.)

11.2.1 Magnetic fields for fusion reactors

For fusion reactors, in contrast to magnets for nuclear magnetic resonance (NMR) where a high magnetic field is crucial, there is no benefit (and it is also not even possible) in increasing the magnetic field above 13–15 T; this range of magnetic field is accessible to Nb_3Sn.

In a tokamak design, the factor of merit $\xi = R^2 B_t^3$ is a key driver of the machine's performance in terms of fusion power and the amplification factor of energy Q. R is the major radius of the tokamak and B_t is the magnetic field at the plasma centre. The selection of the two parameters (R, B_t) has to be made in close conjunction with the questions of cost, available technology and accessibility to the plasma through the ports.

At first glance, it could be assumed that it would be advantageous to select a very high value of B_t and a small value of R to minimize the size and thereby the cost of the machine. With a more in-depth approach, however, it is necessary to consider the successive schematic layers of the reactor in the equatorial plane, as presented in Fig. 11.5.

The major radius of the machine must be sufficiently large to accommodate the following within the zone from the plasma to the centre of the machine:

- The plasma–magnet distance Δ_{int} (typically 1.9 m for a 1000 MWe reactor),
- The toroidal field (TF) inner leg,
- The central solenoid (CS), which must be sufficiently large to provide the magnetic flux to increase the plasma current up to the nominal value.

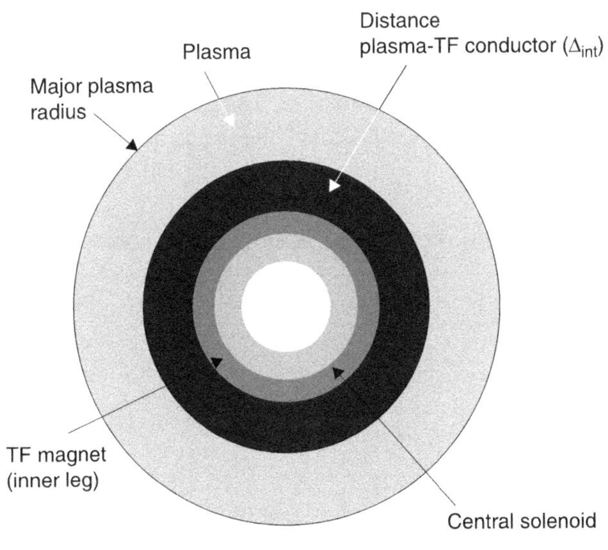

11.5 Tokamak reactor radial build in the equatorial plane.

The maximum dimensioning value B_{tmax} for the TF magnet system can be derived from B_t using the formula[4]

$$B_{tmax} = \frac{kB_t}{1 - \frac{a}{R} - \frac{\Delta_{int}}{R}} = 1.08 \qquad [11.1]$$

where a is the minor plasma radius and Δ_{int} is the distance between magnet and plasma (blanket, shielding). Typical values of these parameters are given in Table 11.1 for ITER and DEMO. From Table 11.1 it can be seen that a typical value of B_{tmax} for the DEMO TF magnet is 12.8 T.

The same conclusion is reached in Sagara *et al.*[2], where different LHD reactor designs are presented with a maximum magnetic field on the magnet system in the same range. For this reason, in the following sections, a maximum magnetic field in the range of 13–15 T will be considered as a working hypothesis.

11.2.2 Overall current density in the magnet system

The overall current density in the ITER TF is in the range of 10 A/mm^2. This overall current density is dictated by the structures and not by the superconductor, as illustrated in Table 11.2 for TF systems of different tokamaks. The Lorentz forces appearing in the TF system, which consist of a centring force and vertical force acting on each coil, require a large amount of steel, taking into account a typical yield stress of 1000 MPa for A316LN, which is the envisaged structural material.

For this reason, as in ITER, the overall current density of the TF magnet for DEMO is expected to be in the range of 10 A/mm^2, which indicates that the cable current density and the non-copper current density can be kept at moderate levels without affecting the overall current density. The cable current density of the ITER TF cable is in the range of 50 A/mm^2, the largest proportion of the cable

Table 11.1 Comparison of characteristics of ITER and DEMO

	ITER	DEMO
Pfus (MW)	500	2 600
R (m)	6.2	8.5
a (m)	2.46	2.79
Δ_{int} (m)	1.23	1.9
B_t (T)	5.3	5.49
ξ	5 722	11 955
TF (m)	2.16 m < r < 2.98 m	2.5 m < r < 3.86 m
B_{tmax} (T)	11.8	12.8

Table 11.2 Current densities in different tokamaks

Tokamak Maximum magnetic field on conductor	TF cable Current density including cooling (A/mm^2)	Average current density in the inner TF leg (A/mm^2)
TORE SUPRA: superconducting TF 9T, steady state	39	19
JET upgrade: copper TF 8 T, pulsed operation 10 s	20	17
ITER: superconducting TF 11.8 T, steady state	52	11

being occupied by the copper section, which plays a major role in keeping the conductor temperature lower than 250 K during the fast safety discharge that follows a quench of the magnet. For DEMO, a target current density value in the HTS material of 300 A/mm^2 can therefore be selected as a basis for a discussion about possible superconducting materials. It is comparable to the non-copper current density in the ITER TF system.

11.3 Operating temperature of the magnet system

11.3.1 Circulating power in a fusion reactor

Like all reactors, fusion reactors need to circulate a proportion of their electrical production P_{ecir} in order to operate their auxiliaries[5]. In the case of fusion reactors, the proportion is particularly large. The main components of the circulating power are associated with:

- power injection for heating and current drive (H and CD), P_{inj},
- pumping power to circulate helium in the blankets, P_{Hepump},
- cryogenics for the superconducting magnets, $P_{cryomagnets}$,
- cryogenics for cryopumps, $P_{cryopump}$.

This is presented schematically in Fig. 11.6. To find P_{inj}

$$P_{inj} = P_{fus} / Q\eta_{inj} \qquad [11.2]$$

where Q is the amplification factor of energy (10 in the case of ITER) and η_{inj} is the efficiency of the reactor heating systems. P_{ecir} is given by

$$P_{ecir} = P_{fus} / Q\eta_{inj} + P_{Hepump} + P_{cryopump} + P_{cryomagnets} \qquad [11.3]$$

Table 11.3 gives values for these parameters for DEMO to give practical figures, but similar conclusions can be drawn for helical reactors and stellarators. As can

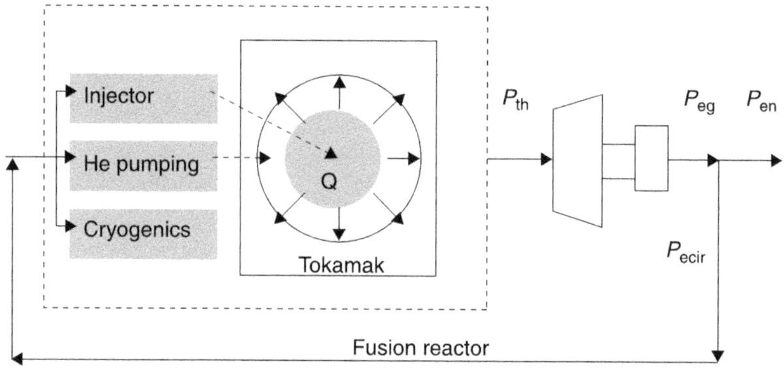

11.6 Illustration of the different components of the circulating power in a tokamak.

Table 11.3 Main components of circulating electrical power for DEMO

Parameter	DEMO
Major, minor radii *R,a* (m)	7.5, 2.5
Fusion power (MW)	2 401
Total produced electrical power P_{eg} (MW)	1 320
Net electric power P_{en} (MW)	700
Heating power CD P_{inj} (MW)	303
He pumping power P_{Hepump} (MW)	282
Magnets cryogenics power $P_{cryomagnets}$ (MW)	25
Cryopumps power $P_{cryopump}$ (MW)	7.5
Total circulating power P_{ecir} (MW)	620

be seen in Table 11.3, the circulating power is large for DEMO and the dominating parameters in the circulating power are P_{inj} and P_{Hepump}.

11.3.2 Cryogenic power of the magnets

The magnet circulating cryogenic power $P_{cryomagnets}$ has two origins: P_{cryo1} and P_{cryo2}. These terms correspond to the electrical power needed to operate the refrigerator of the magnet system. P_{cryo1} is the cryogenic losses around 5 K (as is the case for ITER) and P_{cryo2} is the cryogenic power taken by the thermal shields of the magnets around 77 K. It is possible, starting from the estimations regarding ITER, to extrapolate the values from ITER to DEMO, taking into account that the dimensions of the two reactors are not so very different from each other. In a first approach, the operating temperature of the DEMO magnet system is assumed to be 5 K. It is important to distinguish between the two components of the circulating power, P_{cryo1} and P_{cryo2}, as the introduction of HTS material will only have an impact on the component relating to losses at 5 K, i.e. P_{cryo1}.

Cryogenic losses at 5 K

The cryogenic losses at 5 K are presented in Table 11.4. It is assumed that pulsed losses are negligible in DEMO as the reactor is steady-state. The thermal static losses are proportional to the surface of the TF first wall S_{TF}:

$$S_{TF} \sim (1+\kappa^2)Ra \qquad [11.4]$$

where κ represents plasma elongation.

Cryogenic losses at 77 K

The cryogenic losses at 77 K are presented in Table 11.5. The thermal static losses are proportional to the surface of the TF first wall S_{TF}.

Estimation of P_{cryo1} and P_{cryo2}

P_{cryo1} and P_{cryo2} are respectively estimated from $P_{4.5\ K}$ and $P_{80\ K}$ taking into account the Carnot efficiency of the refrigerator and the efficiency f of the refrigerator, as presented in the following equations.

$$P_{cryo} = P_{cryo1} + P_{cryo2} \qquad [11.5]$$

Table 11.4 Main components of cryogenic losses at 5 K for DEMO and ITER

	ITER	DEMO
R (m)	6.2	7.5
a (m)	2	2.46
κ	1.7	1.9
Pulsed loads (kW)	13.5	0.
Steady static load (kW)	8.1	14.3
Heat load from circulating pumps and cold compressors (kW)	11.4	10
Current leads (kW)	4	8
Total $P_{5\ K}$ (kW)	37	42

Table 11.5 Cryogenic losses at 77 K for DEMO and ITER

	ITER	DEMO
R (m)	6.2	7.5
a (m)	2	2.46
κ	1.7	1.9
First wall surface (m^2)	682	1104
$(1+\kappa^2)Ra$	48.2	85.1
Static heat loads $P_{77\ K}$ (kW)	800	1412

$$P_{cryo1} = P_{5\,K}(T_1 - T_2)/fT_2 \qquad [11.6]$$

where $T_1 = 300$ K, $T_2 = 5$ K, $f = 0.25$, $P_{cryo1} \sim 250\, P_{5\,K}$

$$P_{cryo2} = P_{80K}(T_1 - T_2)/fT_2 \qquad [11.7]$$

where $T_1 = 300$ K, $T_2 = 80$ K, $f = 0.25$, $P_{cryo2} \sim 10\, P_{80\,K}$

The results of the estimations are presented in Table 11.6. It can be seen that, due to the large dimensions of the machine, P_{cryo2} is unexpectedly slightly larger than P_{cryo1}.

Impact of HTS materials on circulating power of future fusion reactor

As can be seen from the previous subsection, only P_{cryo1} can be affected by the introduction of HTS materials in the fusion conductors. P_{cryo2} corresponds to the radiation power and is not affected by the introduction of HTS materials; i.e. this part of the circulating power remains the same whatever the operating temperature of the magnet.

Consider now, without entering into details, the situation that the magnet system is operating at a temperature T with $5\text{ K} < T < 77\text{ K}$; cryogenic losses are still $P_{5\,K}$

$$P_{cryo1} = P_{5\,K}(T_1 - T)/fT. \qquad [11.8]$$

Due to the improvement of the Carnot efficiency, P_{cryo1} decreases as temperature increases, as seen in Fig. 11.7. Some 80% of the improvement (from 10 MW to 2 MW) is already achieved by around 20 K, which demonstrates that there is only limited gain (in terms of circulating power) in further increasing the temperature above 20 K.

Supposing that operation of the reactor magnet system at around 70–80 K is possible, which is currently unlikely, the cryostat design will be slightly affected. Total suppression of the thermal shield, leading to a simplification of the magnet system, is also unlikely as it will always be preferable to remove the radiation

Table 11.6 Cryogenic circulating power for DEMO and ITER

	ITER	DEMO
R (m)	6.2	7.5
a (m)	2	2.46
κ_{95}	1.7	1.9
Cryogenic power 5 K $P_{5\,K}$ (kW)	37	42
Cryogenic power 80 K $P_{80\,K}$ (kW)	800	1422
P_{cryo1} (MW)	9.25	10.5
P_{cryo2} (MW)	8	14.2
$P_{cryomagnets}$ (MW)	17.25	24.7

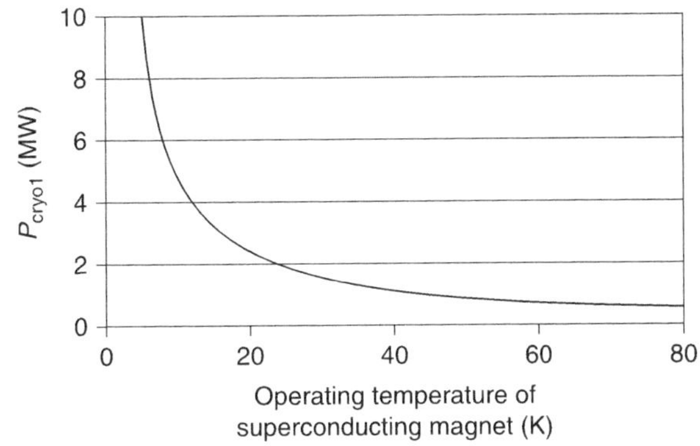

11.7 Impact of magnet operating temperature on P_{cryo1}.

losses on a separate cryogenic shield, thermally separated from the magnet and thermohydraulically optimized.

11.4 Indications for the cooling system of the fusion reactor magnet system

As shown in Section 11.3, there is an advantage in increasing the operating temperature of the conductor to 20–30 K. Within this range of temperature, the magnet operation appears feasible, taking into account the current state of the art of HTS materials such as YBCO or BSCCO.

11.4.1 Cable-in-conduit conductors

For operation in the temperature range 20–30 K, it is not clear that CICC is the best adapted form of conductor for fusion magnets. CICC, as shown in Fig. 11.3, has been developed for tokamak magnets to make use of the very high volumetric heat capacity of helium (about 500 times the volumetric heat capacity of metallic materials at 5 K), to limit the temperature excursion during fast energy deposition. This occurs in tokamaks after the very fast decrease to zero of the plasma current in cases of plasma disruption. In this situation, a fast magnetic field variation (typically within 100 ms in the case of ITER) affects the whole coil over lengths well over several metres, creating losses in the superconducting strands. CICC represents an adequate solution to this problem by providing:

- a local high-enthalpy helium reservoir (30% void fraction in the cable),
- a very high wetted perimeter (~4 m in total for the ITER TF cable) of the superconducting strands, achieved as a result of the subdivision that allows access to the helium reservoir of the cable,

- low AC losses for the conductor by controlling the coupling current time constant through the contact resistance between strands.

The use of this technology is essential for ITER, a pulsed machine that can also be subject to plasma disruptions. However, there are drawbacks to the use of such technology, making it difficult to use in a large fusion reactor when it is associated with a LTS material like Nb_3Sn:

- Decrease of the critical non-copper current density in Nb_3Sn by more than 50% due to strain. This strain is induced by the jacket during cool-down from the reaction temperature to 5 K, and by bending strain due to Lorentz force. This results in a very low temperature margin in the ITER TF system (0.7 K).
- Large pressure drop for helium circulation (8 g/s) resulting in friction losses estimated to about 25% of P_{cryo1} (see Section 11.3).
- Increased complexity of the cryogenic system in comparison with a bath-cooled magnet: helium circulator, specific joints and helium inlets are all potential sources of leaks and weakness with regard to high voltage in cases of fast safety discharge.

In Table 11.7, the role of strain in the degradation of the critical properties of Nb_3Sn is highlighted for three TF systems, respectively ITER at 11.8 T and two versions of DEMO at 13 T and 14 T respectively. Starting with a reference design for the temperature margin of 0.8 K, a gain of more than 2 K can be envisaged if it is possible to reduce the strain from –0.74% (CICC) to a strain of –0.25% which can be reached in a Rutherford cable, with a low void fraction.

It is therefore likely that the CICC is not the best adapted conductor for future fusion reactors and most of the institutes involved in these studies have started to envisage other types of conductors. Implicitly, this requires that a different type of cooling is also envisaged. This is possible if the power density in the winding is sufficiently low. This is the case for FFHRs but also for tokamak reactors, which are supposed to operate in steady-state conditions.

Table 11.7 Influence of Nb_3Sn strain on temperature margin in ITER and DEMO

	ITER TF system (B_{max} = 11.8 T)	DEMO TF system (B_{max} = 13 T)	DEMO TF system (B_{max} = 14 T)
Snoncu @ 68 kA (mm^2)	235	34	505
ΔT (ε = –0.74 %) (K)	0.78	0.78	0.78
ΔT (ε = –0.25 %)	2.16	3.02	3.08
$J_{noncu}(T_{op}, \varepsilon$ = –0.74 %) (A/mm^2)	373	265	143
$J_{noncu}(T_{op}, \varepsilon$ = –0.25 %) (A/mm^2)	680	542	444

11.4.2 Other types of solution: the indirect cooling concept

Magnets cooled by forced flow helium, whether using CICCs or not, in reality represent only a minority in applied superconductivity. In fact, most of the superconducting magnets in the world are bath-cooled magnets. This is the case for the tokamak Tore Supra where the conductors of the TF system are directly in contact with a superfluid static helium bath. This is also the case for MRI magnets where the winding is dry but the magnet is globally submerged in a liquid helium bath that indirectly cools the superconducting winding by conduction. In the case of MRI magnets, these are DC magnets subject to nearly zero heat deposition.

In the case of accelerator magnets such as the beam magnets of the Large Hadron Collider (LHC), the amount of helium in contact with the conductor is so small that it can also be said that the magnets are indirectly cooled by conduction.

For future fusion reactors such as DEMO or FFHR, the density of power deposition within the magnet is certainly very low, as these machines are expected to be steady-state systems and not pulsed systems like ITER. It is recommended that the nuclear radiation is shielded with a sufficient shielding thickness (to be included in Δ_{int} for tokamaks) in order to moderate the losses of this category of reactor.

In view of such a system, the teams of the National Institute for Fusion Science (NIFS) in Japan have started a development programme and have already presented an attractive conceptual design for the winding pack of a FFHR cooled by conduction, as shown in Fig. 11.8[3,6]. This concept is very specific to FFHRs and cannot be directly extrapolated to tokamak reactors; however, the concept includes general characteristics that are certainly common to both concepts, for

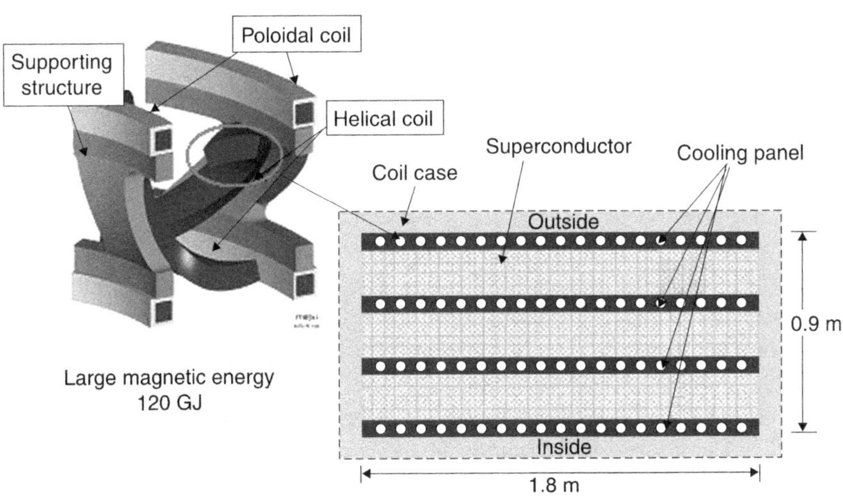

11.8 Indirect cooling in the superconducting FFHR system. (Courtesy of NIFS.)

example a stored energy of 120 GJ, a maximum magnetic field of 13 T and a conductor carrying a current of 100 kA. In fact, there is definitely a need to develop a high-current conductor in the range of 100 kA (compared with 68 kA in ITER TF), to keep the inductance of the winding and the associated voltage at a low level in situations of fast safety discharge.

The cable in this concept is made of several elementary components (ribbons or strands). It is situated within a structural jacket with a high filling factor. The heat generated is transferred by conduction to panels in which coolant is included.

This concept cannot be used for LTS materials such as Nb_3Sn or Nb_3Al because the steady-state heat load, even when maintained at the lowest level, must be removed by thermal conduction. If the temperature of the panel is kept at $T_0 = 4.5$ K, thanks to the channels cooled by two-phase helium, a temperature gradient due to thermal conduction will raise the conductor temperature to $T_0 + \Delta T$. This probably necessitates a conductor jacket made of aluminium alloy to keep ΔT moderate.

The concept is certainly more suited to HTS materials, and solutions with BSCCO materials or with YBCO materials as well may be possible. The cooling panels can be cooled to a temperature around 20 K, giving a temperature of 20 K $+ \Delta T$ for the conductor, but in this case ΔT can be in the range of several kelvin.

11.4.3 Stability: HTS versus LTS materials

A simplified approach to the stability problem for fusion magnets is presented in the two equations below[15].

$$A_{\mathrm{comp}} C_{\mathrm{comp}} \frac{\mathrm{d}T_{\mathrm{comp}}}{\mathrm{d}t} = A_{\mathrm{comp}} Q_{\mathrm{p}} - hP(T_{\mathrm{comp}} - T_{\mathrm{He}}) \qquad [11.9]$$

$$A_{\mathrm{He}} C_{\mathrm{He}} \frac{\mathrm{d}T_{\mathrm{He}}}{\mathrm{d}t} = hP(T_{\mathrm{comp}} - T_{\mathrm{He}}) \qquad [11.10]$$

where A_{comp} = composite section, A_{He} = helium section, C_{comp} = composite volumetric heat capacity, C_{He} = helium volumetric heat capacity, $Q_{\mathrm{p}}(t)$ = disturbance power density (W/m^3), P = wetted perimeter.

In a zero D approach, neglecting the spatial temperature variations in the composites and in the helium, the CICC is presumed to be infinite and to be at uniform temperature T_{comp}, exchanging with the surrounding helium by convection at uniform temperature T_{He}. A disturbance power density Q_{p} is deposited in the cable within a certain time constant. The cable is stable if the observed maximum cable temperature as a function of time during the disturbance is always under T_{cs}. There is no joule heating in the composite in this approach to the stability problem.

As noted in Section 11.4.1, at 4 K, disruptions or vertical displacement events are the fast external disturbances in tokamaks like ITER. As a result of coupling losses, they can deposit several tens of J/m^3 in the conductor on a long length, typically within 100 ms; this cannot be dealt with acceptably without the help of helium, the only substance with high enthalpy in this temperature range.

The thermal time constant of the helium reservoir,

$$\tau_{He} = C_{He} A_{He} / hP \qquad [11.11]$$

plays a major role in this case. The CICC concept with its temperature margin (0.7 K in ITER TF), ensures a large wetted perimeter and a low thermal time constant consistent with the disturbance time deposition. As seen in Table 11.8, this is no longer the case at temperatures around 20 K, where the material's heat capacity is even greater than that of helium. The stability can be envisaged adiabatically for HTS materials and is proportional to the temperature margin. This is a considerable advantage compared with LTS materials, leading to substantial simplification of the conductor relative to the CICC concept.

The downside of this stability is the extremely slow propagation (mm/s for HTS materials rather than the m/s for LTS materials) in the case of a quench, making quench detection extremely difficult. This problem is the subject of further discussion by researchers.

The theoretical adiabatic energy margin is

$$e_h = \int_{Top}^{Tcs} C_{comp} \, dT \qquad [11.12]$$

Depending on the temperature margin, it is possible to estimate this energy margin using Table 11.8; the energy margin is larger than the value that can be expected in the TF system.

Taking into account the actual concept of the ITER TF winding pack, as shown in Fig. 11.9, the conductor (in this case a CICC) is embedded in structural plates. It can be envisaged that this concept could be adapted to indirect cooling using another type of conductor, embedding cooling channels into the existing plates, which would have to be redesigned.

Table 11.8 Heat capacities in superconducting magnets

Material	*C*(*T*) (J/m^3K)		
	4 K	10 K	20 K
Copper–Nb_3Sn	900	8 000	66 000
Silver	1500	20 000	200 000
NbTi	4400	26 000	167 000
YBCO	700		~100 000
Stainless steel	16 000	42 000	105 000
Helium*	340 000	105 000	45 000

* Constant volume, supercritical, 5 bar.

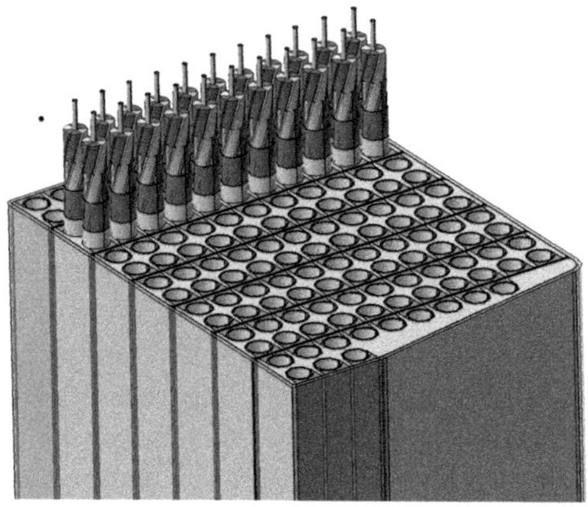

11.9 ITER TF winding plates showing the CICCs embedded in the radial plates. (Courtesy of ITER.)

11.5 Potential HTS conductors and superconducting material for future fusion reactors

As has been shown in the preceding section, it is not likely that the CICC concept, such as the one used in ITER, can be further developed for fusion reactors. The high void fraction, which is intrinsic to this type of conductor, is very detrimental to Nb_3Sn due to bending strain. This limitation will also be relevant for the HTS materials that can currently be envisaged for use in future fusion reactors, i.e. BSCCO and YBCO; these materials also being sensitive to strain. As highlighted in the preceding sections, the conductor must fulfil several requirements such as:

- High current in the range of 100 kA,
- Magnetic field in the range of 13–15 T,
- Some transposition of the elementary components (ribbons or strands).

11.5.1 Conductors with high filling factor

Transposed cables

A long-term project has been underway at Karlsruhe Institute of Technology (KIT)[7,8] since 2006 investigating the development of high-current cables from YBCO tapes. While most of the effort has been aimed at the development of cables for electrical machinery at 77 K, 50 Hz and low field, many aspects of the development are still valid for magnets. Owing to the frequency of operation of the cable, it is not possible to avoid transposition, which would minimize losses and ensure a uniform current distribution inside the conductor.

A first promising result was achieved by producing a cable of ROEBEL assembled coated conductors (RACC) with a critical current of 2.5 kA in self field at 77 K. This was achieved by manufacturing a ROEBEL cable made of 15 strands, each made of three YBCO tapes (see Fig. 11.10, 11.11).

With the aim of further increasing the current capability of the cable, a new design has recently been unveiled at KIT: the coated conductor Rutherford cable (CCRC). In this concept the strands of the Rutherford cable are themselves made of several punched tapes, such as those presented in Fig. 11.10. In order to prove the feasibility of the concept, a sub-size CCRC demonstrator is in preparation, which includes 10 strands of Rutherford type, each being made up of 10 strands.

Solutions with simple stacks of tapes

A completely different concept has been explored at NIFS, focussed on DC conductors at high magnetic field and at temperatures in the range 20–30 K. If the magnet system operates in steady-state it is not clear to what extent full transposition is still necessary at high temperature. While not eliminating the possibility of transposition, the experimental activity has concentrated until now on conductors made of simple stacks of HTS tapes.

A sub-size conductor made of Bi-2223 has been successfully tested showing a critical current of 10 kA at 8 T and 20 K.[9] Recently, a conductor concept has been developed using YBCO tapes and having two alternative layouts[10] (Fig. 11.12). The

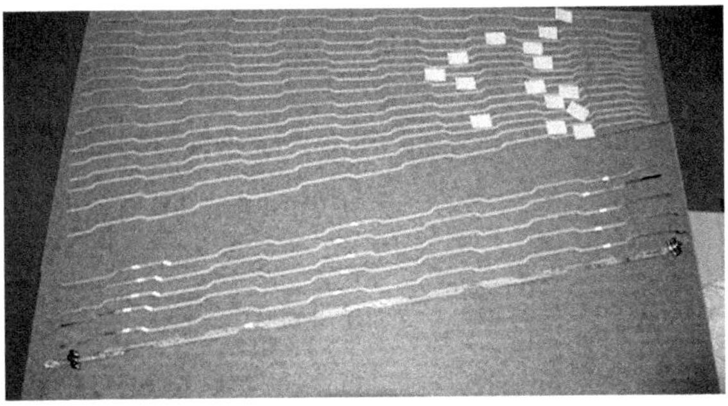

11.10 Punched coated conductors (15 samples top), assembled strands made of three YBCO tapes (five strands middle) and one-third of an assembled RACC cable (front). (Courtesy of KIT.)

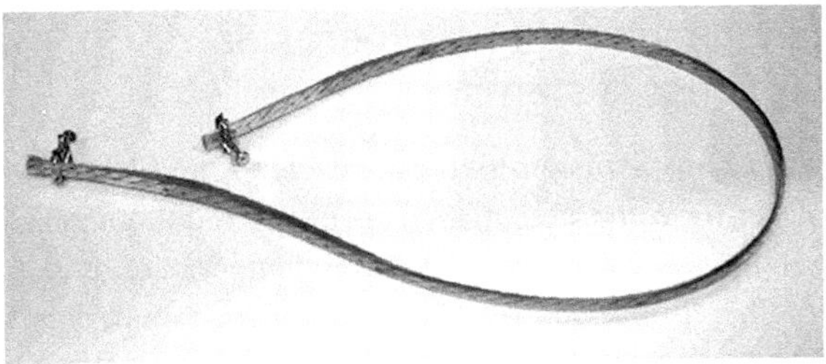

11.11 Assembled RACC cable sample (1.11 m long). (Courtesy of KIT.)

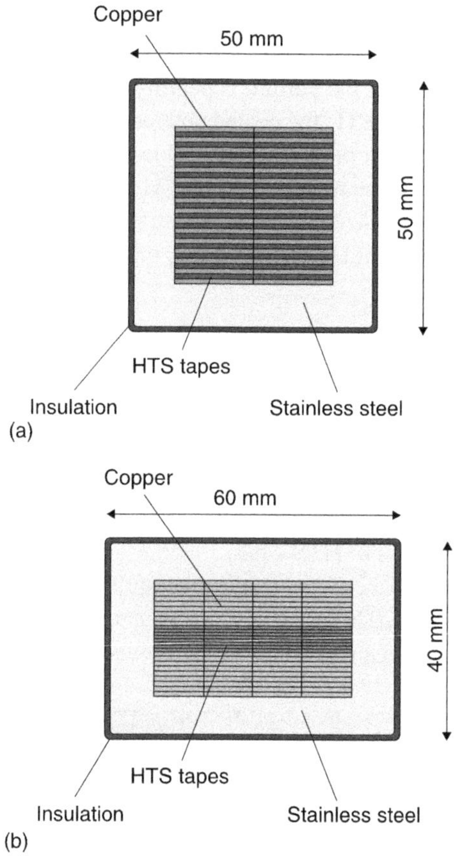

11.12 Cross-sections of the proposed conductors for the FFHR helical coils. Two designs are shown. (Courtesy of NIFS.)

two layouts differ in the shape and the respective location of the stabilizing copper and of the YBCO tapes. Layout (b) in Fig. 11.12 aims at keeping the YBCO tapes as near as possible to the neutral axis in order to mitigate bending strain. Experiments carried out on sub-size samples have confirmed the advantage of layout (b).

11.5.2 Superconducting materials for fusion reactor magnet systems

Focusing on the TF system of fusion tokamak reactors or FFHRs, YBCO and BSCCO HTS materials have the potential to meet the requirements regarding current density, depending on the selected operating temperature. It is difficult to gain an overview of the current status of superconducting materials since the technology is moving and improving rapidly. Taking a target value for current density of 300 A/mm^2, as presented in Section 11.2, YBCO and BSCCO materials are compared below, as a function of temperature, using experimental observations recently obtained at CEA Saclay[11,12].

HTS YBCO ribbons

The main characteristics of the two tapes studied, AMSC 344 and SP SCS 5050, are presented in Table 11.9. The AMSC 344 tape was supplied by American Superconductor (http://www.amsc.com) and the SP SCS 5050 tape was supplied by SuperPower Inc. (http://www.superpower-inc.com).

The critical current density J_c (in spite of the fact that it is not the real critical current density circulating in the YBCO) is calculated by dividing the current by the overall cross-section of the tape or strand, taking into account the fact that the amount of copper section is small in the tapes. The results for the critical current density are presented in Fig. 11.13 and 11.14. It can be seen that for both materials, in the magnetic field range 12–14 T, the target current density is reached at a temperature of 30 K. Note that the magnetic field is applied parallel to the tapes, which is favourable.

BSCCO–2212 ribbon and strands

Special attention has been given to Bi2212 as it is the only cuprate that can be processed in round wires – Bi2223 has to be rolled into a tape to align the crystals. The critical properties of the HTS material in both tape and round wire form have been compared using their critical current densities J_c. The characteristics of the two samples are presented in Table 11.10. It can be seen in Fig 11.15 and 11.16,

Table 11.9 Characteristics of two YBCO tested tapes

	SP SCS 4050	AMSC 344C
Width (mm)	4	4.3
Thickness (mm)	0.1	0.2
Tape section (mm^2)	0.4	0.86

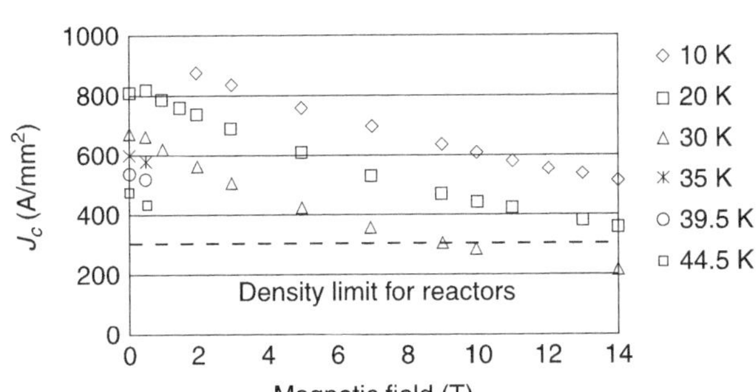

11.13 Available overall current density in AMSC 344C tape.

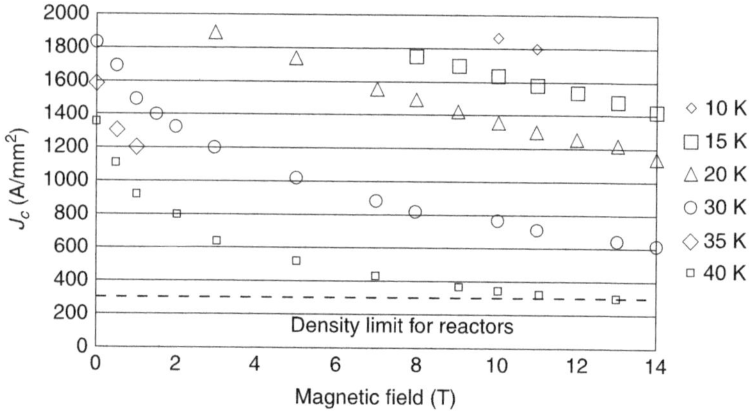

11.14 Available overall current density in SP SCS 4050 tape.

Table 11.10 Characteristics of two Bi2212 tested samples

	Nexans Bi2212 tape	Nexans Bi2212 round wire
Width (mm)	4	
Thickness (mm)	0.21	
Diameter (mm)		0.8
Superconducting section (SC) (mm^2)	0.21	0.053
SC/nonSC ratio (%)	25	10.5

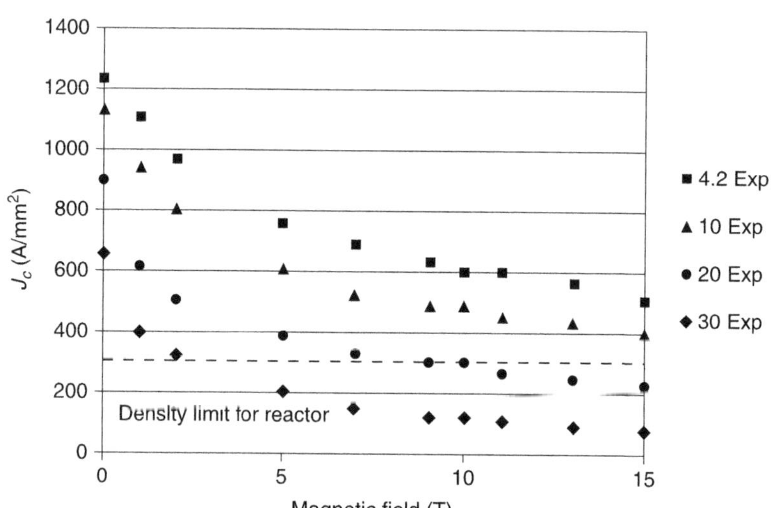

11.15 Available critical current density BSCCO–2212 tape.

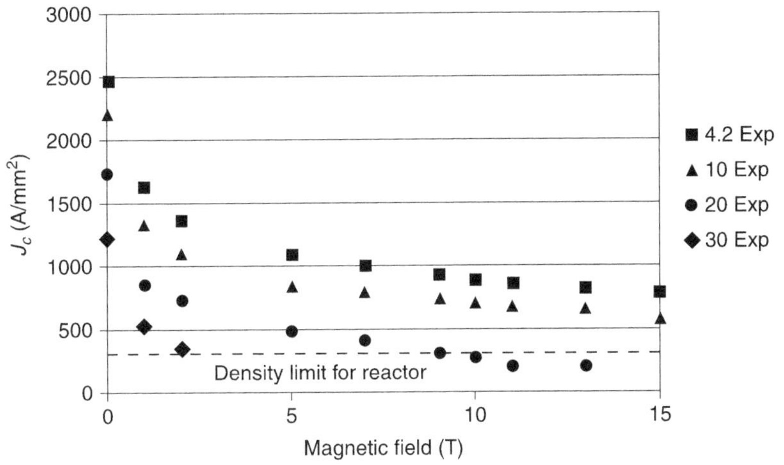

11.16 Available critical current density BSCCO–2212 round wire.

that the available critical current density is in this case lower than for the YBCO tapes, allowing a maximum temperature of operation of 25 K at the required current density of 300 A/mm^2.

11.6 HTS current leads

An HTS material (i.e. Bi2223) is already present in the current leads of existing superconducting fusion machines. Current leads using Bi2223 have been selected for ITER, W7-X and JT–60SA, three superconducting fusion machines that are presently in construction. These leads have also been in operation in the EAST tokamak in China for the last two years.

HTS current leads are an ideal 'natural' application for HTS materials at high temperature because the magnetic field is very low in this region: 0.05 T external magnetic field in the ITER TF system perpendicular to the tape, in addition to the self field. Even though HTS current leads have been developed for a long time within the framework of the LHC programme, there was a specific difficulty in fusion linked to the very high current of the magnets (68 kA for the ITER TF system). Previously, a 70 kA HTS current lead demonstrator had been developed and built by KIT with the collaboration of the Centre en Recherches en Physique des Plasmas (CRPP),[13,14] demonstrating that this technology was available for ITER. The overall length of the current leads for ITER is typically in the range of 3 m. HTS current leads are made in two parts, as shown schematically in Fig. 11.17:

1. A resistive module. The behaviour of this part is very similar to that of conventional current leads. Available cold helium at 50 K is used to actively cool the module.

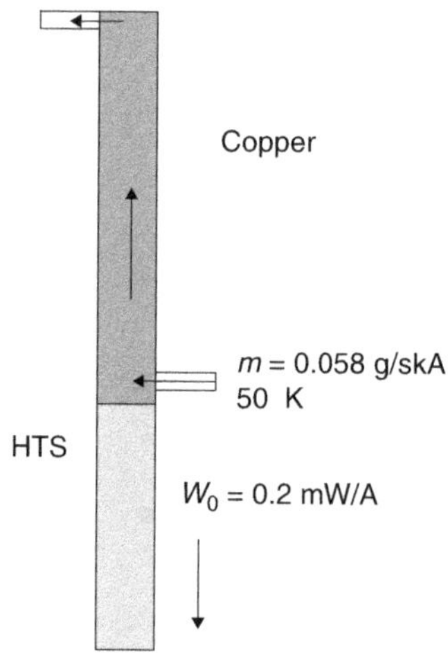

11.17 Schematic of an HTS current lead.

Table 11.11 Influence of HTS current leads in ITER electrical consumption

ITER option for current leads	Conventional current leads	HTS current leads
Total number of pairs of current leads	30	30
Cryogenic power linked to current leaks (kW)	13*	4
Total cryogenic power in ITER at 5 K (kW)	46	37
Total ITER electrical power consumption at 5 K (MW)	11.5	9.25

* 130 g/s.

2. An HTS module using BSCCO–2223/AgAu tapes embedded in stainless steel. The HTS module is cooled by heat conduction from 4.5 K up to 65 K, the transition temperature between the HTS module and the resistive module. At 4.5 K the current lead is connected to the superconducting overlapping clamped joint.The current leads in a tokamak represent a substantial amount of the refrigeration power at low temperature (see Table 11.11 for ITER). Owing to the reduction of the range of operation of the resistive part (50–300 K versus 5–300 K for conventional current leads), the HTS current leads typically enable the power needed for cooling the current leads to be divided by a factor of three.

11.7 Conclusion

HTS materials are already present in existing superconducting machines. In ITER, HTS current leads enable savings of about 20% of the cryogenic refrigeration power of the reactor at 5 K, which is quite substantial. HTS materials have the potential to play a major role in future fusion reactors but development is only just beginning.

By enabling the operation of reactor magnet systems at higher temperatures, the HTS materials could achieve large savings in cryogenic power requirements in comparison with classical operation at 5 K. The potential gain, as demonstrated, is related to the ratio of the new operating temperature and 5 K; it is already clear that an operating temperature of 20–40 K could allow substantial reductions in power requirements; the progression to even higher temperatures would then bring only marginal gains. The critical properties of existing superconducting HTS materials (BSCCO and YBCO) have the potential to allow operation in this range of temperatures and at high relevant magnetic field, while the long-term goal of operation in liquid nitrogen (65–77 K) at high magnetic field is probably still very far in the future.

It is true that operation in liquid nitrogen would probably result in reductions in investment costs because of the simplification of the cryostat and the thermal shield; this, however, is still to be fully investigated.

The most significant potential savings for HTS materials are perhaps more likely to be found in the potential simplification of the cryogenics of these very large systems, rather than in the decrease of the circulating power of future fusion reactors. The NIFS has started this exploration, by introducing the concept of indirect cooling for their FFHR. In addition, it is quite clear that the complexity of the forced flow concept of ITER cryogenics cannot be further extrapolated for robust and reliable fusion reactors.

Owing to the unavoidable degradation of Nb_3Sn strands in CICCs due to strain, the temperature margin of the conductor is automatically affected and reduced, which is probably not acceptable for reactors. Indirect cooling, which is the traditional way of cooling superconducting magnets, can certainly be updated for the magnets of fusion reactors.

The main questions to be answered are therefore linked not only to the choice of the superconducting material but also to the choice of the conductor associated with the optimal magnet concept. It is therefore urgent to launch a development programme associated with these objectives, bearing in mind that it has taken over 20 years to develop the ITER Nb_3Sn CICC. Moreover, it is clear that this concept, quite adapted for the production of large magnetic fields in large volumes, will also find other beneficial applications.

11.8 References

1. A. Li Puma, J. Bonnemason, L. Cachon, J.L. Duchateau, F. Gabriel (2006), 'Consistent integration in preparing the helium cooled lithium lead DEMO reactor', *Fusion Engineering and Design*, **81**, 2703.
2. A. Sagara, O. Mitarai, S. Imagawa, T. Morisaki, T. Tanaka, (2006), 'Conceptual design activities and key issues on LHD-type reactor FFHR', *Fusion Engineering and Design*, **81**, 2703.

3. K. Takahata, T. Mito, H. Tamara, S. Imagawa, A. Sagara (2007), 'Conceptual design for an indirect cooled superconducting magnet for an LHD-type fusion reactor FFHR', *Fusion Engineering and Design*, **82**, 1487.
4. J.L. Duchateau, P. Hertout (2008), 'Which superconducting material for the toroidal field system of the fusion DEMO reactor?', *J. Phys.: Conf. Ser.*, **97**, 012038.
5. J.L. Duchateau, J.Y. Journeaux, F. Millet (2006), 'Estimation of recycled power associated with cryogenic refrigeration power of a fusion reactor based on TORE SUPRA experiment and ITER design', *Nucl. Fusion*, **46**, S94–S99.
6. H. Tamura, K. Takahata, T. Mito, S. Imagawa, A. Sagara (2010), 'Conceptual design and development of an indirect-cooled superconducting helical coil in the FFHR', *Plasma and fusion research*, **5**, 1035.
7. W. Goldacker (2009), 'Improvement of superconducting properties in ROEBEL assembled coated conductors (RACC)', *IEEE Trans. Appl. Sup.*, **19**, 3, 3098.
8. S. I. Schlachter, W. Goldacker, F. Grilli, R. Heller, A. Kudymow (2010), 'Coated conductor Rutherford cables (CCRC) for high current applications: concept and properties', Presented at the Applied Superconducting Conference 2010 Washington, to be published.
9. G. Bansal, N. Yanagi, T. Hemmi, K. Takahata, T. Mito (2008), 'Experimental results of large current capacity HTS conductors', *IEEE Trans. Appl. Sup.*, **18**, 1151.
10. R. Champailler, N. Yanagi, G. Bansal, H. Tamura, T. Mito *et al.* (2010), 'Experiments on bending strain on reduced scale conductors for fusion energy reactors', *IEEE Trans. Applied Sup.*, **30**, 3, 1565.
11. T. Lécrevisse (2010), 'Critical current measurements in HTS YBaCuO ribbons', Presented at the Applied Superconducting Conference 2010 Washington, to be published.
12. J.M. Rey, A. Allais, J.L. Duchateau, P. Fazilleau, J.M. Gheller *et al.* (2009), 'Critical currents measurement in HTS Bi2212 ribbons and round wires', *IEEE Trans. Appl. Sup.*, **19**, 3, 3088.
13. R. Wesche, R. Heller, P. Bruzzone, W.H. Fietz, R. Lietzow *et al.* (2007), 'Design of high temperature superconductor current leads for ITER', *Fusion engineering and Design*, **82**, 1385.
14. R. Heller, S.M. Darweschsad, G. Dittrich, W.H. Fietz, S. Fink *et al.* (2005), 'Experimental results of a 70 kA high temperature superconductor current lead demonstrator for the ITER magnet system', *IEEE Trans. Appl. Sup.*, **15**, 1, 1496.
15. J.L. Duchateau (2009), 'New considerations about stability margins of NbTi CICCs', *IEEE Trans. Appl. Sup.*, **19**, 2, 55.

12

High temperature superconductors (HTS) in accelerator systems

N. ZANGENBERG, Danish Technological Institute, Denmark

Abstract: The first major-scale industrial application of high temperature superconductor (HTS) materials was in accelerators as current leads for low temperature superconductor (LTS) magnets. This chapter highlights the challenges concerning the use of HTS materials for various accelerator elements such as dipole accelerator magnets, focusing elements and insertion devices. HTSs on the market are evaluated in the context of accelerator technology. The most promising application is as inserts in high-field dipole magnets and as focusing elements in high radiation environments.

Key words: particle accelerators, dipole magnet, quadrupole magnet, insertion devices, current leads.

12.1 Introduction

The accelerator community has been the first large-scale industry to adopt both low temperature and high temperature superconductivity. Low temperature superconductivity is an enabling technology in the construction of large-scale particle accelerators that pioneer studies into the basics of the laws of nature.

In the Large Hadron Collider (LHC) at CERN, Geneva, Switzerland/France, HTS materials have been used for current leads to power the LTS magnets (Fig. 12.1) that transfer up to 13 000 A (Ballarino, 2002). This has been the first large-scale industrial application of HTS with an order volume of 31 km of Bi–2223 tape (Ballarino *et al.*, 2007).

Superconductors are employed in accelerators for two main reasons: performance and cost. Take, for example, the case of dipole magnets used for bending the particle beams in circular accelerators. Conventional resistive iron-based magnets are limited in magnetic field by the iron saturation at 1.8 T, while a superconducting magnet can achieve at least ten times this value (Devred *et al.*, 2006). The higher field enables stronger particle bending and focusing, leading to smaller machines and more well-defined interaction zones. Consider the savings in construction costs from building a circular accelerator with a circumference of 27 km with dipoles of 8.3 T (LHC) as compared to an imaginary ring equipped with 1.8 T dipoles which would then need to have a circumference of 125 km! Typically, superconducting magnets will also use less power in operation, which is becoming an issue also for accelerators. For example, the European Spallation Source in Lund, Sweden aims to be the first CO_2 neutral accelerator.

12.1 A 13 kA HTS current lead for LHC, CERN. (Copyright CERN.)

To discuss the use of HTS materials in accelerators, some perspective may be gained from studying the story of LTSs.

Since 1983, NbTi has been employed in the Tevatron at Fermilab in Batavia, IL, US, the Hadron Elektron Ring Anlage (HERA), at DESY in Hamburg, Germany, the Relativistic Heavy Ion Collider (RHIC) at BNL in Upton, NY, US and the LHC at CERN in Geneva, Switzerland for the dipole and quadrupole magnets. Also, several so-called compact synchrotrons aimed at the need for x-ray lithography were constructed with LTS bending magnets (Weihreter, 1996).

The large research infrastructure facilities such as Fermilab in the US, CERN, and also the International Thermonuclear Experimental Reactor (ITER) have been industry drivers in important areas such as cost reduction and performance reliability for LTS technology (NbTi and Nb_3Sn) and for HTSs through current lead technology. The main research and development centres occupied with design studies of HTS magnets in accelerators quite logically originate from the activities related to the large research infrastructure facilities.

But for HTS technology to be fully developed for use in accelerators, a 'flagpole' project basing its operation on the successful employment of HTS magnets is probably required. There are on-going initiatives constructing prototype magnets as a first step towards these applications and, in this chapter, we will investigate the current actions and future possibilities for HTS accelerator magnets.

12.2 Fundamental issues for HTS in accelerators

The special requirements for the magnets used in accelerators are presented. The crucial challenges of field homogeneity and field ramping are illuminated and then translated into desirable properties of superconducting wires.

12.2.1 Main challenges

The main challenge for an accelerator magnet used in a circular accelerator is the field homogeneity – both under steady-state operation and during ramping. This is because the accelerated particles typically will pass the same magnet millions of times and the effect of even small inhomogeneities will be added in a resonant way and lead to full or partial loss of the particle beam.

Another serious issue for accelerator magnets is ramping or quasi-DC operation where the current in the magnet coils is increased from one value to another, thus, changing the magnetic field to match the increasing energy of the particles. For a synchrotron, the ramping is synchronised with the particle acceleration gained from a radio frequency source. The accelerated particles often have a limited lifetime and will experience losses over time, so the ramping rate should generally be very fast. This is a challenge for superconductors due to heat dissipation and induced screening currents which makes the homogeneity requirements even more difficult to comply with.

The challenges of homogeneity and ramping put special requirements on superconducting wires, which are discussed below.

12.2.2 Filament size

When a superconducting filament is exposed to a magnetic field, shielding persisting currents build up from the outside of the filament to shield the inside as shown in Fig. 12.2. The magnetization is found by integrating the current in the shaded area in Fig. 12.2:

$$M = -\left(\frac{4}{3\pi}\right)\mu_0 J_c a\left(1-\frac{b^2}{a^2}\right) \quad [12.1]$$

At the 'penetration field', the peak magnetization occurs with a value of:

$$\left|M_{peak}\right| = \left(\frac{4}{3\pi}\right)\mu_0 J_c a \quad [12.2]$$

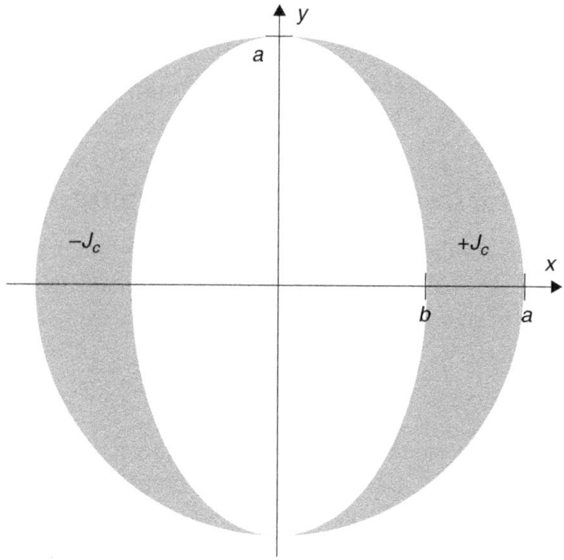

12.2 Magnetization of a superconducting filament during up-ramp from a non-magnetized state.

The magnetization will add to the magnetic field from the coils and cause a distortion to the field experienced by the particle beam. According to the Bean critical model, the shielding currents will be larger at lower fields, so the distortion is largest in this case. Also, the filaments in different positions will experience different magnetic fields and, thus, be magnetised differently making it difficult to maintain a high homogeneity during the ramp.

Clearly, these complicated effects are best minimised by reducing the magnetization which, as seen by the formula above, can be done by reducing the filament diameter, a. Smaller filaments also reduce the magnetic instabilities (flux jumps).

12.2.3 Filament twisting

When a multi-filamentary wire is subjected to a time-varying magnetic field, current loops are generated between filaments. If the filaments are straight, large loops are generated, which leads to large losses. Furthermore, if the strands are magnetically coupled, the effective filament size is larger, leading to more severe flux jumps.

It is advantageous if the filaments in the wires are twisted to reduce interfilament coupling and AC losses. This also prevents the current from flowing, for example, mainly on one side of a wire.

12.2.4 Wire transposing

It is very desirable to bundle wires connected in parallel into cables to enable operation at lower voltages. There are also other advantages from bundling: reduction of the wire piece length, reduction of number of turns to ease the winding process, smaller coil inductance, and current redistribution in case of a defect or a quench in one strand.

However, to prevent an uneven current distribution in the cables, the wires need to be transposed. For LTSs this is done by making Rutherford-type cables as depicted in Fig. 12.3. The transposition also reduces the inter-strand coupling currents (as for inter-filament coupling) which minimises the losses and field distortions and provides better mechanical stability.

Round HTS wires (Bi-2212, MgB_2) can potentially be made into Rutherford cables (Hasegawa *et al.*, 2001) while another type of transposing is necessary for tapes (coated conductor, Bi-2223). For example, that used for the Roebel cable (Goldacker *et al.*, 2006; Goldacker *et al.*, 2009; Long *et al.*, 2008) is shown in Fig. 12.4.

12.2.5 Current carrying ability

A high critical current for in-field operation is of course essential for the applicability of HTSs in accelerators and it is the main argument in favour of using LTSs where superconductors are required. The engineering current density, J_e, is the most relevant figure of merit since the contribution to the central magnetic field inside a beam aperture from a current flowing in a distance r from the centre

12.3 Rutherford cable of NbTi wires each with several thousand filaments. (Copyright CERN.)

12.4 Roebel cable made from YBCO tapes. Courtesy of Industrial Research Ltd., New Zealand.

decreases with $1/r$. In other words, the closer to the aperture the wires can be placed, the more effectively the current is used.

The use of a current-carrying matrix material is desirable since this reduces the risk of catastrophic failure that would destroy the superconductor in the case of a quench. A proper material will allow current to pass and also hold a significant heat capacity to mitigate a runaway local heat deposition in the system.

12.2.6 Tolerances

Low production tolerances are critical for good field homogeneity. For accelerator magnets, a homogeneity requirement of $dB/B < 10^{-4}$ (in some cases 10^{-3}) is used, which requires a conductor tolerance of 10–50 μm. Random variations in the wire geometry or inner superconductor distribution will to some extent even out in stacked structures such as a coil.

12.2.7 Insulation

Generally, accelerator magnets are operated in DC or quasi-DC and the minimum breakdown voltage is not required to be very high. The specifications are normally defined by the designed quench ramp-down of the magnet.

However, finding the proper insulation is a significant challenge because of the need for:

- Strict tolerance requirements as described above,
- Maintaining a high J_e,
- Compatibility with epoxy impregnation. Insulation with thin Kapton foils is often used but this technology is generally not compatible with impregnation. In these cases, an epoxy-based system is more desirable.

12.2.8 Competing with LTS

These challenges have led to the development of multi-filament LTS wires with small filament size, twisted filaments and transposition of wires in cables. NbTi wires have been developed during the last 20 years to address exactly these challenges and extraordinary advances have been achieved regarding production tolerances, conductor/filament development and winding techniques.

To illustrate the maturity of LTSs for accelerator purposes, consider that the LHC uses 240 000 km of NbTi wire for the magnets and inter-magnet bus-bars. The wire used for the outer dipole layers have a diameter of 0.825 mm and consist of 6425 6 μm NbTi filaments (Adam *et al.*, 2002). The filaments are twisted inside the wire with a pitch of 15 mm. Thirty-six wires were bundled in Rutherford cables where each wire is transposed with a pitch of 100 mm. These Rutherford cables can carry 13 kA in a field of 9 T when cooled to 1.9 K.

With these figures in mind and given the current HTS wire/tape specifications, the substitution of LTSs by HTSs is not foreseen anytime soon. Instead, HTSs could have an important role to play in singular applications where special conditions are present.

Some issues do, however, remain a challenge for LTS magnets: The dipole magnets based on LTS materials so far are slow-ramping devices compared to synchrotron accelerators using resistive magnets. For example, the ramp rate for RHIC and LHC are 0.042 T/s and 0.007 T/s (Kaugerts *et al.*, 2007), respectively, while for a 3 GeV synchrotron based on resistive magnets, such as for example the Australian Light Source, the rate may be 3 T/s (Weber *et al.*, 2007). Development is in progress to produce fast ramping 1 T/s (Kaugerts *et al.*, 2007) and 4 T/s (Fischer *et al.*, 2010) LTS magnets by adding resistive elements into the cable matrix to relax the induced flux (persistent current) faster.

Inserts in high-field superconducting magnets could be an area where HTS may actually become the enabling technology. The critical current of the LTSs used presently (NbTi and Nb_3Sn) decrease rapidly at high magnetic fields, but this is not the case for YBCO and Bi–2212 as shown in Fig. 12.5.

Another niche application for HTS magnets is the use as focusing elements in radiation-hard environments where the HTS can survive temperature increases of several degrees whereas LTS will typically only operate within tenths of degrees.

Looking at the quench scenarios for LTS and HTS also reveals an advantage for HTS. For LTS, flux flow will usually result in flux jumps leading to rapid quenching, whereas the quench for HTS is a more gentle process involving an intermediate step of thermally induced flux motion where the larger heat capacity (as compared to LTSs) will in part mitigate a quench avalanche. Instead, for HTSs, the challenge is to detect the quench early and prevent extreme hot-spots from forming.

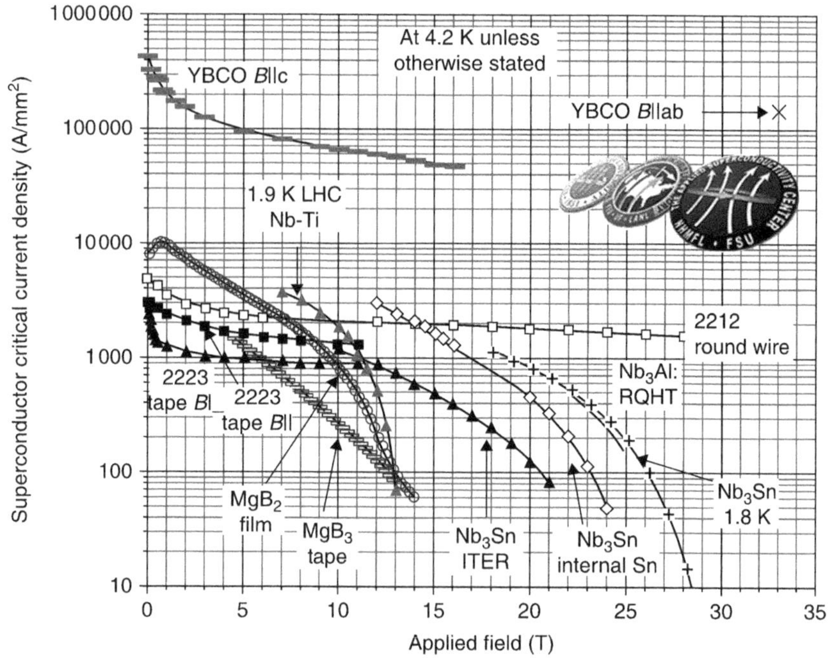

12.5 Critical current versus magnetic field demonstrating the potential for HTS coils as inserts in high-field magnets. (Courtesy of Peter J. Lee, National High Magnetic Field Laboratory.)

12.3 Types, properties and development of HTS for accelerators

In this section, we will discuss the potential of each specific HTS material to be used for different accelerator applications. We will devote attention to Bi-2212, Bi-2223, coated conductors and MgB_2. The conductors will be evaluated based on their potential to meet the criteria mapped out in Section 12.2.

12.3.1 Bi-2212

Bi-2212 is made in multi-filament round wires and it is thus the HTS most similar in structure to the industrially available NbTi and Nb_3Sn LTS wires employed in accelerator magnets. These similarities make the comparison of LTS by Bi-2212 more straightforward than for tapes since both geometry and the theory for magnetisation and losses are the same. Furthermore, the Bi-2212 wire is electromagnetically isotropic.

However, the wire faces many challenges before the switch from LTS to Bi-2212 technology can be justified. The wire has poor mechanical properties, and practical usage requires a complex wind and react technique involving amongst others a critical annealing step at 890 °C in pure oxygen (Kahn *et al.*, 2009; Shen *et al.*, 2008).

The most intense research into HTS for accelerator magnets has been into Bi-2212 wires, particularly as coil inserts into high-field magnets to boost the magnetic field (Weijers *et al.*, 2004; McIntyre and Sattarov, 2005). The wire architecture is being developed to include more filaments and to twist them. There are also on-going activities to make Rutherford cables but, so far, this has lead to a significant reduction in the critical current of the individual conductors (Scanlan *et al.*, 1999).

12.3.2 Bi-2223

Being the workhorse technology since HTS was introduced to the market, Bi-2223 has developed into a mature technology when it comes to product specifications and reliability which, as stated in the previous section, are important parameters for accelerator uses. Bi-2223 was also used for the LHC current leads and still represents the most reliable and cost-effective solution for current leads.

However, the architecture and lack of mechanical flexibility present severe limitations for uses in accelerator magnets and it is questionable whether Bi-2223 has the potential in the long term to compete against Bi-2212, which is more suited for accelerator uses due to the shape, and the coated conductors, which offer higher strength and more flexibility. Activities to twist the filaments are on-going but so far with loss of current-carrying ability (Knoopers *et al.*, 2002).

12.3.3 Coated conductors and YBCO

The ability to produce long-length, high-performance YBCO and rare-earth-element-BCO thin-film-based HTS (coated conductors) opens up whole new concepts for the accelerator industry. The technology cannot easily be evaluated on the scale used for multi-filamentary LTS/HTS materials. Also, simulation models, e.g. for losses based on filament conductors, cannot be directly used for thin films. The fact that the superconducting layer is only 1 μm thick allows for new non-filamentary architectures for the accelerator industry that have not been explored yet. The tapes are non-isotropic, which adds complexity to the design phase by challenging the modelling of current distribution. Mechanically, tapes are robust and extremely flexible compared to Bi-based HTS.

The thin superconducting film can be deposited on very ductile substrates and this gives the technology an advantage over Bi-2223 and Bi-2212 HTS. Based on the flexibility and robustness of the tape, the technology to 'twist' conductors into the so-called Roebel cable has been developed with good results (Goldacker *et al.*, 2006; Goldacker *et al.*, 2009; Long *et al.*, 2008). This also adds to the complexity of the system and, relating to accelerator technology, more studies are currently required (Turenne *et al.*, 2010).

An important question regarding the use of coated conductors in accelerator magnets is current distribution across the thin film and what effect this will have on the field homogeneity. To achieve a dipole magnet field homogeneity of $\partial B/_B < 10^{-4}$ requires a production tolerance of at least 50 μm. Modelling the current distribution across, for example, the 4 mm width of a standard coated conductor is required to examine the potential in detail.

A few programmes aiming to build YBCO prototypes for accelerator uses are in progress. Brookhaven National Laboratory and Fermilab in the US are pioneering the use of YBCO magnets/inserts for particle accelerators (Gupta *et al.*, 2010) using racetrack coils. Danfysik A/S has made the first non-flat coils for a small-scale prototype magnet targeting gantry magnets in particle therapy (Danfysik, 2011).

The potential of substantially higher critical current offered by, for example, multiple stacked thin-films in coated conductors (Foltyn *et al.*, 2005) makes it realistic to expect that coated conductors will even be able to compete with LTS in the foreseeable future.

12.3.4 MgB_2

With a critical temperature of 39 K, it is debatable if MgB_2 can be considered 'HTS'. For accelerators, MgB_2 is considered among the HTS conductors offering more cost-efficient performance and more stable operation at temperatures where cryocoolers may be employed. The isotropic nature and ductility of the wire together with rapid advances of the production technology and the scale-up potential (Grasso *et al.*, 2008) stresses the fact that MgB_2 should be considered seriously for accelerator magnet technology.

Recent investigations have gone into the possible organisation of MgB_2 filaments in wires and cables (Hušek *et al.*, 2009; Ballarino, 2010). HTS accelerator magnets based on MgB_2 has been studied (Cooley *et al.*, 2001) and an iron-based magnet with the copper conductors replaced by MgB_2, a so-called superferric magnet, has also been constructed (Musenich *et al.*, 2008). However, no niche area for MgB_2 accelerator magnets has been identified and the potential is still uncertain.

12.3.5 Bulk HTS

Bulk HTS in accelerators will typically be bench-marked against permanent magnets. The advances in the permanent magnet (for example, NdFeB) manufacturing technology driven by the market pull from several industrial sectors, such as hard-drive technology and bearings, has been more pronounced than the advances in bulk HTS technology and it is unlikely that any bulk HTS will find its way into accelerators for several years.

In the long term, however, the technology may revolutionise accelerators by enabling strong magnetic fields and gradients over short distances (Masuzawa *et al.*, 2006). Next step is to identify relevant demonstrator cases in which reliability and accuracy issues must be addressed.

12.4 Applications of HTS in accelerators

An accelerator consists of several high-technology components, specification level depending on the purpose and design of each individual machine. As mentioned, specific programmes aim at developing HTSs to be used in the most promising enabling areas such as current leads, strongly focusing elements and inserts for dipole magnets. Prototypes have been developed and tested but, with the exception of current leads, the long-term reliability of HTS technologies in a real accelerator or storage ring is yet to be tested. In this section, an overview will be given on the accelerator components where HTSs are or should be considered.

12.4.1 Bending magnets

An accelerator dipole magnet has a purely vertical magnetic field that will deflect the beam ions or electrons inwards towards the centre of the accelerator. Superconducting dipole accelerator magnets represent a large fraction of the overall cost of superconducting circular accelerators. In a conventional iron magnet, the field shape and homogeneity is defined by the pole shape, which is machined to a high precision of up to 0.05 mm, translating into a high field homogeneity of up to $\partial B/B < 10^{-4}$. The strength of the magnet (maximum magnetic flux) is, however, limited by the iron saturation at a flux of 1.5–1.8 T.

When using superconductors in magnets, the field is mainly defined by the current flowing in the superconducting wires and coils. Placing these with a high precision is a challenge, and superconducting magnets always face challenges from the material (superconductor and matrix) production tolerances, thermal contraction during cooling and material movement caused by Lorentz forces acting on the magnet. Special attention is also given to higher order multipole errors in the superconducting magnets, which will affect the beam by modifying critical accelerator parameters such as orbit, tune, coupling and chromaticity.

Iron is used outside the coils in part to shield the surroundings from the field, and in part to increase the central field – an increase of up to 30% may be achieved in this way.

A hybrid of the conventional and superconducting technologies is the so-called superferric magnet, where the iron yoke is maintained, but superconductors replace the copper conductors. This allows for a reduction in magnet size and weight (Kalimov *et al.*, 2004; Fischer *et al.*, 2010; Piekarz *et al.*, 2008).

To create a homogeneous horizontal magnetic field, the conductor coils need to be placed very accurately around the beam aperture. The coil distribution can be illustrated from two basic approaches: the cos θ and the elliptical.

The so-called cos θ distribution is seen in Fig. 12.6. The direction of the current is inwards on the left hand side and outwards on the right hand side and the current is confined to the circle circumference. The magnitude of the current at a point on the circumference is $I = I_0 \cos \theta$. From this current distribution, the field may be found to be a pure dipole field with the magnetic field pointing upwards (Wilson, 1987):

$$B_y = -\frac{\mu_0 I_0}{2a} \text{ and } B_x = 0 \qquad [12.3]$$

An alternative and quite elegant approach to determining the coil distribution is to take two ellipses with equal but opposite current flowing in them and intersect them as shown in Fig. 12.7. Where they overlap, the net current is zero and the combined field from the ellipses are (Wilson, 1987):

$$B_y = -\frac{\mu_0 I_0 SC}{(b+c)} \text{ and } B_x = 0 \qquad [12.4]$$

The actual coils should then be placed to best cover the shaded area in Fig. 12.7. In both cases, the task of the magnet designer is to place coils mainly at the magnet midplane as shown in Fig. 12.8 for the dual-bore LHC dipoles. For the LHC cos θ layout, the coil structure consists of two layers of Rutherford cables, where the inner layer is made with a 28-strand Rutherford cable that passes each side of the aperture 15 times, as seen in Fig. 12.8.

The inclination of the Rutherford cable is defined by the radial position and the azimuth angle together with the requirement that the cable is bent rigidly around the aperture in a complex 3D shape. The cos θ coil placement solves the problem of getting the cable away from the aperture in the magnet ends. It does, however,

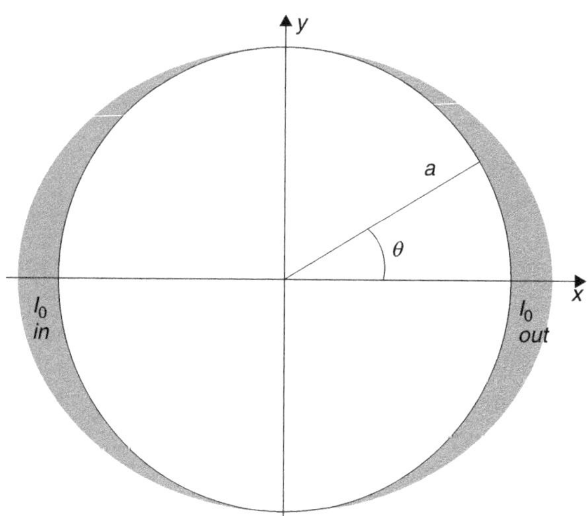

12.6 Cos θ current distribution leading to a pure vertical magnetic field.

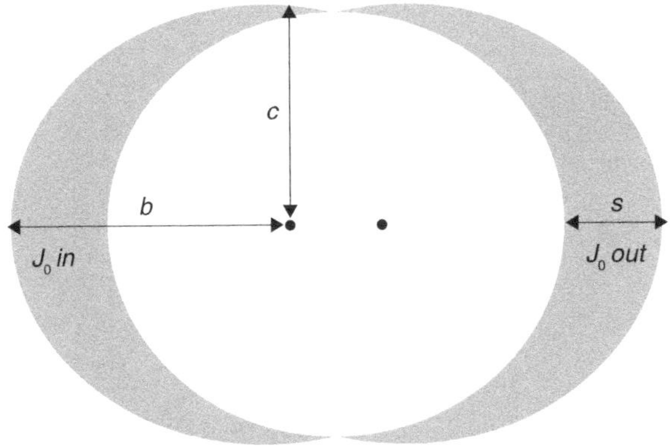

12.7 The ellipse design made by intersecting two identical ellipses carrying a uniform current but in opposite directions; *s* is the off-set of the ellipse centres, *b* and *c* are the semi-major and semi–minor-axis of the ellipse, respectively.

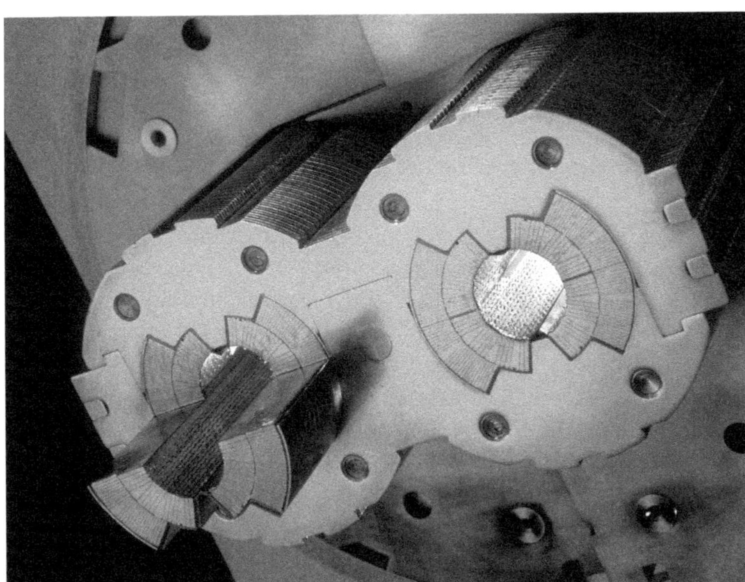

12.8 The LHC twin-aperture dipole magnets. (Copyright CERN.)

pose a challenge for bending magnets that are not straight. The concept of bending cos θ magnets is being explored in the SIS300 magnets that are to be bent with a sagitta of 114 mm (Fabbricatore *et al.*, 2008).

A coil layout less effective than the cos θ layout but better suited for magnets that are to be bent is called the ellipse structure, where coils for example may be stacked as shown in Fig. 12.9 for the proposed Next European Dipole (Toral *et al.*,

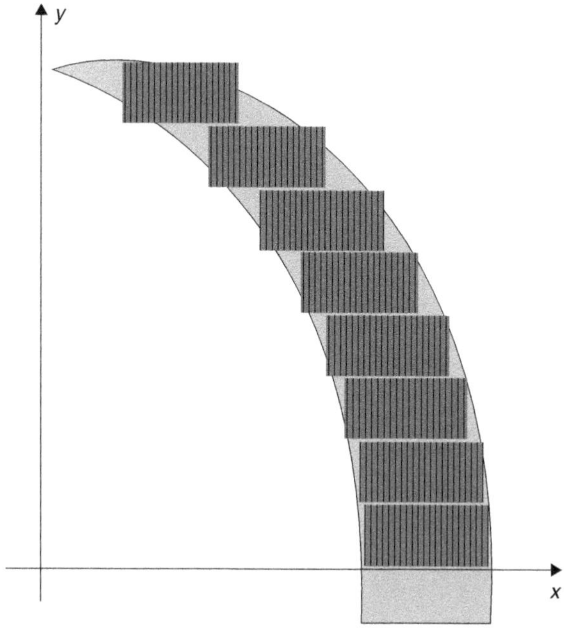

12.9 The ellipse coil layout built up, essentially by filling the shaded area in Fig. 12.7. See for example the proposed designs for the Next European Dipole (Toral *et al.*, 2007).

2007). This design is well-suited for HTS tapes and it has been used in prototype HTS magnets (Danfysik, 2011). The challenge of this design lies instead in the end design where the conductor needs to be wound away from the aperture.

Bending magnets with a field strength higher than 15 T cannot be realised with NbTi technology since the conductor H_{c2} limit is 10–12 T. Nb_3Sn is also a LTS and can go up to 20–24 T but the technology is not as mature as NbTi and the wind and react technology is required. Bi-2212 has the ability to operate at much higher fields but the brittle nature of the material again requires a wind-and-react approach. Several laboratories are investigating the possibility of using Bi-2212 inserts into Nb_3Sn coils to boost the field (Weijers *et al.*, 2004; McIntyre and Sattarov, 2005) as shown in Fig. 12.10.

YBCO conductors are much easier to handle and also offer the advantage of being able to endure high fields while maintaining reasonable critical current. Here, the main challenges concern the winding process and limited degrees of freedom imposed by the tape shape.

Separator magnets are also dipole magnets and are sometimes used at accelerator facilities on the experiment beamlines. The homogeneity requirements are generally less demanding and an HTS separator magnet is currently in use at

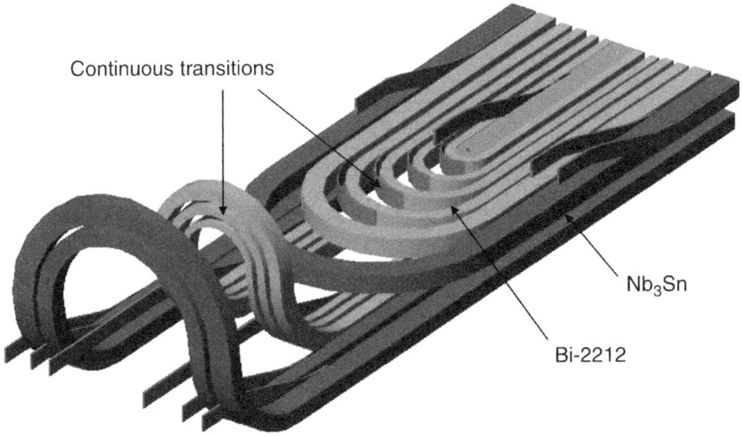

12.10 Block-coil Nb_3Sn dipole coil geometry with coil inserts of Bi-2212 to boost the magnetic field developed by the Accelerator Research Laboratory at Texas A&M University. (Courtesy Peter McIntyre.)

the Berlin electron synchrotron facility (BESSY), Hahn-Meitner Institute, Germany (Pooke *et al.*, 2009).

12.4.2 Focusing elements (solenoids and quadrupole magnets)

All accelerators require magnetic focusing of the particle beam. Quadrupole magnets have a focusing effect in one direction and a defocusing in the other, but if these are combined in pairs focusing in perpendicular directions, an overall focusing of the particle beam can be achieved. An alternative, especially used in linear accelerators, are solenoid magnets.

For accelerators designed to collide particles, the optimal design is to have the final focusing elements before the collision as close to the interaction point as possible, and the element must be able to operate under heavy irradiation from the collision products. Since the temperature margin for operation of a LTS-based magnet is very narrow, the risk of a quench from the irradiation-induced heat generation is high. However, HTSs generally can operate over a larger temperature margin and the specific heat capacity of matrix materials is larger at, for example, 20 K as compared to 4 K. Indeed, studies have shown that coated conductors can operate under irradiation (Greene *et al.*, 2009; Shiroyanagi *et al.*, 2011).

For the Facility for Rare Isotope Beams (FRIB) to be constructed in Michigan, US, a prototype large aperture Bi-2223 based quadrupole magnet has been built

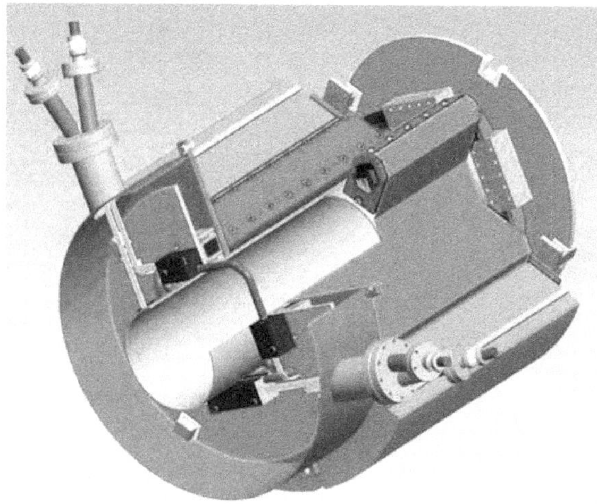

12.11 Design drawing of an HTS quadrupole magnet for FRIB. (Courtesy of Brookhaven National Laboratory.)

(Gupta *et al.*, 2008). The magnet is constructed at Brookhaven National Laboratory and an upgrade using YBCO tapes (Fig. 12.11) is scheduled for delivery in 2012 (Gupta *et al.*, 2010). This will be a breakthrough for HTS technology (and particular YBCO) since it will be the first HTS magnet to operate in a large accelerator facility.

Also in the US, the design of a 45–50 T solenoid magnet to be used for muon cooling in TeV muon colliders is being designed considering both Bi-2212 (McInturff *et al.*, 2007), Bi-2223 (Kahn *et al.*, 2007), and YBCO (Kahn *et al.*, 2010). In general, most recent initiatives tend to favour the use of YBCO (Gupta *et al.*, 2010; Kahn *et al.*, 2010).

A helical muon cooling channel magnet, combining solenoid, dipole and gradient magnetic fields, has also been proposed (Derbenev and Johnson, 2005), and is being designed (Zlobin *et al.*, 2010).

12.4.3 Insertion devices (undulators and wigglers)

An insertion device is an element inserted in a straight section of a synchrotron or free-electron laser where a periodically varying magnetic field causes the electrons to 'wiggle' and emit high-brilliant x-ray radiation in the forward direction. In conventional insertion devices, the magnetic field is defined by an array of permanent magnets mounted above and below the electron beam along the path with opposite polarity and changing along the beam path with a periodicity of λ_u as shown in Fig. 12.12.

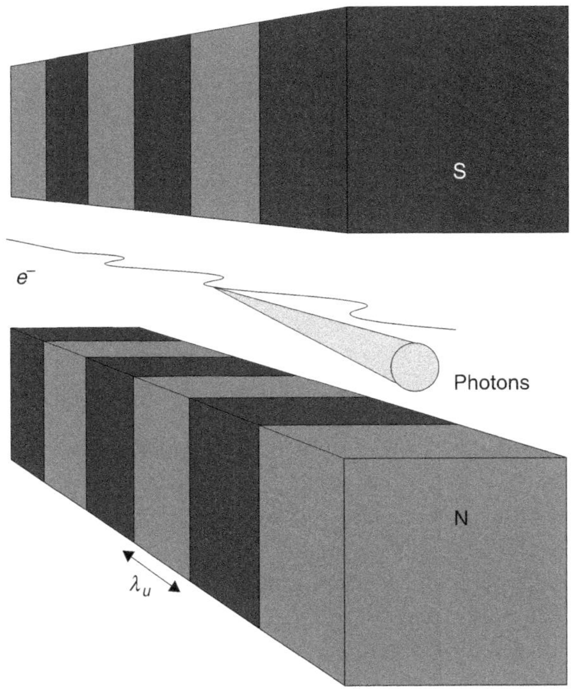

12.12 Insertion device. Bending of an electron beam in an array of magnets leads to the generation of free electron laser or synchrotron light.

When the array of magnets act on the electron beam the same way as a series of bending magnets, the emitted radiation is broadband and the insertion device is called a wiggler. Alternatively, the device is called an undulator when the magnetic fields are tuned so that the emitted radiation experiences constructive interference with the electron beam. The undulator emits light in a narrow bandwidth defined by the parameters of the incoming electron beam, the periodicity and the magnetic field strength.

The use of superconductors can increase the magnetic field strength and enable higher radiation brightness. Decreasing the undulator period will allow shorter wavelengths to be probed. It may also add to the radiation brightness by allowing more periods in the same space, typically defined by available space and accelerator optics.

Following the massive interest in large synchrotron accelerator and free-electron laser projects in recent years, where radiation generation is based on undulator/ wiggler technologies, prototypes of superconducting (NbTi wire) and in-vacuum permanent-magnet cryogenic undulators have been developed (Tanaka *et al.*,

2007). New ideas and proof of principle constructions of undulators employing HTS technologies are also being investigated. One such design is to add bulk HTS rings to a permanent magnet cryo-undulator (Tanaka *et al.*, 2004; Tanaka *et al.*, 2005; Kii *et al.*, 2009). Another concept takes advantage of HTS thin film technology and proposes to make short-period free-electron laser undulators by forcing particular current paths in a thin-film HTS-coated conductor using lithographic techniques (Prestemon *et al.*, 2009).

Currently, however, no breakthrough technology for HTS-based undulators has been identified and the proposed solutions have to compete with already existing technologies based on LTS and permanent magnets.

12.4.4 Current leads and magnet interlinks

The use of HTS current leads to bring high currents from the room-temperature magnet power supplies to the cryogenic LTS magnets was suggested shortly after the discovery of high temperature superconductivity (Mumford, 1989; Matrone *et al.*, 1989; Hull, 1989). With an order volume of 31 km of Bi-2223 for the LHC, current leads represent a breakthrough for HTSs in the accelerator industry. The leads were used to power the LTS magnets.

The LHC design cooled the resistive part of the lead with 20 K, 1.3 bar helium gas, recovered from the LHC beam screen cooling line, while the HTS element operates in self-cooling conditions between an intermediate temperature (T_{HTS}) and the 4.5 K liquid helium bath.

HTS current leads are now standard components that may be acquired commercially from several providers and they are used in all applications where an LTS/HTS magnet is operated. Both wire/tape and bulk-based solutions are on the market and the can be made from Bi-2212, Bi-2223 and YBCO.

12.4.5 Other applications

In a few other areas relevant to the accelerator industry, the use of HTS is also debated. One of these is high-field ion sources – the review by Zhao (2010) describes the current status of LTS ion sources. Design studies and prototype ion sources based on HTS materials have also been made (Rodrigues *et al.*, 2008).

In a recent study, the application of HTS thin films to accelerator cavities is also discussed (Gustafsson *et al.*, 2010). Cavities currently represent a huge market owing to the imminent construction of several linear accelerators.

12.5 Challenges and future trends

HTSs as accelerator magnets are currently struggling to find niche applications in between standard copper magnets and LTSs. The LTS breakthrough was fuelled by the advances in cryotechnology; setting up a cryogenic liquid helium plant is now considered a standard solution for modern facilities such as the International Thermo-Nuclear Reactor (ITER), the European Free-Electron Laser (XFEL) and the European Spallation Source (ESS). A similar development is required either for cryocooler technology or for the development of HTS magnets based on liquid nitrogen operation.

Several case studies and practical tests have shown promising results but HTS accelerator magnets are still lacking in performance measured against LTS and perceived reliability against copper-based magnets.

The main goals for HTS magnets for the accelerator community are:

- Develop a reliable design and construction platform for HTS magnets demonstrating the achievement of production tolerance requirements,
- Better-performing and/or cheaper HTS conductors,
- Cheap and reliable cryocooler technology,
- Availability of cheaper low-current (100–200 A) power supplies.

Overall, the product has to be able to compete both in price (lifetime cost) and reliability with water-cooled room-temperature copper magnets.

With the continuing increasing performance of YBCO tape, we will see several new projects emerge over the coming years. The market for HTS accelerator magnets already exists but other well-established solutions using copper, permanent magnets and LTSs are currently favoured due to price-competiveness and reliability. Once price and performance competiveness have been demonstrated, we will see HTS accelerator magnets as inserts in high-field magnets and in high-radiation environments. The absence of a 'flagship' project requiring HTS technology as an enabling technology will delay their introduction.

12.6 Conclusions

In this chapter, HTSs used for accelerators were discussed. The maturation and development of the NbTi low temperature superconductor for accelerators starting with the Tevatron (1983) was used as an illustration of the incredible industrial drive associated with enabling technologies. HTS current leads are being used to transfer current to the LTS magnets in accelerators but, apart from this, no niche application has been found.

Within a few years, we will, however, probably see HTS accelerator magnets as high-field inserts in dipole magnets and as focusing magnets in high-radiation environments – areas where LTS cannot compete.

Table 12.1 summarises the different types of HTS evaluated for potential applications.

Table 12.1 Overview of the match between applications and types of HTS

	Bi-2212	Bi-2223	YBCO	MgB2	Bulk
Bending magnets	Potential as high-field insert. Performance needs to improve.	Little potential due to low J_e and poor mechanical properties.	Most promising HTS candidate. Uncertainty regarding field homogeneity and ramping effects.	Has long-term potential, but more research and development is required.	Still at research and development level.
Focusing magnets	Not promising compared to Bi-2223 and YBCO.	Large potential.	Prototype under construction.		
Insertion devices		Cannot yet compete with permanent magnets and LTS.	Cannot yet compete with permanent magnets and LTS.		Has potential, but present technology not mature.
Current leads		Preferred solution due to performance, reliability and cost considerations.	Will be used in the future due to the flexibility and durability of the tapes.		Available commercially.

12.7 Sources of further information and advice

- Martin N. Wilson, *Superconducting Magnets*, Monographs on Cryogenics, Oxford University Press, New edition (1987), ISBN 978–0–19–854810–2.
- E. Weihreter, 'Compact Synchrotron Light Sources', *Series on Synchrotron Radiation Techniques and Applications – Vol. 3*, World Scientific Publishing Co. Pte. Ltd. (1996), ISBN 9810224591.
- K. H. Mess, P. Schmuser, S. Wolff, *Superconducting Accelerator Magnets*, World Scientific Publishing Co. Inc. (1996), ISBN 9810227906.
- Fred M. Asner, *High Field Superconducting Magnets*, Oxford University Press (1999), ISBN 0–19–851764–5.
- Y. Iwasa, *Case Studies in Superconducting Magnets*, Plenum Press New York (1994), ISBN 0–306–44881–5.
- US Particle Accelerator School, course material available at http://uspas.fnal.gov/
- CERN Accelerator School, course material available at http://cas.web.cern.ch/cas/

12.8 References

Adam J D, Boutboul T, Cavallari G, Charifoulline Z, Denarie C-H *et al.* (2002), 'Status of the LHC superconducting cable mass production', *IEEE Trans Appl Supercond*, **12**, 1056–1062.

Ballarino A (2002), 'Current leads for the LHC magnet system', *IEEE Trans Appl Supercond*, **12**, 1275–1280.

Ballarino A (2010), 'Design of an MgB_2 feeder system to connect groups of superconducting magnets to remote power converters', *J Phys Conf Ser*, **234**, 032003.

Ballarino A, Martini L, Mathot S, Taylor T and Brambilla R (2007), 'Large scale assembly and characterization of Bi–2223 HTS Conductors', IEEE Trans Appl Supercond, **17**, 3121–3124.

Cooley L D, Eom C B, Hellstrom E E and Larbalestier D C (2001), 'Potential application of magnesium diboride for accelerator magnet applications', *Proceedings of the 2001 Particle Accelerator Conference*, Chicago, 203–207. Available from: http://accelconf.web.cern.ch/AccelConf/p01/PAPERS/TOAB008.pdf [Accessed 30 April 2011].

Danfysik (2011), 'Developing a High Temperature Superconducting Dipole Magnet', *DANFYSIK Newsletter 2010*, 4. Available from: http://www.danfysik.com/_root/media/41520_News%202010.pdf [Accessed 30 April 2011].

Derbenev Y and Johnson R (2005), 'Six-dimensional muon beam cooling using a homogeneous absorber: Concepts, beam dynamics, cooling decrements, and equilibrium emittances in a helical dipole channel', *Phys Rev ST AB*, **8**, 041002.

Devred A, Baudouy B, Baynham D E, Boucboul T, Canfer S *et al.* (2006), 'Overview and status of the next European dipole joint research activity', *Supercond Sci Technol*, **19**, S67–S83.

Fabbricatore P, Farinon S, Musenich R, Alessandria F, Bellomo G *et al.* (2008), 'R&D activities aimed at developing a curved fast ramped superconducting dipole for FAIR SIS300', *Proceedings of EPAC08*, Genoa, Italy, WEOBM01, 1950–1952. Available from: http://accelconf.web.cern.ch/AccelConf/e08/papers/weobm01.pdf [Accessed 30 April 2011].

Fischer E, Schnizer P, Akishin P, Kurnyshov R, Mierau A *et al.* (2010), 'Superconducting SIS100 prototype magnets test results and final design issues', *IEEE Trans Appl Supercond*, **20**, 218–221.

Foltyn S R, Wang H, Civale L, Jia Q X, Arendt P N *et al.* (2005), 'Overcoming the barrier to 1000 A/cm width superconducting coatings', *Appl Phys Lett*, **87**, 162505–162507.

Goldacker W, Frank A, Kudymow A, Heller R, Kling A *et al.* (2009), 'Status of high transport current ROEBEL assembled coated conductor cables', *Supercond Sci Technol*, **22**, 034003.

Goldacker W, Nast R, Kotzyba G, Schlachter S I, Frank A *et al.* (2006), 'High current DyBCO-ROEBEL assembled coated conductor (RACC)', *J Phys Conf Ser*, **43**, 901–904.

Grasso G, Brisigotti S, Pietranera D, Tumino A, Berta S *et al.* (2008), 'Current MgB_2 wire performance and their industrial development', *WAMSDO Proceedings*, Workshop on Accelerator Magnet Superconductors, Design and Optimization, CERN, Geneva, Switzerland, 19–23 May 2008, 78. Available from: http://cdsweb.cern.ch/record/1163719/files/ [Accessed 30 April 2011].

Greene G, Gupta R and Sampson W (2009), 'The effect of proton irradiation on the critical current of commercially produced YBCO conductors', *IEEE Trans Appl Supercond*, **19**, 3164–3167.

Gupta R, Anerella M, Cozzolino J, Ganetis G, Ghosh A *et al.* (2010), 'Second generation HTS quadrupole for FRIB', *Proceedings for the Applied Superconductivity Conference 2010*, Washington DC, USA, 4LZ–05. Available from: http://www.bnl.gov/magnets/staff/gupta/Publications/asc2010/ASC2010-frib-rev-g.pdf [Accessed 30 April 2011].

Gupta R, Anerella M, Ghosh A, Schmalzle J and Sampson W (2008), 'Design, construction and test results of a warm iron HTS quadrupole for the facility for rare isotope beams', *IEEE Trans Appl Supercond*, **18**, 236–239.

Gustafsson A, Calatroni S, Vollenberg W and Seviour R (2010), 'New methods for thin film deposition and first investigations of the use of high temperature superconductors for thin film cavities', *Proceedings of IPAC'10*, Kyoto, Japan, WEPEC047, 2995–2997. Available from: http://accelconf.web.cern.ch/AccelConf/IPAC10/papers/wepec047.pdf [Accessed 30 April 2011].

Hasegawa T, Ohtani N, Koizumi T, Aoki Y, Nagaya S *et al.* (2001), 'Improvement of superconducting properties of Bi–2212 round wire and primary test results of large capacity Rutherford cable', *IEEE Trans Appl Supercond*, **11**, 3034–3037.

Hull J R (1989), 'High temperature superconducting current leads for cryogenic apparatus', *Cryogenics*, **29**, 1116–1123.

Hušek I, Kovả P, Melišek T and Kopera L (2009), 'Transport current densities of MgB2 wire, cable and continually transposed conductor', *Cryogenics*, **49**, 366–370.

Kahn S A, Alsharo' A M, Johnson R P, Kuchnir M, Gupta R C *et al.* (2007), 'A high-field HTS solenoid for muon cooling', *Proceedings of PAC07*, Albuquerque, New Mexico, USA, MOPAN118, 446–448. Available from: http://accelconf.web.cern.ch/AccelConf/p07/PAPERS/MOPAN118.pdf [Accessed 30 April 2011].

Kahn S A, Flanagan G, Johnson R P, Turenne M, Hunte F *et al.* (2010), 'YBCO conductor technology for high field muon cooling magnets', *Proceedings of IPAC'10*, Kyoto, Japan, MOPEB055, 394–396. Available from: http://accelconf.web.cern.ch/AccelConf/IPAC10/papers/mopeb055.pdf [Accessed 30 April 2011].

Kahn S A, Turenne M, Johnson R P, Hunte F and Schwartz J (2009), 'HTS development for 30–50 T final muon cooling solenoids', *Proceedings of PAC09*, Vancouver, BC, Canada, MO6PFP071, 295–297. Available from: http://accelconf.web.cern.ch/AccelConf/PAC2009/papers/mo6pfp071.pdf [Accessed 30 April 2011].

Kalimov A, Moritz G and Zeller A (2004), 'Design of a superferric dipole magnet with high field quality in the aperture', *IEEE Trans Appl Supercond*, **14**, 271–274.

Kaugerts J, Moritz G, Wilson M N, Ghosh A, Den Ouden A *et al.* (2007), 'Cable Design for FAIR SIS 300', *IEEE Trans Appl Supercond*, **17**, 1477–1480.

Kii T, Kinjo R, Bakr M A, Sonobe T, Higashimura K *et al.* (2009), 'Generation of periodic magnetic field using bulk high-*Tc* superconductor', *Proceedings of PAC09*, Vancouver, BC, Canada, WE5RFP074, 2435–2437. Available from: http://accelconf.web.cern.ch/AccelConf/PAC2009/papers/we5rfp074.pdf [Accessed 30 April 2011].

Knoopers H G, Rabbers J J, Ten Haken B and Ten Kate H H J (2002), 'Magnetisation loss in twisted multifilament Bi–2223 tape conductors', *Phys C Supercond*, **372–376**, 1784–1787.

Long N J, Badcock R, Beck P, Mulholl M, Ross N *et al.* (2008), 'Narrow strand YBCO Roebel cable for lowered AC loss', *J Phys Conf Ser*, **97**, 012280.

Masuzawa M, Egawa K and Tsuchiya K (2006), 'Feasibility study of a permanent magnet made from high-*Tc* bulk superconductor', *Proceedings of EPAC 2006*, Edinburgh, Scotland, WEPLS089, 2580–2582. Available from: http://accelconf.web.cern.ch/AccelConf/e06/PAPERS/WEPLS089.pdf [Accessed 30 April 2011].

Matrone A, Rosatelli G and Vaccarone R (1989), 'Current leads with high *Tc* superconducting bus bars', *IEEE Trans Magn*, **25**, 1742–1745.

McInturff A, McIntyre P and Sattarov A (2007), '50 Tesla superconducting solenoid for fast muon cooling ring', *Proceedings of PAC07*, Albuquerque, New Mexico, USA, TUPAS049, 1757–1759. Available from: http://accelconf.web.cern.ch/AccelConf/p07/PAPERS/TUPAS049.pdf [Accessed 30 April 2011].

McIntyre P and Sattarov A (2005), 'On the feasibility of a tripler upgrade for LHC', *Proceedings of 2005 Particle Accelerator Conference*, Knoxville, Tennessee, 634–636. Available from: http://accelconf.web.cern.ch/AccelConf/p05/PAPERS/MOPA008.pdf [Accessed 30 April 2011].

Mumford F J (1989), 'Superconducting current-leads made from high *Tc* superconductor and normal metal conductor', *Cryogenics*, **29**, 206–207.

Musenich R, Sorbi M, Tavilla G, Volpini G, Marabotto R *et al.* (2008), 'A MgB2 superferric racetrack magnet', *Supercond Sci Technol*, **21**, 105014.

Piekarz H, Hays S, Huang Y and Shiltsev V (2008), 'Design considerations of fast-cycling synchrotrons using superconducting transmission line magnets', *Proceedings of EPAC08*, Genoa, Italy, MOPC101, 301–303. Available from: http://accelconf.web.cern.ch/AccelConf/e08/papers/mopc101.pdf [Accessed 30 April 2011].

Pooke D M, Chamritski V, Fee M, Gibson S, King B T *et al.* (2009), 'HTS 5 Tesla synchrotron and neutron beamline magnets', *IEEE Trans Appl Supercond*, **19**, 1372.

Prestemon S, Dietderich D, Madur A, Marks S and Schlueter R D (2009), 'High performance short-period undulators using high temperature superconductor tapes', *Proceedings of PAC09*, Vancouver, BC, Canada, WE5RFP075, 2438–2440. Available from: http://accelconf.web.cern.ch/AccelConf/PAC2009/papers/we5rfp075.pdf [Accessed 30 April 2011].

Rodrigues G, Lakshmy P S, Mathur Y, Rao U K, Dutt R N *et al.* (2008), 'Operational experience with the 18 GHz HTS-ECRIS PKDELIS', *Proceedings of ECRIS08*, Chicago, IL USA, MOPO–17, 107–110. Available from: http://accelconf.web.cern.ch/AccelConf/ecris08/papers/mopo-17.pdf [Accessed 30 April 2011].

Scanlan R M, Dietderich D R, Higley H C and Marken K R (1999), 'Fabrication and test results for Rutherford-type cables made from BSCCO strands', *IEEE Trans Appl Supercond*, **9**, 130–133.

Shen T, Liu X, Trociewitz U P and Schwartz J (2008), 'Electromechanical behavior of $Bi_2Sr_2CaCu_2O_x$ conductor using a split melt process for react-wind-sinter magnet fabrication', *IEEE Trans Appl Supercond*, **18**, 520–524.

Shiroyanagi Y, Greene G, Gupta R and Sampson W (2011), 'Influence of proton irradiation on angular dependence of second generation (2G) HTS', *Proceedings of 2011 Particle Accelerator Conference*, New York, NY, USA, TUP171. Available from: http://www.bnl.gov/magnets/staff/gupta/Publications/pac2011/tup171.pdf [Accessed 30 April 2011].

Tanaka T, Hara T, Kitamura H, Tsuru R, Bizen T *et al.* (2004), 'Application of high-temperature superconducting permanent magnets to synchrotron radiation sources', *Phys Rev ST AB*, **7**, 090704.

Tanaka T, Tsuru R and Kitamura H (2005), 'Pure-type superconducting permanent-magnet undulator', *J Synchrotron Rad*, **12**, 442–447.

Tanaka T, Tsuru R, Nakajima T and Kitamura H (2007), 'Magnetic characterization for cryogenic permanent-magnet undulators: a first result', *J Synchrotron Rad*, **14**, 416–420.

Toral F, Devred A, Felice H, Fessia P, Loveridge P *et al.* (2007), 'Comparison of 2-D magnetic designs of selected coil configurations for the next European dipole (NED)', *IEEE Trans Appl Supercond*, **17**, 1117–1121.

Turenne M, Johnson R P, Hunte F, Schwartz J and Song H (2010), 'Roebel cable for high-field low-loss accelerator magnets', *Proceedings of IPAC'10*, Kyoto, Japan, MOPEB057, 397–399. Available from: http://accelconf.web.cern.ch/AccelConf/IPAC10/papers/mopeb057.pdf [Accessed 30 April 2011].

Weber S V, Bach H, Bødker F, Hauge N, Kristensen J *et al.* (2007), 'Final commissioning results from the injection system for the Australian Synchrotron Project', *Proceedings of PAC07*, Albuquerque, New Mexico, USA, TUPMN007, 926–928. Available from: http://accelconf.web.cern.ch/AccelConf/p07/PAPERS/TUPMN007.pdf [Accessed 30 April 2011].

Weihreter E (1996), 'Compact Synchrotron Light Sources', *Series on Synchrotron Radiation Techniques and Applications – Vol. 3*, World Scientific Publishing Co. Pte. Ltd., ISBN 9810224591.

Weijers H W, Hascicek Y S, Marken K, Mbaruku A, Meinesz M *et al.* (2004), 'The generation of 25.05 T using a 5.11 T $Bi_2Sr_2CaCu_2O_x$ superconducting insert magnet', *Supercond Sci Techn*, **17**, 636–644.

Wilson M N (1987), *Superconducting Magnets*, Monographs on Cryogenics, Oxford University Press, New edition, ISBN 978–0–19–854810–2.

Zhao H W (2010), 'World-wide development of intense highly charged superconducting ECT ion sources', *Proceedings of IPAC'10*, Kyoto, Japan, MOZMH02, 31–35. Available from: http://accelconf.web.cern.ch/AccelConf/IPAC10/papers/mozmh02.pdf [Accessed 30 April 2011].

Zlobin A V, Barzi E, Kashikhin V S, Lamm M J, Lombardo V *et al.* (2010), 'Modeling the high-field section of a muon helical cooling channel', *Proceedings of IPAC'10*, Kyoto, Japan, MOPEB054, 391–393. Available from: http://accelconf.web.cern.ch/AccelConf/IPAC10/papers/mopeb054.pdf [Accessed 30 April 2011].

Index

9780857090126